Lecture Notes in Artificial Intelligence 4923

Edited by J. G. Carbonell and J. Siekmann

Subseries of Lecture Notes in Computer Science

Sadok Ben Yahia Engelbert Mephu Nguifo
Radim Belohlavek (Eds.)

Concept Lattices and Their Applications

Fourth International Conference, CLA 2006
Tunis, Tunisia, October 30–November 1, 2006
Selected Papers

Springer

Series Editors

Jaime G. Carbonell, Carnegie Mellon University, Pittsburgh, PA, USA
Jörg Siekmann, University of Saarland, Saarbrücken, Germany

Volume Editors

Sadok Ben Yahia
El Manar University of Tunis
Department of Computer Science
Campus Universitaire, 1060 Tunis, Tunisia
E-mail: sadok.benyahia@fst.rnu.tn

Engelbert Mephu Nguifo
Université d'Artois–IUT de Lens
Lens Computer Science Research Centre, CRIL CNRS FRE 2499
Rue de l'Université SP 16, 62307 Lens Cedex, France
E-mail: mephu@cril.univ-artois.fr

Radim Belohlavek
State University of New York at Binghamton
P.O. Box 6000, Binghamton, NY 13902–6000, USA
E-mail: rbelohla@binghamton.edu
and
Palacký University
Department of Computer Science
Tomkova 40, 779 00 Olomouc, Czech Republic

Library of Congress Control Number: 2008923361

CR Subject Classification (1998): H.3.1, H.3.3, H.2.8

LNCS Sublibrary: SL 7 – Artificial Intelligence

ISSN 0302-9743
ISBN-10 3-540-78920-0 Springer Berlin Heidelberg New York
ISBN-13 978-3-540-78920-8 Springer Berlin Heidelberg New York

Springer is a part of Springer Science+Business Media

springer.com

Printed in Germany

Typesetting: Camera-ready by author, data conversion by Scientific Publishing Services, Chennai, India
Printed on acid-free paper SPIN: 12250271 06/3180 5 4 3 2 1 0

Preface

This volume contains selected papers from CLA 2006, the 4th International Conference on Concept Lattices and Their Applications. CLA 2006 was held in Hammamet, Tunisia, from October 30 to November 1, 2006, and was organized jointly by the El-Manar University (Computer Science Department, Faculty of Sciences), Tunis, and the Université Centrale, Tunis. The main areas of interest relevant to CLA include formal concept analysis (FCA), foundations of FCA, mathematical structures related to FCA, relationship of FCA to other methods of data analysis, visualization of data in FCA, and applications of FCA.

The conference received 41 submitted papers. This volume contains 18 papers (13 long, 5 short) selected from the submitted papers which were accepted and presented at the conference (selection rate 0.44). Contributions to CLA 2006 were refereed by at least three reviewers on the basis of their originality, quality, significance, and presentation. When one of the Program Chairs was involved in a paper, the reviewing process of this paper was managed independently by the other chair. When both of the Program Chairs were co-authors, Radim Belohlavek managed the reviewing process of those papers.

The program of CLA 2006 also included four invited talks by Rudolf Wille (TU-Darmstadt, Germany), Claudio Carpineto (FUB, Rome, Italy), Peter Eklund (University of Wollongong, Australia), Amedeo Napoli (LORIA, Nancy, France), and a tutorial by Radim Belohlavek (Palacky University, Olomouc, Czech Republic). Three papers based on the invited talks are a part of this volume.

We would like to express our thanks to the authors who submitted their papers to CLA 2006, to the invited speakers, to the members of Program Committee who managed the review of papers, to the additional reviewers, to the members of the Organization Committee, as well as to the conference attendees, who all helped make CLA 2006 a successful event.

November 2007

Sadok Ben Yahia
Engelbert Mephu Nguifo
Radim Belohlavek

Preface

Organization

CLA 2006 was organized by Faculté des Sciences de Tunis of El Manar University and by Université Centrale de Tunis.

Steering Committee

Radim Belohlavek	State University of New York at Binghamton, USA
Sadok Ben Yahia	Faculté des Sciences de Tunis, Tunisia
Engelbert Mephu Nguifo	CRIL CNRS FRE 2499 - IUT de Lens, France
Václav Snášel	VSB-TU Ostrava, Czech Republic

Program Chairs

Sadok Ben Yahia	Faculté des Sciences de Tunis, Tunisia
Engelbert Mephu Nguifo	CRIL CNRS FRE 2499 - IUT de Lens, France

Program Committee

Radim Belohlavek	State University of New York at Binghamton, USA
Anne Berry	LIMOS, Université de Clermont Ferrand, France
Laurent Chaudron	Onera-CERT, France
Claudio Carpineto	Fondazione Ugo Bordoni, Rome, Italy
Richard J. Cole	University of Queensland, Brisbane, Australia
Jean Diatta	Université de la Réunion, France
Vincent Duquenne	Université Pierre et Marie Curie, Paris, France
Peter Eklund	University of Wollongong, Australia
Samir Elloumi	Faculté des Sciences de Tunis, Tunisia
Mohamed M. Gammoudi	ISG, Kairouan, Tunisia
Gemma C. Garriga	Technical University of Catalonia, Spain
Ali Jaoua	Qatar University, Qatar
Stanislav Krajči	UPJS Kosice, Slovakia
Marzena Kryszkiewicz	Warsaw Institute of Technology, Poland
Sergei Kuznetsov	VINITI and RSUH Moscow, Russia
Michel Liquière	LIRMM, Montpellier, France
Mondher Maddouri	INSAT, Tunis, Tunisia
Rokia Missaoui	UQO, Gatineau, Canada
Amedeo Napoli	LORIA, Nancy, France

Lhouari Nourine	LIMOS, Université de Clermont Ferrand, France
Sergei Obiedkov	University of Pretoria, South Africa
Habib Ounelli	Faculty of Sciences, Tunis, Tunisia
Uta Priss	Napier University, Edinburgh, UK
Olivier Raynaud	LIMOS, Clermont-Ferrand, France
Yahya Slimani	Faculty of Sciences, Tunis, Tunisia
Václav Snášel	VSB-TU Ostrava, Czech Republic
Henry Soldano	LIPN, Paris 13, France
Petko Valtchev	DIRO, Université de Montréal, Canada
Vilem Vychodil	State University of New York at Binghamton, USA
Ezzeddine Zagrouba	ISI, Tunis, Tunisia
Mohammed Zaki	Rensselaer Polytechnic Institute, NY, USA

Organization Committee

Samir Elloumi (Chair)	Faculté des Sciences de Tunis, Tunisia
Khedija Arour	INSAT, Tunis, Tunisia
Olivier Couturier	CRIL CNRS FRE 2499 - IUT de Lens, France
Helmi El Kamel	Université Centrale, Tunis, Tunisia
Ghada Gasmi	Faculté des Sciences de Tunis, Tunisia
Tarek Hamrouni	Faculté des Sciences de Tunis, Tunisia
Tienté Hsu	LGI2A - IUT de Lens, France
Nicolas Huicq	IUT de Lens, France
Chiraz Latiri	Ecole Supérieure de Commerce, Manouba
Mondher Maddouri	INSAT, Tunis, Tunisia
Laurent Masse	IUT de Lens, France
Franck Paszkowski	IUT de Lens, France
Moncef Temani	ISI, Tunis, Tunisia
Sami Zghal	Faculté des Sciences de Tunis, Tunisia

Additional Reviewers

Khedija Arour	INSAT, Tunis, Tunisia
Sondess Ben Tekaya	Faculté des Sciences de Tunis, Tunisia
Chaima Ben Youssef	Faculté des Sciences de Tunis, Tunisia
Karell Bertet	L2I, La Rochelle, France
Vicky Choi	Virginia Tech, USA
Richard Emilion	MAPMO, Université d'Orléans, France
Sébastien Ferré	IRISA, Rennes, France
Céline Fiot	LIRMM, Montpellier, France
Huaiguo Fu	University College Dublin, Ireland
Ghada Gasmi	Faculté des Sciences de Tunis, Tunisia
Alain Gély	LIMOS, Clermont Ferrand, France

Tarek Hamrouni	Faculté des Sciences de Tunis, Tunisia
Leonard Kwuida	University of Bern, Swizerland
Raoul Medina	LIMOS, Clermont Ferrand, France
Pascal Poncelet	LGI2P, École des mines d'Alès, France
Francois Rioult	GREYC, Caen, France
Gabriel Semanišin	UPJS Kosice, Slovak Republic

Sponsoring Institutions

Université El-Manar, Tunis
Université Centrale, Tunis
Ambassade de France, Tunis
Centre de Calcul EL KHAWARIZMI
INSAT, Tunis
Institut Supérieur d'Informatique (ISI), Tunis
IUT de Lens, France
UTM, Tunis
VERMEG, Tunis

Table of Contents

Applications

An Intelligent User Interface for Browsing and Searching MPEG-7 Images Using Concept Lattices

Jon Ducrou, Peter Eklund, and Tim Wilson

School of Information Systems and Technology
University of Wollongong
NorthFields Avenue, New South Wales, Australia
{jond,peklund}@uow.edu.au, timwilson1@optusnet.com.au

Abstract. This paper presents the evaluation of a design and architecture for browsing and searching MPEG-7 images. Our approach is novel in that it exploits concept lattices for the representation and navigation of image content. Several concept lattices provide the foundation for the system (called IMAGE-SLEUTH) each representing a different search context, one for image shape, another for color and luminance, and a third for semantic content. This division of information aids in the facilitation of image browsing based on a metadata ontology. The test collection used for our study is a sub-set of MPEG-7 images created from the popular *The Sims 2*^{TM} game. The evaluation of the IMAGE-SLEUTH program is based on usability testing among 29 subjects. The results of the study are used to build an improved second generation program – IMAGE-SLEUTH2– but in themselves indicate that image navigation via a concept lattice is a highly successful interface paradigm. Our results provide general insights for interface design using concept lattices that will be of interest to any applied research and development using concept lattices.

1 Introduction

The objective of this research is to offer a novel way of searching and navigating digital images – images annotated with MPEG-7 multimedia descriptors – and presenting them in a way that is easily understood by humans. The interface we develop presents a new way to graphically represent relationships between images so that navigation across a collection of images occurs in a non-linear or serendipitous way. A suitable method for achieving this is by the application of formal concept analysis. The images (as objects) are organised as a conceptual hierarchy via the formal concept analysis of their image and metadata attributes. The concept lattice that results provides the information space over which the interface navigates. The paper tests the success of this idea through software evaluation in a usability trial.

The images used to test the approach are derived from the popular computer game *The Sims 2*^{TM}. This collection is made interesting by taking into account the properties given to these objects in *The Sims 2*^{TM} game play. The game playing properties are elements such as: how an object addresses *The Sims 2*^{TM} character's needs — like hunger, bladder comfort, tiredness, hygiene, etc; how the object develops *The Sims 2*^{TM} character's skills in different skill areas – e.g. logic, cooking, mechanical skills and creativity.

S. Ben Yahia et al. (Eds.): CLA 2006, LNAI 4923, pp. 1–21, 2008.

In addition to this metadata associated with the game play, MPEG-7 feature descriptors are also used so that the images can be navigated according to their color and shape.

In related work [1] we address the issue of the design theory underlying a Web-based FCA system for browsing and searching MPEG-7 images called IMAGE-SLEUTH. This paper, which emphasises the usability of IMAGE-SLEUTH, is structured as follows. In order to be self-contained, an introduction to formal concept analysis is provided in Section 1. Because the image format we use contains semantic attributes as well as an image signature we give the reader a brief introduction to MPEG-7 in Section 2. We illustrate the idea of image browsing using FCA with a collection of images from the *The Sims 2™* and in Section 3 we present both a synopsis of the *The Sims 2™* game play and details of the MPEG-7 image content. In Section 4 we present our approach to image navigation based on the design of edge traversal in the concept lattice. Our main results are presented in Section 5 where we present the usability test script, the survey instrument and the results of our evaluation for the image browsing software. An important purpose to our usability study was to learn how to improve the performance and design of our software. In Section 6 we show how our findings conditioned the development of IMAGE-SLEUTH2, in particular the way that conceptual scaling is handled and the introduction to IMAGE-SLEUTH2 of distance and similarity metrics for approximate matching. In Section 7 we discuss work in progress on extending our ideas to searching and browsing a dynamic data collection: namely the Amazon catalog.

Formal Concept Analysis Background

Formal Concept Analysis [2] has a long history as a technique of data analysis ([3], [4]) conforming to the idea of Conceptual Knowledge Processing. Data is organized as a table and is modeled mathematically as a many-valued context, (G, M, W, I_w) where G is a set of objects, M is a set of attributes, W is a set of attribute values and I_w is a relation between G, M, and W such that if $(g, m, w_1) \in I_w$ and $(g, m, w_2) \in I_w$ then $w_1 = w_2$. In the table there is one row for each object, one column for each attribute, and each cell is either empty or asserts an attribute value.

A refined organization over the data is achieved via conceptual scales. A conceptual scale maps attribute values to new attributes and is represented by a mathematical entity called a formal context. A formal context is a triple (G, M, I) where G is a set of objects, M is a set of attributes, and I is a binary relation between the objects and the attributes, i.e. $I \subseteq G \times M$. A conceptual scale is defined for a particular attribute of the many-valued context: if $\mathbb{S}_m = (G_m, M_m, I_m)$ is a conceptual scale of $m \in M$ then we define $W_m = \{w \in W | \exists (g, m, w) \in I_w\}$ and require that $W_m \subseteq G_m$. The conceptual scale can be used to produce a summary of data in the many-valued context as a derived context. The context derived by $\mathbb{S}_m = (G_m, M_m, I_m)$ w.r.t plain scaling from data stored in the many-valued context (G, M, W, I_w) is the context (G, M_m, J_m) where for $g \in G$ and $n \in M_m$

$$\begin{aligned} gJ_m n \Leftrightarrow: \exists w \in W : \ & (g, m, w) \in I_w \\ & \text{and } (w, n) \in I_m \end{aligned}$$

Scales for two or more attributes can be combined in a derived context. Consider a set of scales, S_m, where each $m \in M$ gives rise to a different scale. The new attributes supplied by each scale can be combined:

$$N := \bigcup_{m \in M} M_m \times \{m\}$$

Then the formal context derived from combining these scales is:

$$gJ(m,n) \;:\Leftrightarrow\; \exists w \in W : \; (g,m,w) \in I_w \text{ and } (w,n) \in I_m$$

Several general purpose scales exist such as ordinal and nominal scales. A nominal scale defines one formal attribute for each value that a many valued attribute can take. An ordinal scale can be used to interpret an attribute whose values admit a natural ordering, for example the $\leq$ ordering over numbers.

A concept of a formal context (G, M, I) is a pair (A, B) where $A \subseteq G$, $B \subseteq M$, $A = \{g \in G \mid \forall m \in B : (g,m) \in I\}$ and $B = \{m \in M \mid \forall g \in A : (g,m) \in I\}$. For a concept (A, B), A is called the extent and is the set of all objects that have all of the attributes in B, similarly, B is called the intent and is the set of all attributes possessed in common by all the objects in A. As the number of attributes in B increases, the concept becomes more specific, i.e. a specialization ordering is defined over the concepts of a formal context by:

$$(A_1, B_1) \leq (A_2, B_2) :\Leftrightarrow B_2 \subseteq B_1$$

In this representation more specific concepts have larger intents and are considered "less than" ($<$) concepts with smaller intents. The analog is achieved by considering extents, in which case, more specific concepts have smaller extents. The partial ordering over concepts is always a complete lattice [2].

For a given concept $C = (A, B)$ and its set of lower covers $(A_1, B_1)...(A_n, B_n)$ with respect to the above $<$ ordering the object contingent of C is defined as $A - \bigcup_{i=1}^{n} A_i$. We shall refer to the object contingent simply as the contingent in this paper.

2 MPEG-7 Images

Accepted as an ISO standard in 2001, MPEG-7[1] allows the storage of physical and semantic descriptors for use in content management, organization, navigation, and automated processing of images [5]. MPEG-7 is extensible, being based on XML, and can therefore support a broad range of applications.

MPEG-7 comprises Description Tools made up of the metadata elements, along with their structure and relationships, which are used to form Descriptors and Description Schemas. *Descriptions* can then be used by applications for effective and efficient access to multimedia content. These descriptions accommodate a range of abstraction levels, from low-level signal characteristics to high-level semantic information. In this paper, we are interested in both low-level image descriptors, more specifically color descriptors and shape descriptors, as well as high-level semantic metadata by extending the MPEG-7 format to store customised details for each object.

[1] *http://www.chiariglione.org/mpeg/standards/mpeg-7/mpeg-7.htm*

Color descriptors in MPEG-7 consist of seven more specific descriptors: Color space, Color Quantization, Dominant Colors, Scalable Color, Color Layout, Color-Structure, and GoF/GoP Color. Three of the MPEG-7 visual descriptors are used in this research, these are color Layout, Scalable color and Edge Histogram and these are extracted from the image collection as described below.

3 Feature Extraction on the Example Collection

The example collection for our Web-based image browser is based on items from the popular computer game *The Sims 2*^{TM}. Created by Electronic Arts, *The Sims 2*^{TM} is "the sequel to the best-selling PC game of all time"[2]. In brief, the game is a real-life simulation; the player is given control over a suburban neighborhood and the people in it, shaping their careers, friendships, houses, children, and controlling mundane tasks: such as directing them to cook meals, have showers and go to bed. These simulations of the people who populate the neighborhood — the characters of the game — are referred to as *Sims*.

Each Sim has 8 *needs* that affect their well-being. These are hunger, comfort, hygiene, bladder, energy, fun, environment and social. Sims also have 7 types of skills which they can practice and refine. These skills are cleaning, charisma, creativity, body, logic, mechanical and cooking. As well as looking after their needs and lives, a player can build a house for their Sim and purchase different household items to furnish it. These items include furniture, plumbing, appliances, decorations, electronics, plants, lighting and much more. Household items can directly affect a Sim's needs and skills when in use[3]. Hunger, for example, is satisfied to a lesser degree when cooking with a cheap microwave than using an expensive oven. Some of the items can also have a negative impact on a Sim. For example, a coffee machine increases energy, but decreases bladder comfort. A bookcase will allow the Sim to study and increase cooking skills, while an artist's easel will allow the Sim to produce artworks and increase their creativity.

3.1 *The Sims 2*^{TM} Image Collection

Our collection is based on virtual household items that can be bought and sold in the *The Sims 2*^{TM}. The basis for this choice is the dual nature of the items. A household item must aid in successful game playing – as well as have aesthetic appeal – perhaps matching other furnishings already in place. Therefore, a household item has physical properties such as color and shape, as well as properties that effect a Sim's life and well-being .

The color layout descriptor in MPEG-7 breaks the image into an 8×8 grid and represents each grid square by the dominant color in YCbCr format[4]. The scalable color descriptor gives a measure of color distribution over the entire image. The edge

[2] *http://thesims2.ea.com/*

[3] The exception here is the "social" dimension which is only affected by social interactions.

[4] YCbCr is a family of color spaces used in video systems and similar to that used in color television.

```
<VisualDescriptor xsi:type="ColorLayoutType">
   <YDCCoeff>5</YDCCoeff>
   <CbDCCoeff>30</CbDCCoeff>
   <CrDCCoeff>31</CrDCCoeff>
   <YACCoeff63>
      13 23 15 12  5 20  9 14 19 17 16 17 21 18 15 17 18 12 16 11 13
      16 14 15 15 15 17 13 16 15 17 14 20 15 17 16 18 15 16 15 15 12
      14 15 16 15 16 14 16 15 16 16 17 16 15 15 14 15 15 15 16 17 16
   </YACCoeff63>
   <CbACCoeff63>
      16 15 16 16 17 15 16 16 15 15 16 15 15 15 16 15 15 16 15 16 16
      16 16 16 16 16 15 16 16 16 15 16 15 16 15 16 15 16 16 16 16 16
      16 16 15 16 16 16 16 15 16 15 15 16 16 16 16 16 16 16 16 15 15
   </CbACCoeff63>
   <CrACCoeff63>
      16 16 16 16 16 16 15 15 16 16 15 16 16 16 15 15 16 16 16 15 15
      16 15 16 15 15 15 16 16 16 16 16 16 15 16 16 16 15 16 16 16 16
      15 15 16 16 16 15 16 15 16 16 16 16 15 15 16 16 16 16 16 16 16
   </CrACCoeff63>
</VisualDescriptor>
```

Fig. 1. An example of the color Layout extracted values for an image

```
<VisualDescriptor
   xsi:type="ScalableColorType"
   numOfBitplanesDiscarded="0"
   numOfCoeff="64" >
   <Coeff>
      -202  58  40  41  -7  12  20  14   6  13  11  22   1  16  21   9
         0   1   0   2  -1   5   0   0  -9  -2  -2   9 -15   3  -1 -19
         0   0   0   1   0   0   1   2   1   1   1   3   1   2   4   5
         1  -3   2  -2   2  -1  -8  -2   0 -15   0  -4   1  -2  -3 -15
   </Coeff>
</VisualDescriptor>
```

Fig. 2. Scalable color Type extracted values

histogram defines a 4 × 4 grid and gives the strength of the non-homogeneous texture for each grid square in 4 directions and an overall strength.

To extract shape and color information for a household item, a feature extraction tool, Caliph [7], is used to generate color layout [8] and edge descriptors [9]. These are then stored in MPEG-7 using the appropriate tags as shown in Figs. 1 and 2. A secondary feature extraction process, which analyses the resulting descriptors, is then run to produce more user-friendly color descriptors. The secondary color descriptors use a reduced form of the standard HTML color set to assign a meaningful color property. The set of color names have a hierarchy in which parent colors are more general (e.g. Red → Dark Red → Maroon). These secondary descriptors are added into the MPEG-7 datastore using a custom mark-up.

The edge histogram descriptor is a measure of the edge distribution within an image [10]. In a similar method to that used by the color layout descriptor, an image is broken down into a series of non-overlapping square blocks. An edge histogram is then generated on each of these blocks. The descriptor defines 5 values to represent the edge histogram for each block. These 5 values describe the vertical, horizontal, 45 degree and 135 degree edges as well as a non-directional edge. A nondirectional edge is one that has no apparent direction (e.g. a curve).

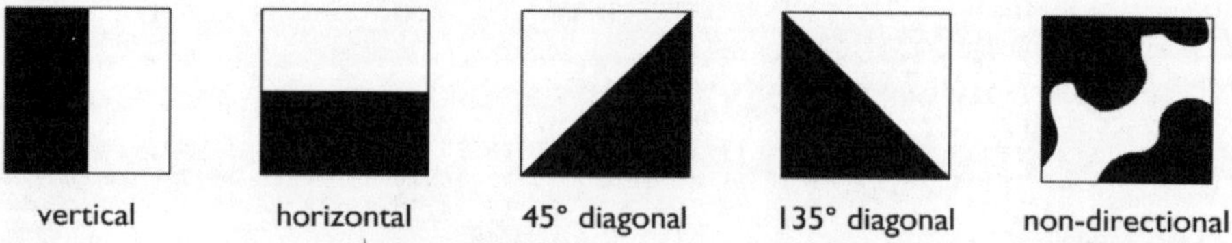

Fig. 3. The 5 edge types used in the edge histogram descriptor

Table 1. A fragment from the 'Item Properties' sub-context

	Price	Needs::Hunger	Needs::Comfort	Needs::Hygiene	Needs::Bladder	Needs::Energy	Needs::Fun	Needs::Environment	Skills	Function	Room Type
4 by 4 Designer Chandelier	§120	0	0	0	0	0	0	1		Lighting	Dining, Living, Bathroom, Bedroom
Absolutely Nothing Special	§850	0	0	0	0	0	0	1		Lighting	Kids, Study, Dining, Living, Bedroom
Ad-a-Quaint Barstool	§285	0	3	0	0	0	0	1		Comfort	Living, Kitchen
Ad-a-Quaint Coffee Table	§140	0	0	0	0	0	0	1		Surfaces	Study, Living
Astrowonder Telescope	§550	0	0	0	0	0	4	0	Logic	Hobbies	Outside
Zenu Meditation Sleeper	§950	0	4	0	0	4	0	2		Comfort	Bedroom

The color and shape descriptors are then complemented with the various semantic metadata derived from the *The Sims 2^{TM}* game information. Needs and skills are an attribute hierarchy where more specific attributes in the hierarchy imply more general. For example, Needs is implied by Fun, Fun is implied by Fun:1 to 5. This gives some level of encapsulation to the attributes because in order to have Fun appear as an attribute, Needs must have also been included. Other item properties such as object price and function and suitability to a given room type are also included. On function, objects are grouped by the game into one of 11 functional categories: electronics, lighting, miscellaneous, comfort, aspiration rewards, career rewards, decorative, plumbing, hobbies, appliances and surfaces. Items are also given a room property based on which room the item would most likely be placed. An item may have one or more values for the room property, meaning that it is suitable in several different rooms, or it may have no room value at all, meaning that it can be put anywhere. The room types are: kids, study, dining room, outside, living room, bathroom, bedroom and kitchen. An fragment of the underlying formal context based on *The Sims 2^{TM}* household objects is shown in Table 1.

3.2 Basis for Selection of *The Sims 2^{TM}* Dataset

In games such as *The Sims 2^{TM}*, where collectible items affect gameplay, much effort is put into the game's design in terms of the balance and distribution of items with respect to item properties. This design principle provides an excellent base for testing ideas that use query-by-example, as for most items there is only a single exact item, and varying cluster types associated with it. For example, there may be only 1 curved, 3-seater, blue couch with Comfort:8, but there is a collection of other blue couches with different comfort levels, a collection of different colored and shaped couches with Comfort:8, and a matching blue coffee table that associates with the couch. These are all acceptable responses to query-by-example for the 3-seater, blue couch with Comfort:8 because in some way they are all household objects of the same grouping: collected based on different facets of the data.

4 Conceptual Design of the Image Browser

4.1 Problem Decomposition

Our approach decomposes the overall lattice – generated from the formal context fragment, a fragment of which is shown in Table 1 – into smaller sub-lattices. These sub-lattices are created by combining attributes compatable meaning. In the case of *The Sims 2*^{TM} data, 3 sub-lattices are evident; color properties (including all attributes regarding the colors used in the MPEG-7 images), Item properties (including all the game play properties) and edge properties (including a human readable form of the MPEG-7's generated EdgeHistogramType classifier). Decomposing the lattice into sub-lattices in this way allows for more overall generality per concept for each concept of each sub-lattice. This is necessary given the unique nature of computer game items used in our image collection, but also allows search via query-by-example.

4.2 Interface Design

At any one time the user will be placed at either a single formal concept of a sub-lattice or at a single object (an image). The formal concept is displayed as a neighbourhood showing the current extent as thumbnailed images (in arbitrary order), and the attributes which allow movement to other formal concepts in the neighbourhood. Movement from the current formal concept or image object can be via either specialisation or generalisation. Specialisation is achieved by adding attributes and moving down in the lattice structure (via an interface control called *include*). Alternatively, generalization is achieved by removing attributes via an upward movement between formal concepts (via an interface control called called *remove*). For *include*, the attributes that can be added are displayed, and for *remove*, the attributes that can be removed are displayed (see Fig. 4 (left)).

Our design philosophy is that the presentation of attributes belonging to the upper and lower neighbour formal concepts allows the current state of the interface to move across the concept lattice in an intuitive way. Further, that this approach to navigation is often preferred to showing a complete list of possible attributes to add, or all attributes that can be removed from the current view. One of the consequences of this design is that it is impossible to navigate to a concept with an empty extent (i.e. no images), because two mutually exclusive attributes can never be selected during navigation via upper and lower concepts in the way described[5]. By using the concept lattice structure

[5] Also, the bottom-most concept is considered inaccessible and the user cannot navigate to it (unless it has an extent size greater than zero).

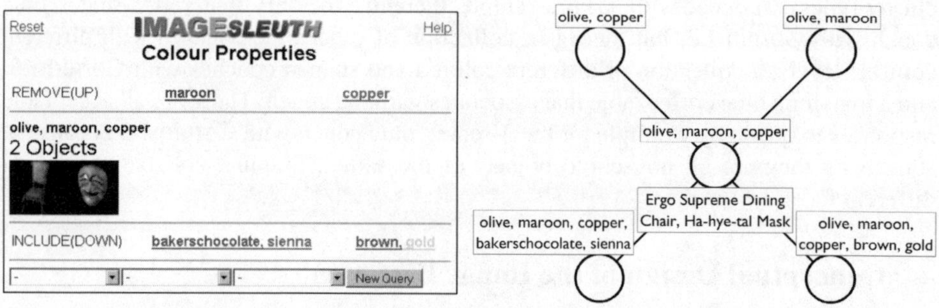

Fig. 4. An example screenshot of *Image Sleuth* and the lattice representation of the corresponding neighbourhood

as the focus for navigation in the interface, the users' perspective is concentrated on conceptual changes that are minimal and incremental. Furthermore, navigation actions in the interface conform directly to the definition of edge traversal in a concept lattice and movment through the information space is therefore directly expressed in the theory of Formal Concept Analysis.

This form of navigation helps to reduced the complexity per 'decision' point, as attributes will often be hidden by others because of implications or attribute hierarchies, whether data-emergent or artificial. For example, given the data has an attribute hierarchy over colors, with more specific color descriptions being children of more general terms, the attributes 'Dark Blue' and 'Light Blue' will not be visible as *include* attributes until the user has included the 'Blue' attribute. Conversely, if the 'Dark Blue' attribute is visible as a *remove* attribute, 'Blue' will be hidden. Complexity is also reduced by attribute equivalence. For example, if 'Dark Green' is the only type of green in the data it will appear in the *include* attributes as a combined pair of attributes 'Green, Dark Green' as there is equivalence between them, (i.e. images with 'Dark Green' will also be 'Green').

As well as navigation via traversal of the concept lattice the IMAGE-SLEUTH interface also provides a traditional query interface that allows direct positioning into the concept lattice. The query interface restricts the user to terms that are attributes of the current sub-lattice. The query interface takes the submitted attributes and finds the most specific concept that has all query terms, namely the query interface performs the equivalent of the double prime operation in FCA. This method ensures that at all times the user is positioned at a formal concept. In the event that the user selects attributes for which no formal concept exists, no images are returned.

IMAGE-SLEUTH also supports the direct selection of an image of interest by clicking on it. In this case the user is presented with the exact set of attributes for the given image and the option of changing sub-lattice or querying-by-example. Any single object can act as a connection point between contexts, and by changing contexts the user is presented with the attributes this object has within the new sub-lattice. Query-by-example uses the current images' attributes to relocate to the most specific concept associated with the image. This will then show all other objects with the same attributes. This means that an object can be found using one sub-lattice, then be used in another

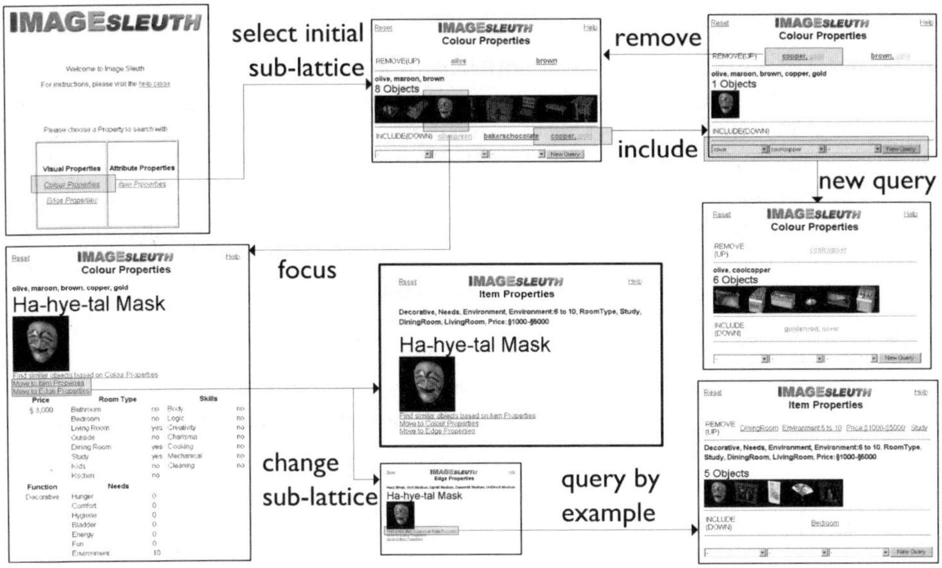

Fig. 5. A navigation overview of the system

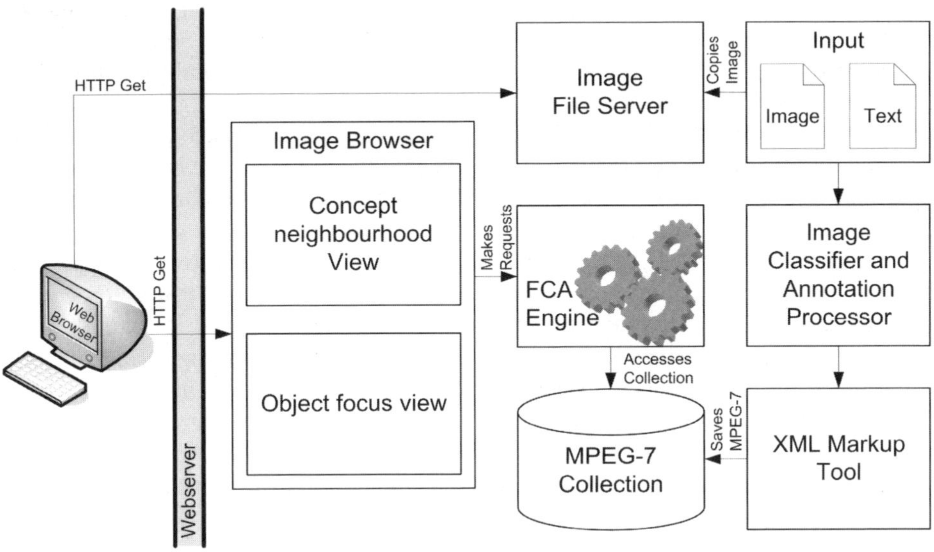

Fig. 6. A technical overview of the system

sub-lattice to find similar objects, but within a different area of interest. For example, using *The Sims 2* collection, a user may find a bed that suits their in-game requirements, then swap sub-lattices to find matching furniture for that bed using color or shape information. The architecture for the IMAGE-SLEUTH system in shown in Fig. 6.

5 Usability

Rozanaski and Haake define several attributes of usability with regards to a user interface [11]: (i) learnability; (ii) efficiency; (iii) memorability; (iv) amount of error; (v) satisfaction. This usability evaluation attempts to measure the success of IMAGE-SLEUTH in meeting these attributes. Usability will be measured through an empirical study that consists of two parts, a test script and a participant survey. The results are also compared with interaction logs which independendly validate that the participant achieved the correct state.

To perform the study, postgraduate and honours students were recruited to participate in the usability tests. Each of the 29 testers were based in one of 6 Faculties at the University of Wollongong in Australia. The students were primarily drawn from Informatics (49%) and Commerce (32%) Faculties, with others from Law, Education, Engineering and Arts. All could be said to exhibit a high degree of computer literacy.

5.1 Test Script

The test script consists of three sections the first two of which returned quantitative results and the third qualitative results. The first section included directions to be followed using Windows Explorer to browse the image collection. Images can be viewed as thumbnails within a folder, and sorted by various criteria such as file name and creation date. Tasks included finding particular items, finding items that matched certain criteria, observations of item groups with certain features and so on. Some tasks in this section – while possible to complete – may have proven time-consuming or difficult to accomplish using Windows Explorer. Windows Explorer is not designed as an image browser, but it can be used to browse images and is used as the base level functionality because all participants are familiar with it. If IMAGE-SLEUTH proved no better than Windows Explorer then this would be a powerful argument to abandon the design.

The second section of the test started with identical tasks to those in the first but now completed using IMAGE-SLEUTH. As the study supervisors were not permitted to assist participants, the steps to be followed using IMAGE-SLEUTH were ordered in such a way as to expose participants to the various functionality of IMAGE-SLEUTH gradually. Designing the test script in this way assists the participant in learning the new navigation style without needing special training. Tasks in the latter half of the second section of the test script were designed to make participants perform more complex interactions in order to solve problems. For example, showing a black and white image of an object in a setting at a different orientation and asking participants to identify its color by finding the corresponding image in IMAGE-SLEUTH. These tasks extended the test script and had no comparable task in the first section of the script. A complete list of the second sections tasks are shown below.

The first and second sections were issued in reverse order to half of the participants. This way, familiarity with the object set did not give an unfair advantage to IMAGE-SLEUTH.

1. Identify how many images have the environment attribute.
2. Identify how many images have a price higher than 5000.

3. Search for all images that are Decorative. Then include the outside room type. How many images are there?
4. Identify how many images can be used in both the Dining Room and Kitchen.
5. Identify how many of the plumbing images also have an environment attribute. What types of objects are these?
6. How many images do not have a price at all?
7. Identify how many images are green using the color properties.
8. There are 2 beds with a high level of horizontal edges. Which are they?
9. Are there more red, navy or green images?
10. Find the name of the chair in the following image:

Shown in color.

11. Find the name of the bed in the following image: (Note - The bedspread need not be the same)

Shown in color.

12. Find the name of the bath/shower in the following image. What color is the curtain around it?

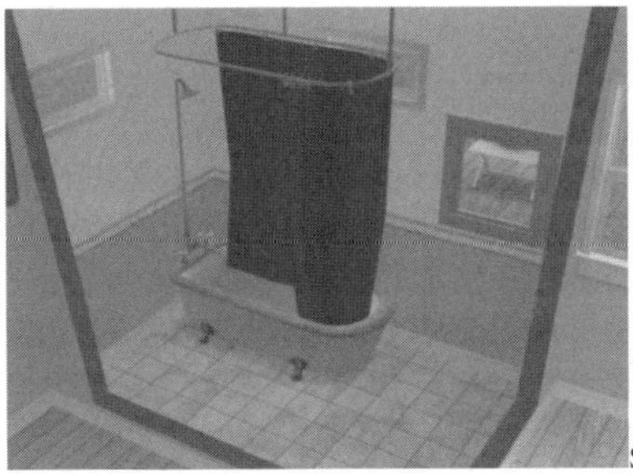
Shown in black and white.

13. This object is not in *The Sims 2™* collection.

Shown in color.

It has attributes of: *Comfort* = 5, *Energy* = 4, *Environment* = 3.
Find one image that is similar in design and one that has similar attributes while not being expensive.
14. Find the object that can be used outside, builds logic and has a price between 1000 to 5000. How many images have similar colors?
15. Aspiration rewards and career rewards are special objects in *The Sims 2™*. Using IMAGE-SLEUTH to browse the objects, what can you say about them, in 50 words or less?

The final section of the test script consisted of a “free exploration” of IMAGE-SLEUTH: encouraging participants to discover features without any particular goal in mind. This allowed participants to gain an understanding of the features without explicit direction. Participants were subsequently asked to provide their positive and negative thoughts regarding the features of the program.

5.2 Survey

The survey asked participants questions on their personal background and experience with IMAGE-SLEUTH. Background information included the faculty of study, experience with other image browsers/viewers and the methods of organisation used for personal image collections. Likert scales were used collect details of their experience with IMAGE-SLEUTH and Windows Explorer. This was followed by a series of questions to assess the participants understanding of IMAGE-SLEUTH and how it worked. Fig. 7 shows a complete list of questions asked in the survey.

5.3 Interpreting the Usability Results

The second section of the test script (where users are first exposed to IMAGE-SLEUTH) proved difficult for participants, but after completion of first few questions, most subjects became acquainted with the user interface and its functionality. On average, once participants had attempted 6 – 7 tasks the number of correct responses increased considerably, even though the tasks became progessively more difficult.

Average correct completion for the Explorer tasks was 70.5%, and 74.5% for the equivalent IMAGE-SLEUTH tasks. The correct completion rate for the entire IMAGE-SLEUTH test script was 73.4%.

Participant Survey

- **Short Response**
  - Which faculty does your university degree belong?
  - Are you color blind? If so, did you experience difficulty in completing the test script?
  - Have you used image management applications before? If so, which?
  - Do you already sort your images based on specific criteria (e.g. date, location, etc.)? If so, what?

- **Likert Scale Statements (0 to 10, disagree to agree)**
  - I am familiar with the PC Game *The Sims 2™*.
  - I found it easy to complete tasks with:
    - Windows Explorer
    - IMAGE-SLEUTH
  - I feel that IMAGE-SLEUTH has a strong advantage over Windows Explorer.
  - I feel that IMAGE-SLEUTH has a strong advantage over other image browsers.
  - IMAGE-SLEUTH allows me to recognise relationships between images that I may not have noticed previously.
  - IMAGE-SLEUTH is a tool that gives more power over searching and browsing catalogs of images.
  - I found that the ______ properties were accurate.
    - color
    - Edge
  - My overall experience with IMAGE-SLEUTH was a positive one.

- **Multiple Choice**
  - What features of IMAGE-SLEUTH did you find assisted most when completing the tasks in the test script? (circle all that apply)
  - What features of IMAGE-SLEUTH made it difficult to complete the tasks in the test script? (circle all that apply)

- **Long Response**
  - Which features of IMAGE-SLEUTH could be used to improve the image browsing experience in the future, and why?
  - In your own words, describe the 4 main components of the IMAGE-SLEUTH interface and what they do. Name the 3 different types of searches and what they do?
  - Do you understand what the Remove(up) and Include(Down) sections mean in IMAGE-SLEUTH?
  - In your own words, please describe what the Include and Remove sections allowed you to do, and comment on whether or not this tool helped you to complete the allocated tasks in the test script.
  - Could you see this application being used in the real world? If so, where?
  - Do you have any other comments?

Fig. 7. Complete list of questions on the survey

Of the respondents, 24 had previously used image management programs. Of these, 17 stated that they sorted their images based on specific criteria and over half the testers had some familiarity with *The Sims 2™*. Not surprisingly, all but 3 testers found that they could complete the test script easier with IMAGE-SLEUTH than with Windows Explorer and 23 testers stated that they felt more comfortable using IMAGE-SLEUTH. All testers believe that IMAGE-SLEUTH was better than Windows Explorer for browsing images and all but 2 testers thought that IMAGE-SLEUTH had advantages over photo browsing applications they had encountered.

Question 6 asked participants to rank the ease of task completion for both Windows Explorer and IMAGE-SLEUTH. It can be seen in Fig. 8 that most subjects found IMAGE-SLEUTH easier to use. On a scale to 10, the average for IMAGE-SLEUTH is 7.3, while Windows Explorer's average is 3.7; almost half that of IMAGE-SLEUTH.

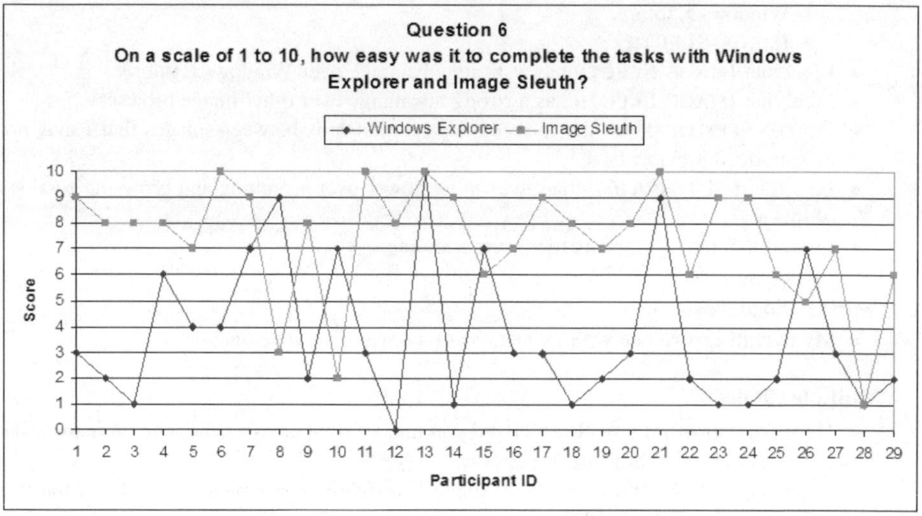

Fig. 8. Comparison of the ease of task completion between Windows Explorer and IMAGE-SLEUTH

Question 12 asked "*What features of* IMAGE-SLEUTH *did you find assisted most when completing the tasks in the test script? (circle all that apply)*". Results (shown in Fig. 9) indicated that the 'include/remove' and attribute search controls were found to be most useful.

Many participants were unhappy with the IMAGE-SLEUTH interface, claiming it appeared primitive and difficult to navigate with the search functions at the bottom of the page (See Fig.10). Another frequently mentioned negative was the accuracy of the color property. Participants did not seem to agree with some of the colors that were returned by IMAGE-SLEUTH for some images and suggested the inclusion of a color palette (or legend) so that testers could identify by label the color they were searching for.

Of the positive comments, the most common involved the include and remove controls, the ability to find specific images quickly and the consistency in design. Testers found it extremely useful to be able to remove single attributes rather than having to perform new searches from scratch.

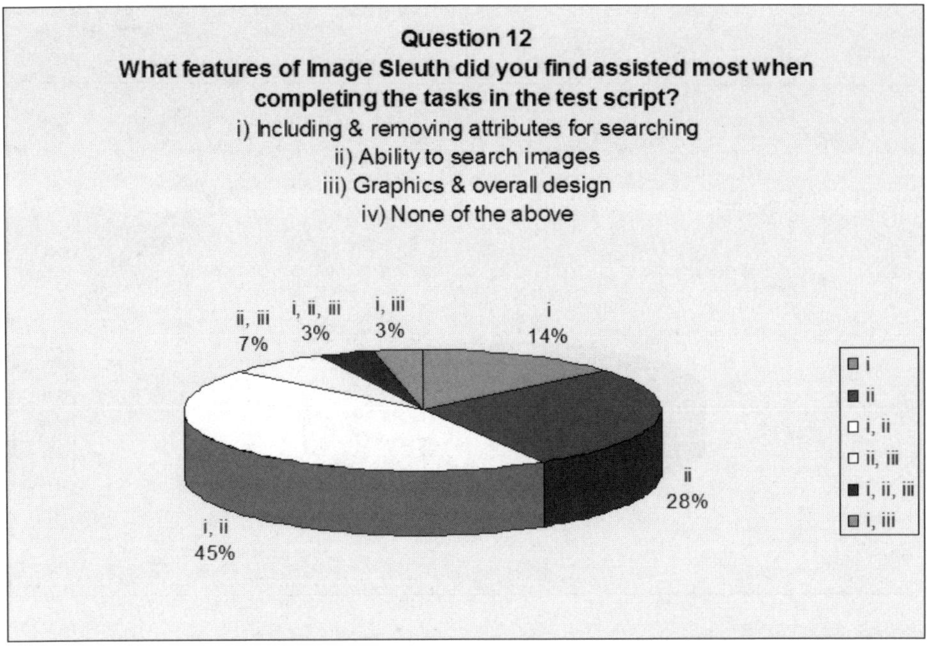

Fig. 9. The most useful features of IMAGE-SLEUTH

Participants for the most part understood the navigation paradigm very well, some even using the terms '*narrow*" and "*move down*" to describe the include control and "*broaden*" and "*move up a level*" to describe the remove control. Other testers felt that the remove was more like the 'Back' button in a Web browser that allowed you to navigate back in multiple ways, for instance, "*...allowed me to reverse a search term without having to go back and redo the search again*" and "*... similar to the function of back in Internet Explorer. The difference is you can choose which step you have before (sic) easily*".

Review of interaction logs found that the most common pattern for participants was to start each task with a term search in the appropriate sub-lattice (color, shape or game play) and then navigate from this concept using the include/remove controls. This method was appropriate but showed an inherent flaw in the term search approach. When attributes are used in a term search and there is no object with all the attributes IMAGE-SLEUTH returns the empty-extent concept. This appeared to leave participants confused, and many reported that the task was unsolvable and moved on to the next task. It is an observation that argues for some form of approximate matching when the result set is empty (discussed below in Section 6).

5.4 Usability Conclusions

The 'include/remove' controls for navigation was very successful, further 'term searching' was liked by participants but sometimes caused empty results which led to con-

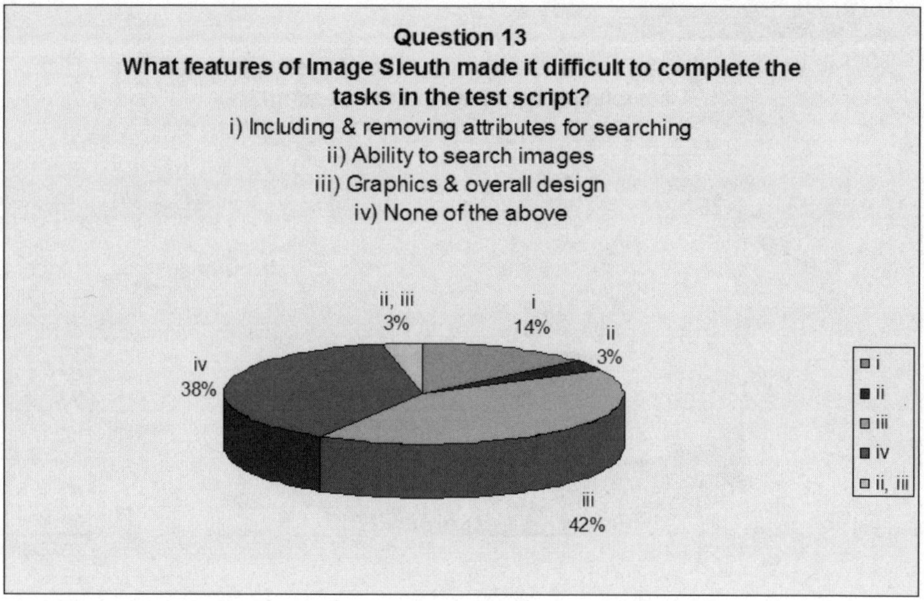

Fig. 10. Features difficult to use in IMAGE-SLEUTH

fusion. Testers responded well to the idea that images were 'grouped' at all times, the groupings being concept extents. Many participants commented on the ease-of-use of IMAGE-SLEUTH, or the fact they had understand the functionality of the software quickly. Most understood or had an idea about how each interaction affected the state of their navigation. This indicates that the navigation paradigm is intuitive.

6 Applying Usability Results

After the analysis of the usability results work on a second version of IMAGE-SLEUTH began. The aims of the second version were to address the problems and issues revealed in the usability testing.

A significant change is to allow overlapping sub-contexts, to combine attributes from color, shape and game play in a more fluid way, so that a more dynamic notion of the sub-lattices that could be created and navigated emerges. In the first version, there were three exclusive contexts concentrating on different facets of the information space. This was changed to one context, with a set of 'perspectives' (conceptual scales) over the formal context. These perspectives can then be used singularly or in combination, and added/removed as necessary during use of the system. This reduces the restrictive nature that separate sub-contexts caused and allows users to see all attributes pertinent to their navigation needs. The original contexts were "Item Properties", "Color Properties" and "Edge Properties", but IMAGE-SLEUTH2 [6] has 10 perspectives;

[6] IMAGE-SLEUTH2 can be trialed by visiting *http://130.130.112.18/jon/test/framebuilder.exe*

- Simplecolors (16 color set.)
- Advancedcolors (216 color set + Simplecolors. Equivalent to "color Properties")
- Needs
- Skills
- Price
- Function
- RoomType
- NeedsAndSkills (combination of Needs and Skills.)
- Gameplay (all game play related attributes. Equivalent to "Item Properties".)
- SimpleGameplay (same as GamePlay, excluding the lowest level of detail.)

This use of a library of concept scales (as perspectives or view over the image collection) means sub-contexts, and concept lattices derived from them, can be drawn from any combination of the scales (e.g. Advancedcolors and SimpleGameplay). A screenshot showing IMAGE-SLEUTH2 is shown in Fig. 11.

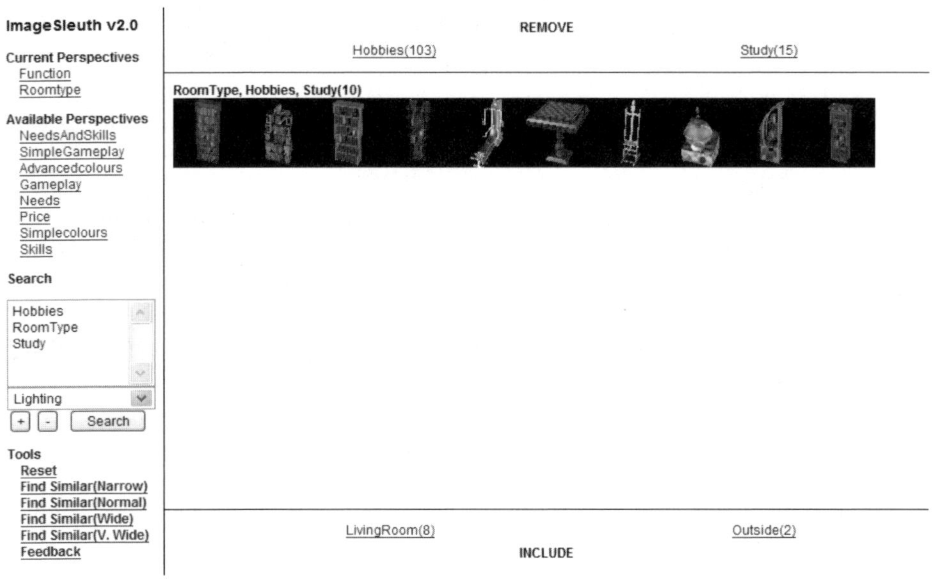

Fig. 11. An example screenshot of IMAGE-SLEUTH2 showing the neighbourhood of the formal concept represented by the images with exactly the attributes Hobbies and Roomtype=study. The scales (or perspectives) that condition for lattice are Function and Roomtype.

The IMAGE-SLEUTH interface received criticism from participants, primarily focusing on poor organisation. To address this IMAGE-SLEUTH2, while still browser-based, has fixed positions for each interface component.

One method for dealing with the return of empty-extents from term search is to provide users with a list of the terms entered so that they can be incrementally removed terms to remove search constraints. Another method is to apply a vector space model of

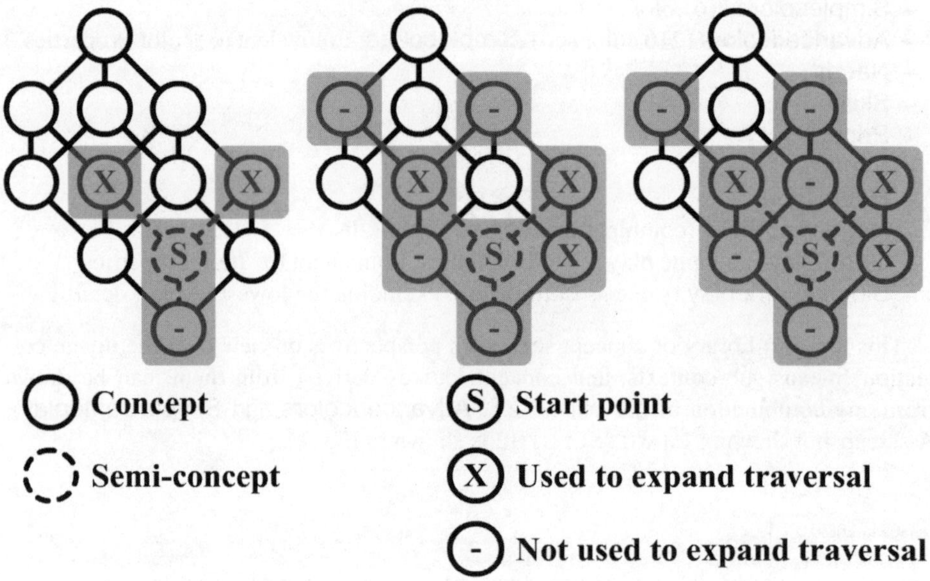

Fig. 12. An example of lattice traversal starting from a semi-concept. The traversal in this example is complete in 3 steps. The shaded area shows the computed concepts at each step.

MPEG-7 images [12] and then apply similarity measures for multi-dimensional feature spaces. IMAGE-SLEUTH2 explores a different approach by using variations on defined distance [13] and similarity [14] metrics in the FCA literature in order to find relevant concepts.

The similarity metric we applied uses the size of the common objects and attributes of the concepts. For two concepts (A, B) and (C, D):

$$similarity((A, B), (C, D)) := \frac{1}{2}\left(\frac{|A \cap C|}{|A \cup C|} + \frac{|B \cap D|}{|B \cup D|}\right).$$

The distance metric uses the size of the total overlap of the intent and extent normalised against the total size of the context. For two concepts (A, B) and (C, D):

$$distance((A, B), (C, D)) := \frac{1}{2}\left(\frac{|A \setminus C| + |C \setminus A|}{|G|} + \frac{|B \setminus D| + |D \setminus B|}{|M|}\right).$$

When a search contains attributes which are not manifest in a single object, IMAGE-SLEUTH2 creates a semi-concept with the searched terms as the intent. This semi-concept is used to prime a traversal of the lattice structure – the traversal applies the distance and similarity metrics to calculate a relevance score. The traversal is bounded by a maximum distance (see Fig. 12). The user is shown the most relevant concepts (with objects as thumbnails) allowing users to decide the concept that best matches their search. This relevance ranking this traversal method is accessible from any concept to find closely matching concepts. A screenshot of results is shown in Fig. 13. This

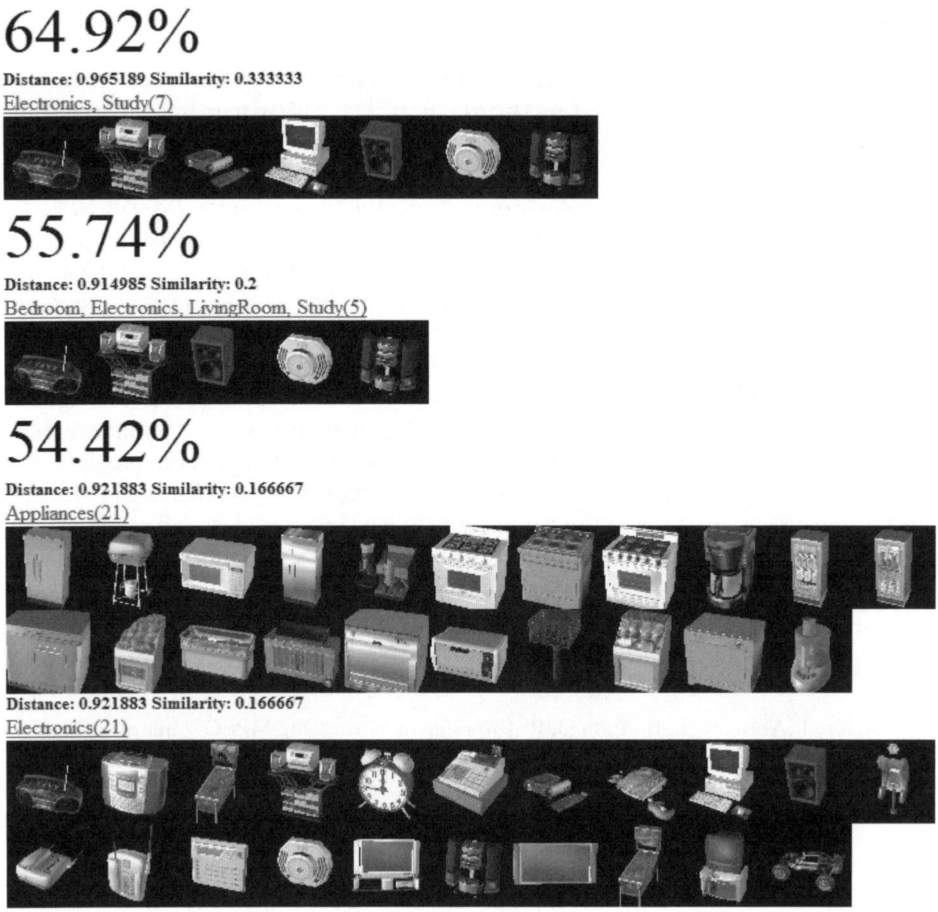

Fig. 13. Results of a concept traversal from the query "*Appliances, Electronics, Study*" using the perspectives "*Function, RoomType*". *Appliances* and *Electronics* are mutually exclusive.

provides a powerful tool for finding similar concepts and objects from a given starting concept.

7 Future Directions

IMAGE-SLEUTH2 is currently being extended to utilise DVD information collected from Amazon's Web store. This allows a DVD to be represented by the front cover of its case, and attributes to be created from the accompanying details (e.g. genre, actor, director, etc). Performing a relevance queries on a DVD allows users to find closely related DVD's based on whichever facets of the data considered important. The architecture of that tool, called DVDSLEUTH, is different from IMAGE-SLEUTH2 because the context is dynamic and grows in various ways depending on the directions taken in the navigation in the Amazon catalog.

8 Conclusion

The design theory underlying a Web-based FCA system for browsing and searching MPEG-7 images was introduced in Ducrou et al. [1]. This paper has presented the evaluation of an architecture and implementation of a browsing and search interface for MPEG-7 images that exploits concept lattices for the representation and navigation of image collections. Sub-contexts provide the foundation for the IMAGE-SLEUTH system, each representing a different search view: one for image shape, another for color and luminance, and a third for semantic content. In this way the initial IMAGE-SLEUTH would navigate over three concept lattices. In the subsequent versions of IMAGE-SLEUTH, a library of conceptual scales (called perpectives) are introduced to allow the more fluid creation of different concept lattices for navigation. The main result of the usability study is it confirms the suitability of the concept lattice as a navigation paradigm for image browsing. We also demonstrate how distance and similarity measures within the concept lattice can be used for approximate matching when search terms do not result in a precise match to a formal concept. The experience with the iterative development of IMAGE-SLEUTH has lead to new insights in search using concept lattices that are being realised for the creation of dynamic contexts and the navigation of Web content.

References

1. Ducrou, J., Vormbrock, B., Eklund, P.: Browsing and Searching MPEG-7 images using Formal Concept Analysis. In: Proceedings of IASTED International Conference on Artificial Intelligence and Applications, ACTA Press (2006)
2. Ganter, B., Wille, R.: Formal Concept Analysis: Mathematical Foundations. Springer, Berlin (1999)
3. Vogt, F., Wille, R.: TOSCANA - a graphical tool for analyzing and exploring data. In: Tamassia, R., Tollis, I(Y.) G. (eds.) GD 1994. LNCS, vol. 894, pp. 226–233. Springer, Heidelberg (1995)
4. Kollewe, W., Skorsky, M., Vogt, F., Wille, R.: TOSCANA - ein werkzeug zur begrifflichen analyse und erkundung von daten. In: Begriffliche Wissensverarbeitung: Grundfragen und Aufgaben, pp. 267–288 (1994)
5. Ohm, J.R.: The MPEG-7 visual description framework - concepts, accuracy, and applications. In: Skarbek, W. (ed.) CAIP 2001. LNCS, vol. 2124, Springer, Heidelberg (2001)
6. Sutanto, D., Leung, C.H.C.: Automatic Index Expansion for Concept-Based Image Query. In: Huijsmans, D.P., Smeulders, A.W.M. (eds.) VISUAL 1999. LNCS, vol. 1614, pp. 399–408. Springer, Heidelberg (1999)
7. Lux, M.: Caliph & emir: Mpeg-7 photo annotation & retrieval (accessed 30/08/05, 2005), `http://caliph-emir.sourceforge.net`
8. Cieplinski, L.: MPEG-7 Color Descriptors and Their Applications. In: Skarbek, W. (ed.) CAIP 2001. LNCS, vol. 2124, Springer, Heidelberg (2001)
9. Won, C.S.: Feature Extraction and Evaluation Using Edge Histogram Descriptor in MPEG-7. In: Aizawa, K., Nakamura, Y., Satoh, S. (eds.) PCM 2004. LNCS, vol. 3333, pp. 583–590. Springer, Heidelberg (2004)
10. Park, D.K., Jeon, Y.S., Won, C.S.: Efficient use of local edge histogram descriptor. In: MULTIMEDIA 2000: Proceedings of the 2000 ACM workshops on Multimedia, pp. 51–54. ACM Press, New York (2000)

11. Rozanski, E.P., Haake, A.R.: The many facets of hci. In: CITC4 2003: Proceedings of the 4th conference on Information technology curriculum, pp. 180–185. ACM Press, New York (2003)
12. Park, K.W., Lee, D.H.: Full-automatic high-level concept extraction from images using ontologies and semantic inference rules. In: Mizoguchi, R., Shi, Z.-Z., Giunchiglia, F. (eds.) ASWC 2006. LNCS, vol. 4185, pp. 307–321. Springer, Heidelberg (2006)
13. Lengnink, K.: Ähnlichkeit als Distanz in Begriffsverbänden. In: Stumme, G., Wille, R. (eds.) Begriffliche Wissensverarbeitung: Methoden und Anwendungen, pp. 57–71. Springer, Heidelberg (2000)
14. Saquer, J., Deogun, J.: Concept approximations based on rough sets and similarity measures. Int. J. Appl. Math. Comput. Sci. 11, 655–674 (2001)

First Elements on Knowledge Discovery Guided by Domain Knowledge (KDDK)

Jean Lieber, Amedeo Napoli, Laszlo Szathmary, and Yannick Toussaint

LORIA (CNRS – INRIA – Universités de Nancy)
Équipe Orpailleur, Bâtiment B, BP 239
F-54506 Vandœuvre-lès-Nancy cedex, France
{''FirstName''.''LastName''}@loria.fr

Abstract. In this paper, we present research trends carried out in the Orpailleur team at LORIA, showing how knowledge discovery and knowledge processing may be combined. The knowledge discovery in databases process (KDD) consists in processing a huge volume of data for extracting significant and reusable knowledge units. From a knowledge representation perspective, the KDD process may take advantage of domain knowledge embedded in ontologies relative to the domain of data, leading to the notion of "knowledge discovery guided by domain knowledge" or KDDK. The KDDK process is based on the classification process (and its multiple forms), e.g. for modeling, representing, reasoning, and discovering. Some applications are detailed, showing how KDDK can be instantiated in an application domain. Finally, an architecture of an integrated KDDK system is proposed and discussed.

1 Introduction

In this presentation, we present research trends carried out within in the Orpailleur team at LORIA, showing multiple aspects of knowledge discovery and knowledge processing. The knowledge discovery in databases process –hereafter KDD– consists in processing a huge volume of data in order to extract knowledge units that are significant and reusable. Assimilating knowledge units to gold nuggets, and databases to lands or rivers to be explored, the KDD process can be likened to the process of searching for gold (in the same way, KDD is compared with archeology in [7]). This explains the name of the research team: the "orpailleur" denotes in French a person who is searching for gold in rivers or mountains. Moreover, the KDD process is iterative, interactive, and generally controlled by an expert of the data domain, called the *analyst*. The analyst selects and interprets a subset of the extracted units for obtaining knowledge units having a certain plausibility. As a person searching for gold and having a certain knowledge of the task and of the location, the analyst may use its own knowledge but also knowledge on the domain of data for improving the KDD process. Indeed, the objective of this paper is to show the role that can be played by domain knowledge within the KDD process.

From an operational point of view, the KDD process is performed within a KDD system including databases, data mining modules, and interfaces for interactions,

S. Ben Yahia et al. (Eds.): CLA 2006, LNAI 4923, pp. 22–41, 2008.

Rough Data, databases
↓ Domain understanding
↓ Data selection (windowing)
Selected data
↓ Cleaning / Preparation
Prepared data
↓ Data mining process (discovering patterns)
↓ Numerical and symbolic KDD methods
Discovered patterns
↓ Post-processing of discovered patterns
↓ Interpretation / Evaluation
Knowledge units for knowledge systems and problem-solving

Fig. 1. From data to knowledge units: the objective of the knowledge discovery process is to select, prepare and extract knowledge units from different data sources. For effective reuse, the extracted knowledge units have to be represented within an adequate knowledge representation formalism.

e.g. editing and visualization. The KDD process is based on three main operations: selection and preparation of the data, data mining, and finally interpretation of the extracted units (see Figure 1).

A way for the KDD process to take advantage of domain knowledge is to be in connection with an *ontology* relative to the domain of data, a step towards the notion of *knowledge discovery guided by domain knowledge* or KDDK. In the KDDK process, knowledge units that are extracted have still a life after the interpretation step: they must be represented in an adequate knowledge representation formalism for being integrated within an ontology and reused for problem-solving needs. In this way, the results of the knowledge discovery process may be reused for enlarging existing ontologies. The KDDK process shows that knowledge representation and knowledge discovery are two complementary tasks: *no effective knowledge discovery without domain knowledge!*

Hereafter, we present various instantiations of the KDDK process that are all based on the idea of *classification*. Classification is a polymorphic process involved in various tasks, e.g. modeling, mining, representing, and reasoning (see also [42,10,53]). Accordingly, a knowledge-based system may be designed, fed up by the KDDK process, and used for problem-solving in application domains, e.g. agronomy, astronomy, biology, chemistry, and medicine (these application domains are studied in the Orpailleur team). A special mention has to be made for Semantic Web activities, involving in particular text mining, content-based document mining, and intelligent information retrieval (see for example [16,8,41]).

The paper is organized as follows. In the next section, symbolic methods for KDD and the CORON platform are introduced. Then, research trends in KDDK are presented and detailed, showing how knowledge can be embedded at each step of the KDD process. In the last section, an architecture for an integrated

KDDK system is described, and the KDD and KDDK processes are studied with respect to this integrated architecture.

2 Methods and Systems for KDD

The KDD process is based on *data mining methods* that are either symbolic or numerical [19,20,14]. The methods that are used in the Orpailleur team are the following (mainly symbolic methods):

- Symbolic methods based on lattice-based classification (concept lattice design or formal concept analysis [18]), frequent itemsets search, and association rule extraction [35]. These symbolic methods are more deeply described in the next subsection.
- Numerical methods based on second-order Hidden Markov Models (HMM2, initially designed for pattern recognition) [30,29]. Hidden Markov Models have good capabilities for locating stationary segments, and are mainly used for mining temporal and spatial data. The CAROTTAGE system[1] is developed in the Orpailleur team for analyzing numerical spatio-temporal data.

In the following, the focus is on symbolic KDD methods. However, an ongoing research work holds on the combination of symbolic and numerical methods, that is discussed in section 3.4.

2.1 Lattice Design, Itemset Search and Association Rule Extraction

Classification problems can be formalized by means of a class of individuals (or objects), a class of properties (or attributes), and a binary correspondence between the two classes, indicating for each individual-property pair whether the property applies to the individual or not [3,18,8]. The properties may be features that are present or absent, or the values of a property that have been transformed into binary variables. Lattice-based classification relies on the analysis of such binary tables and may be considered as a symbolic data mining technique to be used for extracting (from a binary database) a set of concepts organized within a hierarchy (i.e. a partial ordering). The extraction of frequent itemsets, i.e. sets of properties or features of data occurring together with a certain frequency, and of association rules emphasizing correlations between sets of properties with a given confidence, are related activities.

The search for frequent itemsets and association rule extraction are well-known symbolic data mining methods. These processes usually produce a large number of items and rules, leading to the associated problems of "mining the sets of extracted items and rules". Some subsets of itemsets, e.g. frequent closed itemsets (FCIs), allow to find interesting subsets of association rules, e.g. informative association rules. This is why several algorithms are needed for mining data depending on specific applications.

[1] CAROTTAGE is a free software developed in the Orpailleur team, with a GPL license since 2002, see `http://www.loria.fr/~jfmari/App/`

2.2 Rare Itemsets and Rules

Among useful patterns extracted from a database, frequent itemsets are usually thought to unfold "regularities" in the data, i.e. they are the witnesses of recurrent phenomena and they are consistent with the expectations of the domain experts. In some situations however, it may be interesting to search for "rare" itemsets, i.e. itemsets that do not occur frequently in the data (contrasting frequent itemsets). These correspond to unexpected phenomena, possibly contradicting beliefs in the domain. In this way, rare itemsets are related to "exceptions" and thus may convey information of high interest for experts in domains such as biology or medicine. For example, suppose an expert in biology is interested in identifying the cause of cardiovascular diseases (CVD) for a given database of medical records. A frequent itemset such as "{elevated cholesterol level, CVD}" may validate the hypothesis that these two items are frequently associated, leading to the possible interpretation "people having a high cholesterol level are at high risk for CVD". On the other hand, the fact that "{vegetarian, CVD}" is a rare itemset may justify that the association of these two itemsets is rather exceptional, leading to the possible interpretation "vegetarian people are at a low risk for CVD". Moreover, the itemsets {vegetarian} and {CVD} can be both frequent, while the itemset {vegetarian, CVD} is rare.

Rare cases deserve special attention because they represent significant difficulties for data mining algorithms. The underlying mining problems have been studied in detail, with different names, e.g. exceptions, negative rules (see for example [28,43,52,54,37]. These approaches are, most of the time, based on adaptions of the general levelwise Apriori algorithm. These methods typically retrieve large sets of rare itemsets and association rules, but these methods may remain incomplete –rare associations are be discovered– either due to an excessive computational cost or to overly restrictive definitions. Thus, such methods may fail to collect a large number of potentially interesting patterns.

By contrast, a framework is proposed in [45,44,48] is specifically dedicated to the extraction of rare itemsets. It is based on an intuitive yet formal definition of rare itemset. Its goal is to provide a theoretical foundation for rare pattern mining, with definitions of reduced representations and complexity results for mining tasks, as well as to develop an algorithmic tool suite (within the Coron platform, see next subsection) together with the guidelines for its use. The method, for computing all rare itemsets is based on two main steps. The first step thereof is the identification of the *minimal rare itemsets* with an optimized method that limits the exploration to frequent generators only (minimal rare itemsets jointly act as a minimal generation seed for the entire rare itemset family). The second step is performed to restore all rare itemsets from minimal rare itemsets.

2.3 The Coron Platform

The Coron platform[2] is currently developed in the Orpailleur team [46,47]. The platform is composed of three main modules: (i) Coron-base, (ii) AssRuleX,

[2] `http://coron.loria.fr`

(iii) pre-processing and post-processing modules. The CORON-base module is aimed at extracting different kinds of itemsets, e.g. frequent itemsets, frequent closed itemsets, minimal generators, etc. The module contains a collection of important data mining algorithms, such as Apriori, Close, Pascal, Titanic, Charm, Eclat, together with adapted algorithms such as Zart and Eclat-Z (plus some others). This large collection of (efficient) algorithms is one of the main characteristics of the CORON platform. Knowing that each of the algorithms has advantages and disadvantages with respect to the form of the data to be mined, and since there is no universal algorithm for processing any arbitrary dataset, the CORON-base module offers to the user the choice of the algorithm that is the best suited for his needs.

The second module of the system, ASSRULEX (Association Rule eXtractor) generates different sets of association rules, such as informative rules, generic basis, and informative basis.

For supporting the whole life-cycle of a data mining task, the CORON platform proposes modules for cleaning the input dataset and reduce its size if necessary. The module RULEMINER facilitates the interpretation and the filtering of the extracted rules. The association rules can be filtered by (i) attribute, (ii) support, and/or (iii) confidence.

The CORON platform is developed entirely in Java, allowing portability. The system is operational, and has been tested within several research projects within the team [12,31].

2.4 A Data-Mining Methodology with the Coron Platform

A methodology was initially designed for mining biological cohorts, but it can be generalized to any kind of database. It is worth to mention that the whole KDDK process is guided by an analyst. The role of the analyst is important with respect to the following tasks: selecting the data and interpreting the extracted units. This methodology is associated to the CORON platform, that offers various tools necessary for its application in a single platform (nevertheless another platform can be used).

The methodology consists of the following steps: **(1)** Definition of the study framework, **(2)** Iterative step: data preparation and cleaning, pre-processing step, processing step, post-processing step, validation of the results and Generation of new research hypotheses, feedback on the experiment. The life-cycle of the methodology is shown in Figure 2.

Definition of the Study Framework.
The analyst defines a specific field for the analysis (called hereafter "framework"). Thus, he may choose the type of data he wants to work on, e.g. biological data, genetic data, or both, unrelated individuals or families, focus on a special metabolic network or on a particular syndrome.

Iterative Step.
Data preparation and cleaning. Data cleaning is necessary. This step includes the detection and the possible removal of incomplete and out-of-range

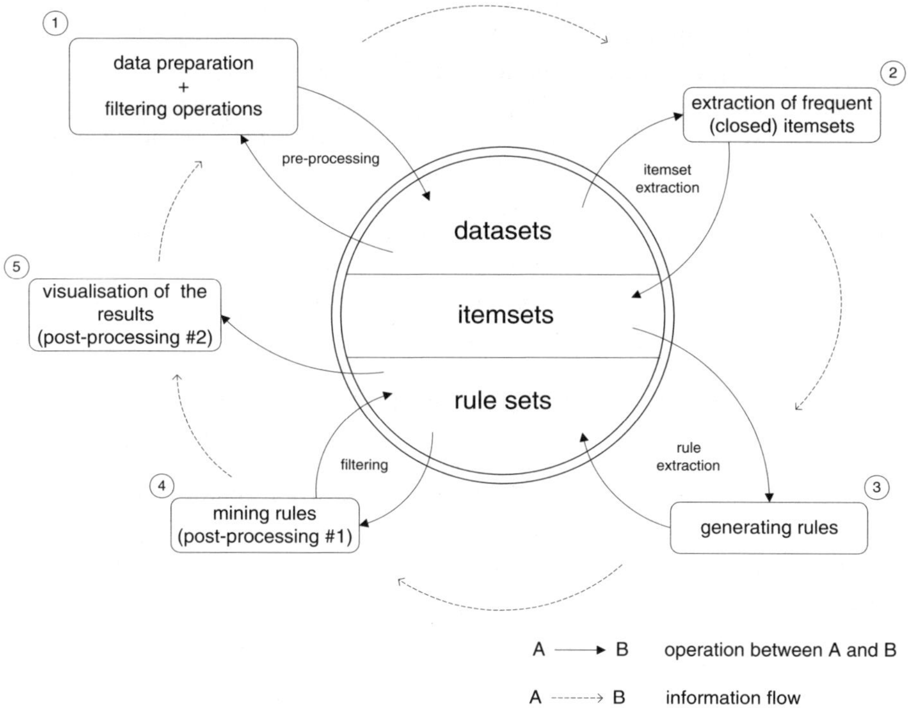

Fig. 2. The life cycle of KDD within Coron

values. Moreover, several actions for converting the data can be done at this step, such as:

(1) *Addition/creation* of new attributes for helping the extraction of association rules by combining attributes (intersection, union and complementary).

(2) *Deletion* of attributes that are not interesting in the chosen biological framework. This option is close to the projections described below.

(3) *Discretization* for transforming continuous data into Boolean values, e.g. by using a threshold defined in the literature, or by separating values of each continuous variable into quartiles.

Data filtering (pre-processing). Several actions can be carried out that correspond to operations in set theory: complement, union and intersection (with operations of additions and projections).

(1) *Apply projections*:

on the rows: i.e. selecting individuals with one or more attributes specified by the expert,

on the columns: i.e. selecting (or deleting) some attributes.

(2) *Consider the complement* of a set of individuals satisfying a rule, defined by the set of individuals that do not satisfy this rule.

The output of the filtering process is considered as a new dataset on which data mining procedures can be applied again.

Applying the data mining procedure. This methodology is related to symbolic data mining methods, as, in particular, frequent itemset search and association rule extraction. With the help of the analyst, the necessary thresholds values can be set for quality measures such as the minimum support and the minimum confidence for generating frequent itemsets and association rules, respectively. As the process is iterative and interactive, the analyst can change these thresholds during a next iteration to carry out new experiments.

Post-processing. After filtering and visualizing the rules, those rules containing the most interesting attributes can be found. If a less relevant attribute is always present in the rules, it can be considered as "noisy", and removed from the input dataset. This means that the dataset is another time modified for a new association rule extraction.

The iterative step can be repeated until the most relevant rules are found. The interpretation of the analyst is mobilized both for rule mining and result visualization.

Rule-mining. In the rule mining step, the analyst has also to make several choices:

- *Choosing rules with a specific form*: e.g. selecting rules that only have one attribute on their left side.
- *Selecting rules with an attribute of interest* from the point of view of the analyst, on the left hand side, on the right hand side, or on both sides.
- *Classifying the extracted rules* in ascending or descending order according to their support or confidence values, or according to other statistical values [9].
  The classification of rules mining step may be dependent on numerical measures, e.g. support and confidence, or on domain knowledge as shown in some experiments [21].
- *Selecting rules* with a support belonging to a given interval $[a, b]$; returning rules with a support less than (or more than) or equal to a given value c. These selections can also be applied with the other statistical measures cited above.

Visualization of the results. A visualization method adapted to symbolic data mining method procedure has to be chosen. For frequent itemset search leading to the extraction of less frequent itemsets, concept lattices may be used beneficially [22].

Validation of the results and generation of new research hypotheses. The evaluation of the rules can be done either by statistical tests, data analysis methods, i.e. automatic classification, component analysis, or with knowledge-based methods, e.g. classification-based reasoning, formal concept analysis. The generated results allow the expert to suggest new directions of research. Accordingly, these new hypotheses are tested by new experiments, for example, managed at the biological level, like genetic epidemiological studies or wet laboratory experiments.

3 Research Directions for KDDK

The principle summarizing KDDK can be read as follows: going "from complex data units to complex knowledge units guided by domain knowledge" (KDDK) or "knowledge with/for knowledge". This principle is discussed below, along research activities such as graph mining, spatio-temporal data mining, text mining and Semantic Web, knowledge discovery in life sciences, combining symbolic and numerical data mining methods for hybrid mining, and finally mining a knowledge base, a kind of "meta-knowledge discovery process". All these research activities share the fact that the mining process is guided and enhanced by domain knowledge (similar ideas are also discussed in [11,53]).

3.1 KDDK and the Mining of Complex Data

Lattice-based classification, formal concept analysis, itemset search and association rule extraction, are suitable paradigms for symbolic KDDK, that may be used for real-sized applications [51]. Global improvements may be carried on the ease of using of the data mining methods, on the efficiency of the methods [24], and on adaptability, i.e. the ability to fit evolving situations with respect to the constraints that may be associated with the KDDK process. Accordingly, the research work presented hereafter is in concern with the extension of symbolic methods to complex data, e.g. objects with multi-valued attributes, relations, graphs, texts, and real world data.

KDDK in databases of chemical reactions.
The mining of chemical chemical reaction databases is an important task for at least two reasons (see also [23]): (i) the first reason is the challenge represented by this task regarding KDDK to be set on, (ii) the second reason lies in the industrial needs that can be met whenever substantial results are obtained. Chemical reactions are complex data, that may be modeled as undirected labeled graphs. They are the main elements on which synthesis in organic chemistry relies, knowing that synthesis —and accordingly chemical reaction databases— are of first importance in chemistry, but also in biology, drug design, and pharmacology. From a problem-solving point of view, synthesis in organic chemistry must be considered at two main levels of abstraction: a strategic level where general synthesis methods are involved –a kind of meta-knowledge– and a tactic level where specific chemical reactions are applied. An objective for improving computer-based synthesis in organic chemistry is aimed at discovering general synthesis methods from currently available chemical reaction databases for designing generic and reusable synthesis plans.

A preliminary research work has been carried on in the Orpailleur team [5], based on frequent levelwise itemset search and association rule extraction, and applied to standard chemical reaction databases. This work has given substantial results for the expert chemists. At the moment, for extending this first work, a graph-mining process is used for extracting knowledge from chemical reaction databases, directly from the molecular structures and the reactions themselves, This research work is currently under development, in collaboration with chemists, and is in accordance with needs of chemical industry [38].

KDDK and the mining of spatio-temporal data.
Temporal and spatial data are complex data to be mined because of their internal structure, that can be considered as multi-dimensional. Indeed, spatial data may involve two or three dimensions for determining a region and complex relations as well for describing the relative positions of regions between each others (as in the RCC-8 theory for example [26,36]). Temporal data may present a linear but also a two-dimensional aspect, when time intervals are taken into account and have to be analyzed (using Allen relations for example). In this way, mining temporal or spatial data are tasks related to KDDK. Spatial and temporal data may be analyzed with numerical methods such as Hidden Markov Models, but also with symbolic methods, such as levelwise search for frequent sequential or spatial patterns.

In the medical domain, the study of chronic diseases is a good example of KDDK process on spatio-temporal data. An experiment for characterizing the patient pathway using the extraction of frequent patterns, sequential and not sequential, from the data of the PMSI[3] system associated with the "Lorraine Region" is currently under investigation. Details on this work are given in [22].

3.2 KDDK, Text Mining and Semantic Web

KDDK and text mining.
The objective of a text mining process is to extract new and useful knowledge units in a large set of texts [21,10,9]. The text mining process shows some specific characteristics due to the fact that texts are complex objects written in natural language. The information in a text is expressed in an informal way, following linguistic rules, making the mining process more complex. To avoid information dispersion, a text mining process has to take into account –as much as possible– paraphrases, ambiguities, specialized vocabulary, and terminology. This is why the preparation of texts for text mining is usually dependent on linguistic resources and methods. In addition, from a KDDK perspective, the text mining process is aimed at extracting new knowledge units from texts with the help of background knowledge. The interpretation of a text relies on knowledge units shared by the authors and the readers. A part of these knowledge units is expressed in the texts and may be extracted by the text mining process. Another part of these knowledge units, background knowledge, is not explicitly expressed in the text and is useful to relate notions present in a text, to guide and to help

[3] For "Programme de Médicalisation des Systèmes d'Informations". This is the name of the information system collecting the administrative data for an hospital.

the text mining process. Background knowledge is encoded in a knowledge base associated to the text mining process. Text mining is especially useful in the context of semantic Web, for manipulating textual documents by their content.

The studies on text mining carried out in the Orpailleur team hold on real-world texts in application domains such as astronomy, biology and medicine, using mainly symbolic data mining methods such as i.e. frequent itemset search and association rule extraction [4]. This is in contrast with text analysis approaches dealing with specific language phenomena. The language in texts is considered as a way for presenting and accessing information, and not as an object to be studied for its own. In this way, the text mining process may be involved in a loop used to enrich and to extend linguistic resources. In turn, linguistic and ontological resources can be exploited to guide a "knowledge-based text mining process".

KDDK within the context of Semantic Web.
Semantic Web constitutes a good platform for experimenting ideas on knowledge discovery –especially text mining–, knowledge representation and reasoning. In particular, the knowledge representation language associated with the Semantic Web is the OWL language, based on description logics (or DLs, see [2]). In OWL, knowledge units are represented within concepts (or classes), with attributes (properties of concepts, or relations, or roles), and individuals. The hierarchical organization of concepts (and relations) relies on a subsumption relation that is a partial ordering. The inference services are based on subsumption, concept and individual classification, two tasks related to "classification-based reasoning". Concept classification is used for inserting a new concept at the right location in the concept hierarchy, searching for its most specific subsumers and its most general subsumees. Individual classification is used for recognizing the concepts an individual may be an instance of. Furthermore, classification-based reasoning may be extended into case-based reasoning (CBR), that relies on three main operations: retrieval, adaptation, and memorization. Given a target problem, retrieval consists in searching for a source (memorized) problem similar to the target problem. Then, the solution of the source problem is adapted to fulfill the constraints attached to the target problem. When there is enough interest, the target problem and its solution may be memorized in the case base to be reused. In the context of a concept hierarchy, retrieval and adaptation may be both based on classification (and "adaptation-guided retrieval" [17]).

In the framework of Semantic Web, the mining of textual documents on the Web, or "Web document mining" [6], can be considered from two main points of view: (i) mining the content of documents, involving text mining, (ii) mining the internal and external –hypertext links– structure of pages, involving information extraction. Web document mining is a major technique for the semi-automatic design of real-scale ontologies, the backbone of Semantic Web. In turn, ontologies are used for annotating the documents, enhancing document retrieval and document mining. In this way, Web document mining improves annotation, retrieval, and the understandability of documents, with respect to their structure and their content. The extracted knowledge units can then be used for completing domain ontologies, that, in turn, guide text mining, and so on.

A research carried on in the team aims at understanding the structure of documents for analyzing and for improving text mining. The design of a system for extracting information units –that have to be turned into knowledge units after interpretation– from Web pages involves a wrapper-based machine learning algorithm combined with a classification-based reasoning process, taking advantage of a domain ontology implemented within the Web Ontology Language (OWL). The elements returned by the process are used as "semantic annotations" for understanding and manipulating the documents with respect to their structure and content [50,49]. The application domain of this research work is the study of research themes in the European Research Community. This study supports the analysis of research themes and detection of research directions.

3.3 KDDK for Life Science: Organizing and Navigating Biological Sources

The application domains that are currently investigated at the moment by the Orpailleur team are related with life sciences, with a particular emphasis on biology (bioinformatics) and medicine. Indeed, there are various reasons explaining why life sciences are a major application domain. In general, life sciences are getting more and more importance as a domain application for computer scientists. In this context, the collaboration between biologists and computer scientists is very active, and the understanding of biological systems provides complex problems for computer scientists. When these problems are solved (at least in part), the solutions bring new ideas not only for biologists but also for computer scientists in their own research work. Thus, advances in research appear on both sides, life and computer sciences.

Knowledge discovery is gaining more and more interest and importance in life sciences for mining either homogeneous databases (DBs) such as protein sequences or structures, heterogeneous DBs for discovering interactions between genes and environment, or between genetic and phenotypic data, especially for public health and pharmacogenomics domains. The latter case appears to be one main challenge in knowledge discovery in biology and involves knowledge discovery from complex data and thus KDDK. The interactions between researchers in biology and researchers in computer science improve not only knowledge about systems in biology, but knowledge about computer science as well. Solving problems for biologists using KDDK methods may involve the design of specific modules that, in turn, leads to adaptations of the KDDK process, especially in the preparation of data and in the interpretation of the extracted units.

A research work carried on in the team is in concern with the search and the access to relevant biological sources (including biological DBs) satisfying a set of given constraints, expressed with respect to concepts lying in a domain ontology –as in the BioRegistry repository [40]. The sources may be described in terms of these concepts, yielding a formal context, from which a concept lattice can be built [32]. Given a specific query, a lattice-based information retrieval process is set on. The classification of the query in the lattice returns a ranked list of relevant sources, according to the characteristics of the sources with respect to the characteristics of the query (see [33]). The next step is to generalize the approach, and to use a "fuzzy concept lattice" and "fuzzy formal concept analysis" (see for example [39]). Moreover, studies hold on complex question answering methods taking into account fuzzy concept lattices, nested queries (intersection, union, and complement), analogical queries, and composition of answers elements. These techniques are still under study.

Another challenge is to extract knowledge from heterogeneous DBs for understanding interactions between clinical, genetic and therapeutic data. For example, a given genotype, i.e. a set of selected gene versions, may explain adverse clinical reactions (e.g. hyperthermy, toxic reaction...) to a given therapeutic treatment. This requires first the integration of both genomic and clinical data into a data warehouse on which KDDK methods have to be applied. This research work is connected with Semantic Web purposes, and in particular with the following elements: (i) data preparation and extracted units interpretation based on domain ontologies, (ii) knowledge edition for building and enriching domain ontologies, (iii) knowledge management for access to knowledge units, querying and reasoning (for problem-solving).

3.4 Combining Symbolic and Numerical Methods for KDDK

Why combining symbolic and numerical methods.
HMM2 have proved to be a valuable tool for extracting knowledge from complex numerical data, e.g. spatio-temporal data. In this way, the CarottAge system has been involved for data mining purposes in two main application domains, namely biology and agronomy. In collaboration with biologists, genome segmentation and interpretation have been investigated [15]. In collaboration with agronomists, spatial and temporal land-use data have been mined for extracting and understanding crop successions, i.e. the way how crops are carried out during a given period of time [25,30]. In these two applications, the effort has focused on two main points, with respect to the questions of the biologists and of the agronomists: (i) the elaboration of a mining process for extracting dependencies in temporal and spatial data involving an unsupervised classification process based on HMM2, (ii) the specification of associated and adequate visualization tools giving a synthetic view of the extraction process results to the experts in charge of interpreting the extracted classes and/or of specifying new experiment directions.

However, some operations remain very difficult to be carried out and could be eased using symbolic methods: (i) the modeling of the HMM2 process for a set of given data, (ii) the interpretation of units extracted by HMM2, (iii) the organization and the visualization of the extracted units for further reuse, e.g. as knowledge units in a knowledge-based system. A proposition is to combine HMM2 with symbolic methods, such as case-based reasoning and concept lattices, for helping the modeling and interpretation process.

A challenge is to set on a methodology for hybrid KDDK, coupling HMM2 and symbolic methods, that can be adapted and reused as a general KDDK method on various data, leading to a multi-functional and multi-purpose KDDK system.

Combining CBR and HMM.
Case-based reasoning seems to be especially interesting since researchers in an application domain often use their own knowledge or knowledge resulting from first experiments to improve steps within the data mining process, e.g. modeling and interpretation. In this way, the elements of the cases within the case-based reasoner can be composed of knowledge units about parameters of the HMM2, and as well of knowledge units on the design –modeling, data preparation–, and the interpretation –relying on ontological knowledge– of the HMM2. In addition, CBR can be of great interest for recording mining strategies that can be adapted and reused in similar situations. Indeed, a study on CBR for guiding mining scenarios in a given situation –with retrieval and adaptation of a similar situation– has not yet been carried on and should give substantial results. More generally, HMM2-based data mining process may take advantage of being coupled with CBR, that can be used at a strategic level for guiding the HMM2-based data mining process.

Combining concept lattices and HMM.
For their part, concept lattices can be used to organize and to visualize the results of the HMM2-based data mining process. The objects resulting of the application of the HMM2 process can be characterized by a set of properties. For example, in a spatio-temporal framework, space regions may be considered as objects and characteristics of the region at a given time can be considered as properties, yielding a kind of formal context. In addition, itemsets and association rules may also be extracted from such a context, offering an easy way of interpreting results of the HMM2 process.

The analysis of complex data in biology also calls for the coupling of symbolic and numerical data mining methods. There are complex data on which HMM2 show a good behavior, for recognizing and extracting regular structures. Such complex data hold on interactions between processes or agents, such as data from transcriptomic biochips –DNA chips or microarrays– experiments (used for extracting knowledge on interactions between plants and microorganisms). Still, an important objective of this kind of study is to investigate and to understand more deeply the modeling of biological systems, at symbolic and numerical levels.

3.5 Meta-knowledge Discovery of Mining Knowledge Bases

The Kasimir system.
The main tasks of the KASIMIR system are decision support and knowledge management for the treatment of cancer. The system is developed within a multidisciplinary research project in which participate researchers from different community (computer science, ergonomics, and oncology). For a given cancer localization, a treatment is based on a protocol similar to a medical guideline. For most of the cases (about 70%), a straightforward application of the protocol

is sufficient and provides a solution, i.e. a treatment, that can be directly reused. A case out of the 30% remaining cases is said to be *out-of-the-protocol*, i.e. either the protocol does not provide a treatment for this medical case, or the proposed solution raises some difficulties, e.g. contraindication, treatment impossibility, etc. For such an out-of-the-protocol case, oncologists try to *adapt* the protocol. In turn, these adaptations can be used to propose *evolutions* of the protocol based on a confrontation with actual cases. The idea is then to make suggestions for protocol evolutions based on frequently performed adaptations.

In knowledge-intensive CBR, the reuse of cases is generally based on adaptation, the goal of which is to solve the target problem by adapting the solution of a source case. The adaptation process is based on adaptation knowledge that –for the main part– is domain-dependent, and thus needs to be acquired for a new application of CBR. Adaptation knowledge plays a key issue in applications, e.g. in knowledge-intensive case-based reasoning systems [1].

In parallel, the Semantic Web technology relies on the availability of large amount of knowledge in various forms [16,41]. The acquisition of ontologies is one of the important issues that is widely explored in the Semantic Web community. Moreover, the acquisition of decision and adaptation knowledge for the Semantic Web has not been so deeply explored, though this kind of knowledge can be useful in numerous situations. For example, given a decision protocol and an adaptation knowledge base, the KASIMIR system can be used to apply and/or to adapt the protocol to specific medical situations.

Semi-automatic acquisition of adaptation knowledge.

The goal of *adaptation knowledge acquisition* (AKA) is to mine a case base, to extract adaptation knowledge units, and to make these units operational. Until now, the research work on CBR in the Orpailleur team has mainly focused on the design of algorithms and knowledge representation formalisms for implementing the adaptation process in a CBR system. A next step is to investigate the AKA process, a research topic that has still not received so much in the CBR community. A parallel research topic is to apply AKA to the extraction of decision knowledge units. Indeed, adaptation knowledge is closely related with decision theory, e.g. the Wald pessimistic criterion is frequently applied when pieces of information about a patient are missing.

Accordingly, the objective of the research work on AKA is to study how KDD techniques can be used for feeding a knowledge server embedded in a semantic portal –as the KASIMIR semantic portal [13]– and thus to instantiate the KDDK process. In the KASIMIR semantic portal, OWL-based formalisms for representing medical ontologies, decision protocols (the case base), and adaptation knowledge, are designed. Web services associated to the CBR process are developed. Several protocols are implemented, with a few of them including adaptation knowledge.

Practically, AKA can be considered from two main points of view. AKA from experts is based on 'manual" analysis of documents related to current problems. The AKA from expert process leads to the elaboration of *adaptation rules*, depending on formal parameters and associated with explanations. The adaptation rules are human-understandable –thanks to explanations– but they need additional knowledge for instantiating the parameters and being applied (more on AKA from experts is given in [27,34]).

Semi-automatic AKA is based on the principles of KDD, and involves data preparation, data mining, and interpretation of the extracted units, under the control of an analyst. The input of the AKA process is a set of adaptations –thus elements at the knowledge level– and the output is a set of adaptation rules. Such an adaptation rule is an operational association rule, that lack explanations. Mixed AKA combines AKA from experts and semi-automatic AKA for supplying operational and human-understandable adaptation knowledge.

In the current experiments within the KASIMIR system, semi-automatic AKA is based on frequent itemset search. A system for AKA, named CABAMAKA–case base mining for AKA, is currently under development within the KASIMIR system and relies on semi-automatic AKA [12]. The CABAMAKA system analyzes a simple representation of the variations $\Delta\mathrm{u}$ between units of knowledge u_1 and u_2, where $\Delta\mathrm{u}$ encodes the substitutions transforming u_1 into u_2. The variations are represented in an expressive DL-based formalism, allowing a high-level expression of the extracted adaptation rules.

Beyond CBR, such a research work can be useful for ontology alignment: an alignment expresses a correspondence between the elements of two ontologies, but it could also express the variations between corresponding elements, within a rich representation formalism for the variations.

4 Towards an Integrated KDDK System

From a global point of view, the research objectives for KDDK can be summarized as follows:

- A methodology for a "knowledge discovery from complex data guided by domain knowledge process" (KDDK), i.e. a process leading from complex data units to complex knowledge units taking advantage of domain knowledge, at each step of the knowledge discovery process.
- A combination of symbolic and numerical data mining methods for setting up a complete and hybrid mining methodology to be applied on various types of data.
- An implementation of the "knowledge discovery from complex data guided by domain knowledge process" within an operational system, to be used on a large set of data types, e.g. textual documents, genomic data, spatio-temporal data, graphs, and even on sets of knowledge units (a kind of meta-knowledge mining), i.e. mining a knowledge base instead of a database.
- Accordingly, the design of a KDDK system, based on the above principles, and involved in application domains such as astronomy, agronomy, biology, chemistry, medicine, for decision support and problem-solving.

From a middle-term perspective, a system for KDDK can be considered as a "decentralized system" the architecture of which is described hereafter.

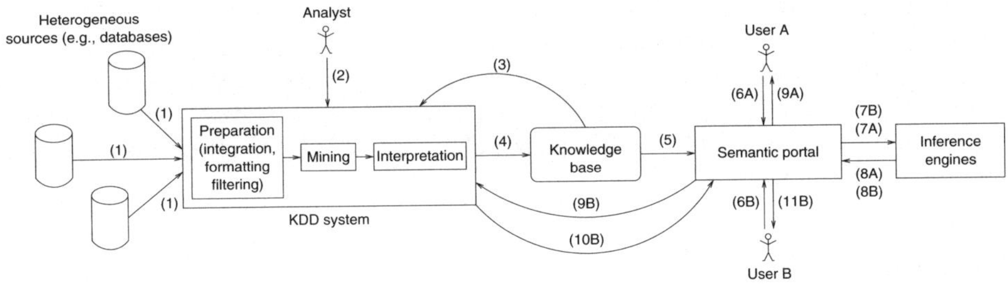

Fig. 3. An architecture for a system aimed at "knowledge discovery (from complex data) guided by domain knowledge process (KDDK)". The classical KDD process can be read from left to right, while, by contrast, the KDDK system can be read from right to left.

- One or several ontologies (knowledge bases) include knowledge from different domains with different points of view, and as well, a case base. A set of services are related through a semantic portal, for knowledge editing, navigating, and visualizing the ontologies.
- An inference engine provides, in association with the knowledge bases, a collection of inference rules for problem-solving purposes, among which subsumption, classification (lattice-based classification, clustering), case-based reasoning. Reasoning services are present for handling concrete datatypes such as strings or numbers (and possibly, for controlling procedural or functional reasoning modes if-needed).
- A set of heterogeneous databases holding on a domain to be mined for providing knowledge units enriching domain ontologies.
- A platform for KDDK proposes a collection of data mining modules –such as the Coron platform– and a set of services for data preparation and extracted unit interpretation.

Moreover, the system has to provide channels for allowing communications with human agents, such as experts and end-users. The resulting KDDK system architecture has to be reusable in any application domain. Accordingly, the integration of such a KDDK system in the framework of the semantic Web can be seen as follows. The data sources, i.e. databases, sets of documents, are explored, navigated, and queried, under the supervision of an analyst, thanks to a KDDK process guided by knowledge bases of the domain. The data are prepared and manipulated by the KDDK process, while the knowledge units are validated by the analyst, and then manipulated by the inference engine.

The figure 3 presents the architecture proposal for a KDDK system, in which different scenarios can be made operational. Heterogeneous sources (e.g. databases) feed the KDD system (1), under the supervision of an analyst (2), using available domain knowledge (3). The KDD system returns new knowledge units for extending and enriching a knowledge base (4), that may be queried through a semantic portal (5) by distant geographically distributed users (users A and B). The users A and B query the portal (6A, 6B), that in turn may use the services of a knowledge base and the associated inference engine (7A, 7B). When the available knowledge provides, with the help of the inference engine, an answer to the request (8A), this answer is transmitted to the user (9A). Otherwise (8B), the request is transferred in a filtering module used by the KDD system (9B) for mining the available data, trying to extract information related to the request. The resulting extracted knowledge units relying on this filter (10B) may provide an answer to the user (11B).

5 Conclusion

In this paper, we have presented the research work carried out in the Orpailleur team at LORIA. Multiple and combined aspects of knowledge discovery and knowledge processing have been introduced and discussed: symbolic KDD methods such as lattice-based classification itemset search, and association rule extraction, and numeric methods such as HMM2. Next, the KDD process has been considered from a knowledge representation perspective, explaining how and why the KDD process may take advantage of domain knowledge embedded in ontologies relative to the domain of data. This perspective leads to the idea of KDDK, for knowledge discovery (from complex data) guided by domain knowledge. The KDDK process is based on classification tasks, for modeling, representing, reasoning, and discovering. Various instantiations of the KDDK process have been described, among which the mining of molecular graphs –for knowledge discovery in chemical reaction databases–, text mining and Semantic Web for designing and enlarging ontologies from documents, knowledge discovery in life sciences, and hybrid knowledge discovery, combining numerical and symbolic methods for data mining. An original experiment has also been introduced and discussed: meta-knowledge mining, or mining a knowledge base instead of a database. This research work has been carried out for the need of adaptation knowledge acquisition (AKA), that is a promising research domain, and that can be reused for mining various kind of strategical knowledge units, e.g. decision knowledge units. At the end of the paper, an architecture of an integrated KDDK system has been proposed and discussed.

References

1. Aamodt, A.: Knowledge-Intensive Case-Based Reasoning and Sustained Learning. In: Aiello, L.C. (ed.) Proc. of the 9th European Conference on Artificial Intelligence (ECAI 1990) (1990)
2. Baader, F., Calvanese, D., McGuinness, D., Nardi, D., Patel-Schneider, P. (eds.): The Description Logic Handbook. Cambridge University Press, Cambridge (2003)
3. Barbut, M., Monjardet, B.: Ordre et classification – Algèbre et combinatoire (2 tomes). Hachette, Paris (1970)
4. Bendaoud, R., Rouane Hacene, M., Toussaint, Y., Delecroix, B., Napoli, A.: Text-based ontology construction using relational concept analysis. In: Flouris, G., d'Aquin, M. (eds.) Proceedings of the International Workshop on Ontology Dynamics, Innsbruck (Austria), pp. 55–68 (2007)

5. Napoli, A., Berasaluce, S., Laurenço, C., Niel, G.: An Experiment on Knowledge Discovery in Chemical Databases. In: Boulicaut, J.-F., Esposito, F., Giannotti, F., Pedreschi, D. (eds.) PKDD 2004. LNCS (LNAI), vol. 3202, pp. 39–51. Springer, Heidelberg (2004)
6. Stumme, G., Berendt, B., Hotho, A.: Towards Semantic Web Mining. In: Horrocks, I., Hendler, J. (eds.) ISWC 2002. LNCS, vol. 2342, Springer, Heidelberg (2002)
7. Brachman, R.J., Selfridge, P.G., Terveen, L.G., Altman, B., Borgida, A., Halper, F., Kirk, T., Lazar, A., McGuinness, D.L., Resnick, L.A.: Knowledge representation support for data archaeology. In: Proceedings of the 1st International Conference on Information and Knowledge Management (CKIM 1992), Baltimore, pp. 457–464 (1992)
8. Carpineto, C., Romano, G.: Concept Data Analysis: Theory and Applications. John Wiley & Sons, Chichester (2004)
9. Cherfi, H., Napoli, A., Toussaint, Y.: Towards a text mining methodology using association rules extraction. Soft Computing 10(5), 431–441 (2006)
10. Cimiano, P., Hotho, A., Staab, S.: Learning Concept Hierarchies from Text Corpora using Formal Concept Analysis. Journal of Artificial Intelligence Research 24, 305–339 (2005)
11. Stumme, G., Hotho, A., Tane, J., Cimiano, P.: Conceptual Knowledge Processing with Formal Concept Analysis and Ontologies. In: Eklund, P.W. (ed.) ICFCA 2004. LNCS (LNAI), vol. 2961, pp. 189–207. Springer, Heidelberg (2004)
12. d'Aquin, M., Badra, F., Lafrogne, S., Lieber, J., Napoli, A., Szathmary, L.: Case base mining for adaptation knowledge acquisition. In: Veloso, M.M. (ed.) IJCAI 2007, Hyderabad, India, pp. 750–755. Morgan Kaufman, San Francisco (2007)
13. d'Aquin, M., Bouthier, C., Brachais, S., Lieber, J., Napoli, A.: Knowledge Edition and Maintenance Tools for a Semantic Portal in Oncology. International Journal on Human–Computer Studies 62(5), 619–638 (2005)
14. Dunham, M.H.: Data Mining – Introductory and Advanced Topics. Prentice Hall, Upper Saddle River (2003)
15. Eng, C., Thibessard, A., Hergalant, S., Mari, J.-F., Leblond, P.: Data mining using hidden markov models (HMM2) to detect heterogeneities into bacteria genomes. In: Journées Ouvertes Biologie, Informatique et Mathématiques – JOBIM 2005, Lyon, France (2005)
16. Fensel, D., Hendler, J., Lieberman, H., Wahlster, W. (eds.): Spinning the Semantic Web. The MIT Press, Cambridge, Massachusetts (2003)
17. Fuchs, B., Lieber, J., Mille, A., Napoli, A.: An Algorithm for Adaptation in Case-based Reasoning. In: Horn, W. (ed.) Proceedings of the 14th European Conference on Artificial Intelligence (ECAI-2000), Berlin, pp. 45–49. IOS Press, Amsterdam (2000)
18. Ganter, B., Wille, R.: Formal Concept Analysis. Springer, Berlin (1999)
19. Han, J., Kamber, M.: Data Mining: Concepts and Techniques. Morgan Kaufmann Publishers, San Francisco (2001)
20. Hand, D., Mannila, H., Smyth, P.: Principles of Data Mining. The MIT Press, Cambridge (2001)
21. Janetzko, D., Cherfi, H., Kennke, R., Napoli, A., Toussaint, Y.: Knowledge-based selection of association rules for text mining. In: de Màntaras, R.L., Saitta, L. (eds.) 16h European Conference on Artificial Intelligence – ECAI 2004, Valencia, Spain, pp. 485–489 (2004)
22. Jay, N., Kohler, F., Napoli, A.: Using formal concept analysis for mining and interpreting patient flows within a healthcare network. In: Ben Yahia, S., Mephu Nguifo, E., Behlohlavek, R. (eds.) CLA 2006. LNCS (LNAI), vol. 4923, pp. 263–268. Springer, Heidelberg (2008) (this volume)

23. Kuznetsov, S.O.: Machine Learning and Formal Concept Analysis. In: Eklund, P.W. (ed.) ICFCA 2004. LNCS (LNAI), vol. 2961, pp. 287–312. Springer, Heidelberg (2004)
24. Kuznetsov, S.O., Obiedkov, S.A.: Comparing performance of algorithms for generating concept lattices. Journal of Theoretical Artificial Intelligence 14(2/3), 189–216 (2002)
25. Le Ber, F., Benoit, M., Schott, C., Mari, J.-F., Mignolet, C.: Studying crop sequences with CarrotAge, a HMM-based data mining software. Ecological Modelling 191(1), 170–185 (2006)
26. Le Ber, F., Napoli, A.: Design and comparison of lattices of topological relations for spatial representation and reasoning. Journal of Experimental & Theoretical Artificial Intelligence 15(3), 331–371 (2003)
27. Lieber, J., d'Aquin, M., Bey, P., Napoli, A., Rios, M., Sauvagnac, C.: Adaptation knowledge acquisition, a study for breast cancer treatment. In: Dojat, M., Keravnou, E.T., Barahona, P. (eds.) AIME 2003. LNCS (LNAI), vol. 2780, pp. 304–313. Springer, Heidelberg (2003)
28. Liu, H., Lu, H., Feng, L., Hussain, F.: Efficient Search of Reliable Exceptions. In: Zhong, N., Zhou, L. (eds.) PAKDD 1999. LNCS (LNAI), vol. 1574, pp. 194–204. Springer, Heidelberg (1999)
29. Mari, J.-F., Haton, J.-P., Kriouile, A.: Automatic Word Recognition Based on Second-Order Hidden Markov Models. IEEE Transactions on Speech and Audio Processing 5, 22–25 (1997)
30. Mari, J.-F., Le Ber, F.: Temporal and spatial data mining with second-order hidden models. Soft Computing 10(5), 406–414 (2006)
31. Maumus, S., Napoli, A., Szathmary, L., Visvikis-Siest, S.: Fouille de données biomédicales complexes: extraction de règles et de profils génétiques dans le cadre de l'étude du syndrome métabolique. In: Journées Ouvertes Biologie Informatique Mathématiques – JOBIM 2005,, pp. 169–173 (2005)
32. Napoli, A., Messai, N., Devignes, M.-D., Smaïl-Tabbone, M.: Querying a Bioinformatic Data Sources Registry with Concept Lattices. In: Dau, F., Mugnier, M.-L., Stumme, G. (eds.) ICCS 2005. LNCS (LNAI), vol. 3596, pp. 323–336. Springer, Heidelberg (2005)
33. Messai, N., Devignes, M.-D., Napoli, A., Smaïl-Tabbone, M.: Br-explorer: An fca-based algorithm for information retrieval. In: Ben Yahia, S., Mephu-Nguifo, E. (eds.) Fourth International Conference on Concept Lattices and their Applications (CLA-2006), Hammamet, Tunisia (2006)
34. Mollo, V.: Usage des ressources, adaptation des savoirs et gestion de l'autonomie dans la décision thérapeutique. Thèse d'Université, Conservatoire National des Arts et Métiers (2004)
35. Napoli, A.: A smooth introduction to symbolic methods for knowledge discovery. In: Cohen, H., Lefebvre, C. (eds.) Handbook of Categorization in Cognitive Science, pp. 913–933. Elsevier, Amsterdam (2005)
36. Napoli, A., Le Ber, F.: The Galois lattice as a hierarchical structure for topological relations. Annals of Mathematics and Artificial Intelligence 49(1–4), 171–190 (2007); Special volume on Knowledge discovery and discrete mathematics and a tribute to the memory of Peter L. Hammer
37. Padmanabhan, B., Tuzhilin, A.: On characterization and discovery of minimal unexpected patterns in rule discovery. IEEE Transactions on Knowledge and Data Engineering 18(2), 202–216 (2006)

38. Pennerath, F., Napoli, A.: La fouille de graphes dans les bases de données réactionnelles au service de la synthèse en chimie organique. In: Ritschard, G., Djeraba, C. (eds.) Extraction et gestion des connaissances (EGC 2006), Lille, pp. 517–528 (2006) RNTI-E-6, Cépaduès-Éditions Toulouse
39. Quan, T.T., Hui, S.C., Fong, A.C.M., Cao, T.H.: Automatic generation of ontology for scholarly semantic web. In: McIlraith, S.A., Plexousakis, D., van Harmelen, F. (eds.) ISWC 2004. LNCS, vol. 3298, pp. 726–740. Springer, Heidelberg (2004)
40. Smaïl-Tabbone, M., Osman, S., Messai, N., Napoli, A., Devignes, M.-D.: Bioregistry: A structured metadata repository for bioinformatic databases. In: R. Berthold, M., Glen, R.C., Diederichs, K., Kohlbacher, O., Fischer, I. (eds.) CompLife 2005. LNCS (LNBI), vol. 3695, pp. 46–56. Springer, Heidelberg (2005)
41. Staab, S., Studer, R. (eds.): Handbook on Ontologies. Springer, Berlin (2004)
42. Stumme, G.: Formal concept analysis on its way from mathematics to computer science. In: Priss, U., Corbett, D.R., Angelova, G. (eds.) ICCS 2002. LNCS (LNAI), vol. 2393, pp. 2–19. Springer, Heidelberg (2002)
43. Suzuki, E.: Undirected Discovery of Interesting Exception Rules. International Journal of Pattern Recognition and Artificial Intelligence (IJPRAI) 16(8), 1065–1086 (2002)
44. Szathmary, L.: Symbolic Data Mining Methods with the Coron Platform.In: Thèse d'informatique, Université Henri Poincaré – Nancy 1, France (2006)
45. Szathmary, L., Maumus, S., Petronin, P., Toussaint, Y., Napoli, A.: Vers l'extraction de motifs rares. In: Ritschard, G., Djeraba, C. (eds.) Extraction et gestion des connaissances (EGC 2006), Lille, pp. 499–510 (2006) RNTI-E-6, Cépaduès-Éditions Toulouse
46. Szathmary, L., Napoli, A.: Coron: A framework for levelwise itemset mining algorithms. In: Ganter, B., Godin, R. (eds.) ICFCA 2005. LNCS (LNAI), vol. 3403, pp. 110–113. Springer, Heidelberg (2005)
47. Szathmary, L., Napoli, A., Kuznetsov, S.O.: Zart: A multifunctional itemset mining algorithm. In: Diatta, J., Eklund, P., Liquière, M. (eds.) Proceedings of the Fifth International Conference on Concept Lattices and their Applications, Montpellier, France, pp. 26–37 (2007)
48. Szathmary, L., Napoli, A., Valtchev, P.: Towards rare itemset mining. In: Proceedings of the IEEE International Conference on Tools with Artificial Intelligence (ICTAI), Patras, Greece, IEEE Computer Society Press, Los Alamitos (2007)
49. Ténier, S., Toussaint, Y., Napoli, A., Polanco, X.: Instantiation of relations for semantic annotation. In: The 2006 IEEE/WIC/ACM International Conference on Web Intelligence - WI 2006, Hong Kong, pp. 463–472. IEEE Computer Society Press, Los Alamitos (2006)
50. Ténier, S., Napoli, A., Polanco, X., Toussaint, Y.: Semantic annotation of webpages. In: Handschuh, S. (ed.) ISWC 2005. LNCS, vol. 3729, Springer, Heidelberg (2005)
51. Valtchev, P., Missaoui, R., Godin, R.: Formal concept analysis for knowledge discovery and data mining: The new challenges. In: Eklund, P.W. (ed.) ICFCA 2004. LNCS (LNAI), vol. 2961, pp. 352–371. Springer, Heidelberg (2004)
52. Weiss, G.M.: Mining with rarity: a unifying framework. SIGKDD Exploration Newsletter 6(1), 7–19 (2004)
53. Wille, R.: Methods of conceptual knowledge processing. In: Missaoui, R., Schmidt, J. (eds.) Formal Concept Analysis. LNCS (LNAI), vol. 3874, pp. 1–29. Springer, Heidelberg (2006)
54. Wu, X., Zhang, C., Zhang, S.: Efficient mining of both positive and negative association rules. ACM Transactions on Information Systems 22(3), 381–405 (2004)

Formal Concept Analysis as Applied Lattice Theory

Rudolf Wille

Technische Universität Darmstadt, Fachbereich Mathematik
wille@mathematik.tu-darmstadt.de

Abstract. *Formal Concept Analysis* is a mathematical theory of concept hierarchies which is based on *Lattice Theory*. It has been developed to support humans in their thought and knowledge. The aim of this paper is to show how successful the lattice-theoretic foundation can be in applying Formal Concept Analysis in a wide range. This is demonstrated in three sections dealing with *representation*, *processing*, and *measurement* of conceptual knowledge. Finally, further relationships between abstract Lattice Theory and Formal Concept Analysis are briefly discussed.

1 Introduction

Formal Concept Analysis has been developed since 1979 as part of applied mathematics based on a mathematization of concept and concept hierarchy. The initial motivation for this development originated from a research seminar at the TU Darmstadt in which mathematicians tried to understand sense and meaning of order and lattice theory for our society. This activity was particularly influenced by the German Scholar of Education, *Hartmut von Hentig*, who demands in [He74] that (from time to time) sciences should be *restructured* to make them better understandable, available, and criticizable (even beyond disciplinary competence). This means in particular that scientists should rethink their theoretical developments in order to integrate and rationalize origins, connections, interpretations, and applications. Generally, abstract developments should be brought back to the common place in perception, thinking, and action. In this sense, *restructuring lattice theory* is understood as an attempt to reinvigorate connections with our general culture by interpreting lattice theory as concretely as possible, and in this way to promote better communication between lattice theorists and potential users of lattice theory (cf. [Wi82]).

The basic connection between ordered structures and concept hierarchies in human thought originates in the *traditional philosophical logic* as, for instance, presented in I. Kant's lectures on logic [Ka88], where he writes in §7:

> "Every concept, as a *partial concept*, is contained *in* the presentation of things; as a *ground of cognition*, i.e. as a *characteristic*, it has these things contained *under it*. In the former regard, every concept has an *intension* [content]; in the latter, it has an *extension*.
> Intension and extension of a concept have an inverse relation to each other. The more a concept contains under it, the less it contains in it."

S. Ben Yahia et al. (Eds.): CLA 2006, LNAI 4923, pp. 42–67, 2008.

In this quotation a concept is viewed as a composition of an extension and an intension where the extension has "things" under it, while the intension has "things" in it; furthermore, the more the concept extension has under it, the less the concept intension has in it. Thus, each collection of concept extensions can be naturally mathematized by an *ordered set*, where the dualization of the ordered set mathematizes the corresponding collection of concept intensions.

The *focus on lattices* was realized when the notion of a (formal) context was introduced as a frame in which concept extensions and concept intensions could be constructed (cf. [Wi82]). Mathematically, a *formal context* is defined as a triple (G, M, I) where G is a set (of "objects"), M is a set (of "attributes"), and I is a binary relation between G and M where gIm (which means $(g, m) \in I$) indicates that the object g has the attribute m. A *formal concept* of the formal context (G, M, I) is defined as a pair (A, B) with $A \subseteq G$, $B \subseteq M$, $A = \{g \in G \mid gIm \text{ for all } m \in B\}$, and $B = \{m \in M \mid gIm \text{ for all } g \in A\}$; A and B are called the *extent* and the *intent* of the formal concept (A, B), respectively. The hierarchical relation *subconcept-superconcept* - expressed by sentences as "the formal concept (A_1, B_1) is a subconcept of the formal concept (A_2, B_2)" - is modelled by the definition:

$$(A_1, B_1) \leq (A_2, B_2) :\Longleftrightarrow A_1 \subseteq A_2 \quad (\Longleftrightarrow B_1 \supseteq B_2)$$

The set of all formal concepts of (G, M, I) with this order relation is a complete lattice, called the *concept lattice* of the formal context (G, M, I) and denoted by $\underline{\mathfrak{B}}(G, M, I)$. This contextual approach yields a close connection between *abstract lattice theory* and the *theory of concept lattices* because each abstract lattice is embeddable in a concept lattice and each complete abstract lattice is even isomorphic to some concept lattice. This connection makes abstract lattice theory more meaningful because it yields the fundamental basis for the many applications of *Formal Concept Analysis*, the theory of constructing, analysing, and applying concept lattices (cf. [GW99]).

It should be mentioned that some other approaches have led to theories which partly overlap with Formal Concept Analysis. Especially, the rich theory about *Galois connections* (cf. [DEW04]) relates in many ways with Formal Concept Analysis. It should at least be mentioned that G. Birkhoff, the father of lattice theory, has already discussed in his first edition of his lattice theory book [Bi40] the Galois connection between objects and attributes with respect to the relation: "an object *has* an attribute". Another approach of activating Galois connections has been founded in the monograph [BM70] where, in particular, a binary relation between two sets is used to construct a so-called *Galois lattice* whose elements are corresponding pairs of subsets of the two given sets.

This paper shall explain in more detail how the mathematization of concept and concept hierarchy leads to the understanding of Formal Concept Analysis as part of applied lattice theory.

First, *Formal Concept Analysis* has been developed as a mathematical theory based on set-theoretical semantics, i.e., all notions of Formal Concept Analysis are mathematically defined in terms of set theory. These notions can especially be assigned to *Lattice Theory*; this becomes substantiated, for instance, by B. A. Davey's and H. A. Priestley's book "Introduction to lattices and order" [DP02] in which the third chapter is fully devoted to Formal Concept Analysis.

Secondly, many notions of Formal Concept Analysis can be understood as mathematization of non-mathematical notions; for instance, a *formal context* mathematizes a context represented by a cross table, an *extent* (*intent*) mathematizes a concept extension (concept intension) and a *formal concept* mathematizes a concept in the understanding of traditional philosophical logic, a *concept lattice* mathematizes a concept hierarchy derived from a context represented by a cross table, etc. All those mathematical notions may serve as bridges to non-mathematical fields and may support these fields with patterns of formal thought; in other words, they make possible the application of mathematical thinking to problems in the real world. This is the reason why Formal Concept Analysis can be understood as part of *Applied Lattice Theory*.

The next three sections discuss applications of Formal Concept Analysis in the field of conceptual knowledge under the headings "Representation", "Representing and Processing", and "Measurement". Supporting the *representation* of conceptual knowledge is basic for all applications of Formal Concept Analysis and therefore should make the represented contents as transparent as possible. The *processing* of conceptual knowledge builts on conceptual knowledge representations and should therefore be developed on top of the conceptual structures of the underlying content representations. *Measurement* of conceptual knowledge structures gaines from both, representation and processing, but it has, moreover, to meaningfully include numeric and algebraic structures.

2 Representation of Conceptual Knowledge

Conceptual knowledge representations are constituted in human thought by semantic structures. Therefore such representations can be comprehended in terms of *Semantology*, the theory and methodology of semantic structures (see [GW06]). The *meaning of semantic structures* in the field of Conceptual Knowlede Representation can be analysed on at least three levels:

- First, there is the meaning on the *concrete level* on which the considered conceptual knowledge originates. This is usually the semantics belonging to the fields whose language and understanding are used to describe that knowledge.
- Second, there is the meaning on the general *philosophic-logical level* on which the semantics is highly abstracted from the semantics of the concrete level, but is still related to actual realities. It is the semantics of the traditional philosophical logic based on the main functions of human thought: concept, judgment, and conclusion (cf. [Ka88]).

– Third, there is the meaning on the *mathematical level* on which the semantics is strongly restricted to the purely abstract: like numbers, ideal geometric figures and, since the twentieth century, set structures (and their generalizations).

	vocalic	sonorant	consonantal	high	obstruent	non-high
i	X	X		X		
e	X	X				X
a	X	X				X
o	X	X				X
u	X	X		X		
p			X		X	X
b			X		X	X
f			X		X	X
v			X		X	X
m		X	X			X
t			X		X	X
d			X		X	X
n		X	X			X
s			X		X	X
z			X		X	X
k			X	X	X	
g			X	X	X	
l		X	X			X

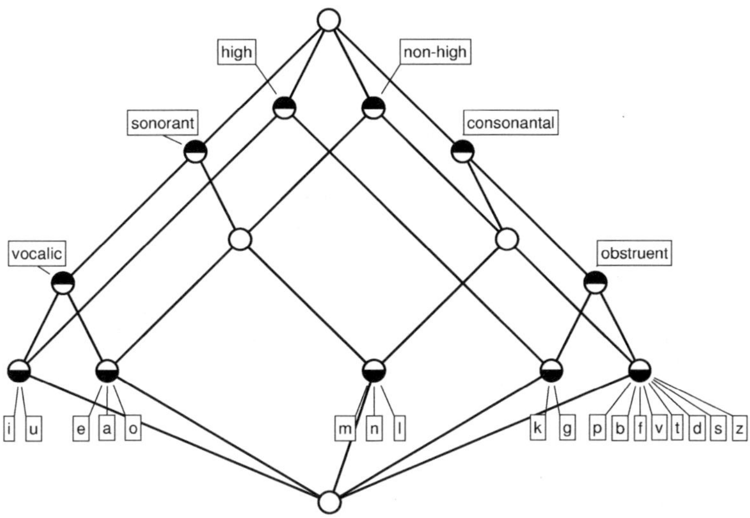

Fig. 1. Context and concept hierarchy about the sound pattern of English

Let us illustrate this three-fold semantics by the example presented in Fig. 1. On the *concrete level*, context and concept hierarchy are understood with respect to the linguistic semantics as presented in [CH68], i.e., the context represents an elementary semantic structure of the English speech sounds and the corresponding concept hierarchy represents the related conceptual semantic structure of these sounds. On the *philosophic-logical level*, the abstract-logical structure of the speech sounds and their concepts comes under consideration; a general discovery is, for example, that there are exactly five atomic concepts classifying the speech sounds. Finally, on the *mathematical level*, the formal context and concept lattice in Fig. 1 represent most abstractly the semantic structure of the linguistic example; relationships in the mathematical structure are good candidates to be interpreted as interesting relationships in the linguistic structure as, for instance, the implication: "high, sonorant → vocalic".

For a comprehensive understanding of the possibilities to apply Formal Concept Analysis, an elaborated mathematical theory of concept lattices has been worked out over more than 25 years so that there exists now a rich store of methods and results for applications in a wide range (cf. [GW99], [SW00], [GSW05]). Many of the research contributions rely on the *Basic Theorem on Concept Lattices*, which shall therefore be cited here and, moreover, it shall also be used for deducing a new theorem for basic applications.

Basic Theorem on Concept Lattices [Wi82]. *Let* $\mathbb{K} := (G, M, I)$ *be a formal context. Then* $\underline{\mathfrak{B}}(\mathbb{K})$ *is a complete lattice, called the* concept lattice *of* (G, M, I)*, for which infimum and supremum can be described as follows:*

$$\bigwedge_{t\in T}(A_t, B_t) = (\bigcap_{t\in T} A_t, (\bigcup_{t\in T} B_t)^{II}),$$

$$\bigvee_{t\in T}(A_t, B_t) = ((\bigcup_{t\in T} A_t)^{II}, \bigcap_{t\in T} B_t).$$

In general, a complete lattice L *is isomorphic to* $\underline{\mathfrak{B}}(\mathbb{K})$ *if and only if there exist mappings* $\tilde{\gamma} : G \longrightarrow L$ *and* $\tilde{\mu} : M \longrightarrow L$ *such that*
1. $\tilde{\gamma}G$ *is* $\bigvee$-dense *in* L *(i.e.* $L = \{\bigvee X \mid X \subseteq \tilde{\gamma}G\}$*),*
2. $\tilde{\mu}M$ *is* $\bigwedge$-dense *in* L *(i.e.* $L = \{\bigwedge X \mid X \subseteq \tilde{\mu}M\}$*),*
3. $gIm \Longleftrightarrow \tilde{\gamma}g \leq \tilde{\mu}m$ *for* $g \in G$ *and* $m \in M$*;*
in particular, $L \cong \underline{\mathfrak{B}}(L, L, \leq)$ *and furthermore:* $L \cong \underline{\mathfrak{B}}(J(L), M(L), \leq)$ *if the set* $J(L)$ *of all* $\bigvee$*-irreducible elements is* $\bigvee$*-dense in* L *and the set* $M(L)$ *of all* $\bigwedge$*-irreducible elements is* $\bigwedge$*-dense in* L*.*

In practice, the basic theorem is most frequently used to examine whether a line diagram really describes the concept lattice of a given formal context. Let us demonstrate this checkup at the line diagram in Fig. 1: For the checkup we assume that the line diagram represents a lattice which we name L; for proving $L \cong \underline{\mathfrak{B}}(\mathbb{K})$, by the basic theorem, we have only to prove the statements 1., 2., and 3. for our example to confirm the claim that the line diagram in Fig. 1 represents the concept lattice of the formal context in Fig. 1. Condition 1 holds because each circle in the diagram is uniquely determined by the set of all letters which can be reached from the circle by a descending path of line segments. Dually, condition 2 holds because each circle in the diagram is uniquely determined by the set of all attribute names which can be reached from the circle by an ascending path of line segments. For justifying condition 3, we have to examine that a speech sound has an attribute if and only if there is a cross in the table of the formal context, the row of which is headed by the name of the speech sound and the column of which is headed by the name of the attribute; this examination gives a positive result for our example.

Unfortunately, our checkup has a weakness, namely: we have to assume that the line diagram represents a (complete) lattice. Although this is often true, there are sometimes hardly idendifiable defects which prevent a lattice structure. To overcome such problems, we need a mathematization of line diagrams of (finite) bounded ordered sets as introduces in [Wi07a]:

In general, a *line diagram of a finite bounded ordered set* $\underline{O} := (O, \leq)$ is mathematically defined as a quadruple $\mathbb{D}_\eta(\underline{O}) := (C_{\underline{O}}, S_{\underline{O}}, T_{\underline{O}}, \eta)$ formed by

– a set $C_{\underline{O}}$ of disjoint little circles of the same radius in the Euclidean plane $\mathbb{R}^2$,
– a set $S_{\underline{O}}$ of straight line segments in $\mathbb{R}^2$ having at most one point in common,

- a ternary relation $T_{\underline{O}} \subseteq C_{\underline{O}} \times S_{\underline{O}} \times C_{\underline{O}}$ which contains for each $s \in S_{\underline{O}}$ exactly one triple (c_1, s, c_2) indicating that the line segment s links up the circles c_1 and c_2 in $\mathbb{R}^2$ and that $c_1 <_2 c_2$ (i.e. for all points $p_i \in c_i$ with $i = 1, 2$, the second coordinate of p_1 is smaller than the second coordinate of p_2).
- a bijection $\eta : O \to C_{\underline{O}}$ which makes explicit that the covering pairs $o_1 \prec o_2$ in $\underline{O}$ are in one-to-one correspondence to the triples $(\eta(o_1), s, \eta(o_2))$ of $T_{\underline{O}}$ (consequently, $|\prec| = |T_{\underline{O}}|$).

The line diagrams $\mathbb{D}_\eta(\underline{O})$ and $\mathbb{D}_{\hat{\eta}}(\hat{\underline{O}})$ of finite bounded ordered sets $\underline{O} := (O, \leq)$ and $\hat{\underline{O}} := (\hat{O}, \leq)$ are called *isomorphic* if and only if there exist bijections $\zeta : C_{\underline{O}} \to C_{\hat{\underline{O}}}$ and $\sigma : S_{\underline{O}} \to S_{\hat{\underline{O}}}$ such that $(c_1, s, c_2) \in T_{\underline{O}} \iff (\zeta(c_1), \sigma(s), \zeta(c_2)) \in T_{\hat{\underline{O}}}$; the corresponding isomorphism is denoted by (ζ, σ).

A cross table which represents a finite context $\mathbb{K} := (G, M, I)$ contains the object names of the objects in G and the attribute names of the attributes in M. Since those names are understood as *proper names* (german: *Eigennamen*), there is a bijection ν mapping each object resp. attribute in $G \dot\cup M$ to its proper name. A line diagram $\mathbb{D}_{\bar{\eta}}(\underline{\mathfrak{B}}(\mathbb{K}))$ together with the bijection ν is called a $(\nu G, \nu M)$-*labelled line diagram* denoted by $\mathbb{D}^\nu_{\bar{\eta}}(\underline{\mathfrak{B}}(\mathbb{K}))$. Analogously, for a finite bounded ordered set $\underline{O}$ and mappings $\check{\gamma} : G \to \underline{O}$ and $\check{\mu} : M \to \underline{O}$, a line diagram $\mathbb{D}_\eta(\underline{O})$ together with the introduced naming bijection ν on $G \dot\cup M$ is called a $(\nu G, \nu M)$-*labelled line diagram* denoted by $\mathbb{D}^\nu_\eta(\underline{O})$.

Basic Theorem on Labelled Line Diagrams of Finite Concept Lattices. [*Wi07a*] *Let $\underline{\mathfrak{B}}(\mathbb{K})$ be the concept lattice of a finite context $\mathbb{K} := (G, M, I)$ and let $\underline{O} := (O, \leq)$ be a finite bounded ordered set with mappings $\check{\gamma} : G \to \underline{O}$ and $\check{\mu} : M \to \underline{O}$. Then, a $(\nu G, \nu M)$-labelled line diagram $\mathbb{D}^\nu_\eta(\underline{O})$ of the ordered set $\underline{O}$ is isomorphic to a $(\nu G, \nu M)$-labelled line diagram $\mathbb{D}^\nu_{\bar{\eta}}(\underline{\mathfrak{B}}(\mathbb{K}))$ of the concept lattice $\underline{\mathfrak{B}}(\mathbb{K})$ if and only if, in $\mathbb{D}^\nu_\eta(\underline{O})$,*

1. *each circle, having exactly one line segment downwards, is labelled (from below) by at least one object name out of νG,*
2. *each circle, having exactly one line segment upwards, is labelled (from above) by at least one attribute name out of νM,*
3. *a circle labelled by an object name out of νG is linked up by an ascending chain of line segments to a circle labelled by an attribute name out of νM, or those labelled circles are identical, if and only if the named object has the named attribute,*
4. *there exists an injection $\zeta : C_{\underline{\mathfrak{B}}(\mathbb{K})} \to C_{\underline{O}}$ such that, for each circle $\bar{c}$ in the diagram $\mathbb{D}^\nu_{\bar{\eta}}(\underline{\mathfrak{B}}(\mathbb{K}))$, $\zeta(\bar{c})$ represents a minimal upper bound of $\{\check{\gamma} g \mid g \in G \text{ with } \gamma g \leq \bar{\eta}^{-1}\bar{c}\}$ which is also a maximal lower bound of $\{\check{\mu} m \mid m \in M \text{ with } \mu m \geq \bar{\eta}^{-1}\bar{c}\}$,*
5. *the number of all circles of $\mathbb{D}^\nu_\eta(\underline{O})$ equals the number of all circles of $\mathbb{D}^\nu_{\bar{\eta}}(\underline{\mathfrak{B}}(\mathbb{K}))$,*
6. *the number of all line segments of $\mathbb{D}^\nu_\eta(\underline{O})$ equals the number of all line segments of $\mathbb{D}^\nu_{\bar{\eta}}(\underline{\mathfrak{B}}(\mathbb{K}))$.*

Now, we test the “Basic Theorem on Labelled Line Diagrams of Finite Concept Lattices” by examining the *line diagram* in Fig. 1. For checking the *conditions 5 and 6*, we need to know the number of elements and the number of covering pairs of elements of the presented lattice. Applying P. Burmeister's program *“ConImp”* [Bu03] to the formal context in Fig. 1, we obtain that the lattice has 15 elements and 23 covering pairs; since the line diagram has 15 circles and 23 straight line segments between those circles, condition 5 and 6 are valid. *Conditions 1 and 2* are obviously true because the five circles directly above the lowest circle are the only circles having exactly one line segment downwards and being labelled by at least one object name of the presented context, and the six circles labelled by an attribute name are the only circles having exactly one line segment upwards. *Condition 3* can be confirmed by a systematic comparison of the crosses in the cross table with the pairs of object circle and attribute circle of which the object circle is linked up to the attribute circle by an ascending chain of line segments. Finally, *condition 4* is checked by confirming for each circle in the line diagram that it represents an element which is a minimal upper bound of elements whose corresponding circles are labelled by an object name and is a maximal lower bound of elements whose corresponding circles are labelled by an attribute name. For our example, this checkup is positive. Thus, the line diagram in Fig. 1 represents the concept lattice of the formal context in Fig. 1.

The readability of line diagrams of concept lattices becomes intricate when the number of concept relationships increases so much that, because of multifarious intersections, the individual line segments cannot be clearly pursued. To overcome this problem, *nested line diagrams* have been invented (see [Wi84]) which allow to draw readable line diagrams with even more than hunderd concept circles. The basic idea of nested line diagrams is to partition a line diagram in such a way that parallel line segments between two classes of the partition can be replaced by one line segment by which those parallel line segments may be still reconstructed. A representation of a concept lattice by a nested line diagram can be deduced from a partition of the set of attributes of the underlying formal context. The basis for this is the following theorem (cf. [GW99], p.77):

Theorem 1. ($\bigvee$-Embedding into Direct Products of Concept Lattices) *Let (G, M, I) be a context and $M = M_1 \cup M_2$. The map*

$$(A, B) \mapsto (((B \cap M_1)', B \cap M_1), ((B \cap M_2)', B \cap M_2))$$

is a $\bigvee$-preserving order embedding of $\underline{\mathfrak{B}}(G, M, I)$ into the direct product of the lattices $\underline{\mathfrak{B}}(G, M_1, I \cap G \times M_1)$ and $\underline{\mathfrak{B}}(G, M_2, I \cap G \times M_2)$. The component maps

$$(A, B) \mapsto ((B \cap M_i)', B \cap M_i)$$

are surjective on $\underline{\mathfrak{B}}(G, M_i, I \cap G \times M_i)$.

Let us exemplify the use of this theorem by the concept lattice presented in Fig. 1. As partition classes we consider the attribute sets
$M_1 := \{vocalic, sonorant, consonantal, obstruent\}$, $M_2 := \{high, non-high\}$ together with the reduced object set $G_r := \{i, e, p, m, k\}$, for which the concept lattice $\underline{\mathfrak{B}}(G_r, M, I \cap G_r \times M)$ is still isomorphic to $\underline{\mathfrak{B}}(G, M, I)$. Fig. 2 shows a *nested line diagram* of the direct product of $\underline{\mathfrak{B}}(G_r, M_1, I \cap G_r \times M_1)$ and $\underline{\mathfrak{B}}(G_r, M_2, I \cap G_r \times M_2)$. A common line diagram of the direct product can be

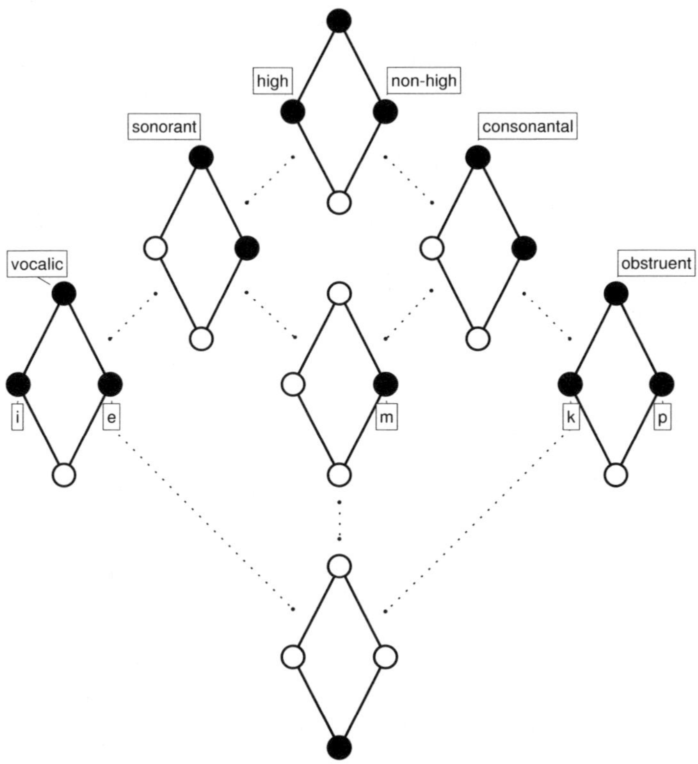

Fig. 2. Direct product of two concept lattices about sound patterns in English

obtained by replacing each line segment between two of the seven rhombuses by four parallel line segments which join the corresponding circles of the two rhombuses. Each rhombus represents a congruence class of the direct product isomorphic to $\underline{\mathfrak{B}}(G_r, M_2, I \cap G_r \times M_2)$, and all those congruence classes together form the elements of a quotient lattice isomorphic to $\underline{\mathfrak{B}}(G_r, M_1, I \cap G_r \times M_1)$. Thus, the nested line diagram of the direct product shown in Fig. 2 can be derived by first drawing a line diagram of $\underline{\mathfrak{B}}(G_r, M_1, I \cap G_r \times M_1)$ and then replacing each of its circles by a copy of a line diagram of $\underline{\mathfrak{B}}(G_r, M_2, I \cap G_r \times M_2)$.

The black circles in the nested line diagram in Fig. 2 represent the concept lattice of the original context (G, M, I) whose lattice order $\leq$ is just the restriction of the lattice order of the direct product. To emphasize such lattice representation, the non-blackened circles are often contracted to a point, respectively. If in the lowest congruence class the lowest circle is black and all other circles are not black, and if the lowest circle in all other circles is not black, respectively, it is common to delete all non-blackened circles and all line segments in the lowest congruence class and to delete the lowest circle with its adjacent line segments in all other congruence classes (cf. the nested line diagram concerning handicapped children in [Wi84]).

The idea of nested line diagrams has been extended in the 1990th to the conception of *TOSCANA-systems* (see [VWW91], [SVW93], [KSVW94], [VW95]). TOSCANA-systems allow to explore data represented by larger formal contexts (G, M, I). Usually, the attribute set M of such a context is divided into a larger number of "meaningful" subsets $M_1, \ldots, M_n$ with $M = M_1 \cup \cdots \cup M_n$. The concept lattice of each subcontext $(G, M_k, I \cap G \times M_k)$ $(k = 1, \ldots, n)$ has to be visualized by a suitable line diagram. A TOSCANA-system for such a family of line diagrams enables the user to present the prepared line diagrams on a computer screen and also nested line diagrams combining two, three ore more of the prepared diagrams. When nestings of line diagrams become difficult to read it is preferrable to restrict to an interesting congruence class and eventually to refine this class by some further nesting. Such procedures makes possible to explore more and more the data coded in (G, M, I). The description of actual software for maintaining and activating TOSCANA-systems can be found in [BH05].

3 Representing and Processing Conceptual Knowledge

In [Wi06], 38 methods of *Conceptual Knowledge Processing* are presented and classified under the twelve headings: 1. Conceptual Knowledge Representation, 2. Determination of Concepts and Contexts, 3. Conceptual Scaling, 4. Conceptual Classification, 5. Analysis of Concept Hierarchies, 6. Aggregation of Concept Hierarchies, 7. Conceptual Identification, 8. Conceptual Knowledge Inferences, 9. Conceptual Knowledge Acquisition, 10. Conceptual Knowledge Retrieval, 11. Conceptual Theory Building, 12. Contextual Logic. In [EW07], those classes of methods are discussed with the focus on *Conceptual Knowledge Representation*, *Conceptual Knowledge Inference*, *Conceptual Knowledge Acquisition*, and *Conceptual Knowledge Communication.* All the discussed methods rely basically on lattice theory. In this section we concentrate on methods of representing and processing conceptual knowledge starting from (quasi-)ordered sets as basic structures.

The representation of knowledge by a formal context is often criticized with the argument that the choice of objects and attributes of the context are too restrictive, so that the given selection of objects and attributes have at least to be justified. This critics has led to develop methods by which formal contexts can be derived from more elementary structures which occur in human thought earlier than concepts (see [Pi70], [SW86], [WW03]). Mathematically, such elementary structures are the *(quasi-)ordered sets* which have been successfully used, especially, as the basic structures for representational measurement theory (see [KLST71]).

A general method of turning an ordered set into a formal context is based on the idea of convergence which allows to create formal objects and formal attributes by converging structures. For instance, the ordered set $(\mathbb{Q}, \leq)$ of all rational numbers gives rise to new numbers like π and e which are represented by infinitely decending number sequences and infinitely ascending number sequences, respectively. In general, the idea of convergence can be substanciated in an ordered set $\underline{P} := (P, \leq)$ by the notions of "filter" and "ideal" where a *filter* of $\underline{P}$ is a non-empty subset F of P, for which $a \in F$ and $a \leq b$ imply $b \in F$ and $a, c \in F$ guarantees the existence of some $d \in F$ with $d \leq a, c$, and an *ideal* of $\underline{P}$ is a non-empty subset I of P, for which $a \in I$ and $a \geq b$ imply $b \in I$ and $a, c \in I$ guarantees the existence of some $d \in I$ with $d \geq a, c$. Filters are of decending nature and ideals are of ascending nature. This is the reason why the filters of $\underline{P}$ are viewed as objects and the ideals of $\underline{P}$ are viewed as attributes. The formal context derived from the ordered set $\underline{P}$ is then defined by $\mathbb{K}(\underline{P}) := (\mathfrak{F}(\underline{P}), \mathfrak{I}(\underline{P}), \Delta)$ where $\mathfrak{F}(\underline{P})$ is the set of all non-empty filters F of $\underline{P}$ and $\mathfrak{I}(\underline{P})$ is the set of all non-empty ideals I of $\underline{P}$ with $F \Delta I :\Longleftrightarrow F \cap I \neq \emptyset$; hence a filter as 'object' has an ideal as 'attribute' if and only if filter and ideal have at least one element in common.

Important are the *ideal-maximal filters* F in $\mathfrak{F}(\underline{C})$ for which an ideal I exists in $\mathfrak{I}(\underline{C})$ such that F is maximal in having the property $F \cap I = \emptyset$; F is named an *I-maximal filter* and, furthermore, if I is a maximal ideal with $F \cap I = \emptyset$ the I is called an *F-opposite*. As dual notions we have *filter-maximal ideals*, *F-maximal ideals*, and *I-opposites*. The set of all ideal-maximal filters is denoted by $\mathfrak{F}_0(\underline{C})$ and the set of all filter-maximal ideals is denoted by $\mathfrak{I}_0(\underline{C})$. The following theorem informs about meaningful structural properties of the concept lattice of $\mathbb{K}(\underline{C})$ (cf. [Ur78], [Ha92], [Wi07b]):

Filter-Ideal-Theorem. *The ordered set $\underline{C}$ is naturally embedded into the concept lattice of the derived context $\mathbb{K}(\underline{C})$ by the map*

$$\iota : x \mapsto (\{F \in \mathfrak{F}(\underline{C}) \mid x \in F\}, \{I \in \mathfrak{I}(\underline{C}) \mid x \in I\})$$

where $\iota(x \wedge y) = \iota(x) \wedge \iota(y)$ if $x \wedge y$ exists in $\underline{C}$ and $\iota(x \vee y) = \iota(x) \vee \iota(y)$ if $x \vee y$ exists in $\underline{C}$. The set $J(\underline{\mathfrak{B}}(\mathbb{K}(\underline{C})))(= \gamma\mathfrak{F}_0(\underline{C}))$ of all $\bigvee$-irreducibles is $\bigvee$-dense and the set $M(\underline{\mathfrak{B}}(\mathbb{K}(\underline{C})))(= \mu\mathfrak{I}_0(\underline{C}))$ of all $\bigwedge$-irreducibles is $\bigwedge$-dense, i.e.,

$$\underline{\mathfrak{B}}(\mathbb{K}(\underline{C})) \cong \underline{\mathfrak{B}}(\mathfrak{F}_0(\underline{C}), \mathfrak{I}_0(\underline{C}), \Delta).$$

For representing knowledge by a formal context derived from an ordered set, the most important result of the Filter-Ideal-Theorem lies in the described reduction of the general context $\mathbb{K}(\underline{C})$ to the reduced context $(\mathfrak{F}_0(\underline{C}), \mathfrak{I}_0(\underline{C}), \Delta)$ whose concept lattice is still isomorphic to the concept lattice of $\mathbb{K}(\underline{C})$. How this general piece of knowledge can be used to clarify human thought shall be demonstrated by treating the question: Does the ordered set $\underline{\mathbb{R}} := (\mathbb{R}, \leq)$ of all real numbers represent a *one-dimensional continuum*?

Real numbers are used to represent time points, so that we can rephrase the continuity question: Is the linear order of time points a one-dimensional continuum? If we follow *Aristotle*, the answer is "no". Aristotle understands time and durations as one-dimensional continua, which means according to his continuum definition that they "are unlimitedly divisible into smaller parts" ([We72], p.431). Therefore, for Aristotle, durations do not consist of time points, but *time points are only limits of durations. Aristotle's conception* of the time and the space continuum yields in general that a continuum does not consists of points, but has as parts only continua again whose nature is to be extensive. In contrast to that, points are in principle of different nature: they are not extensive and can only be understood as limits of extensives.

If we accept that the ordered set $\underline{\mathbb{R}} := (\mathbb{R}, \leq)$ of all real numbers does not represent a one-dimensional continuum, we nevertheless can use the real numbers for describing the *structure of a continuum.* Let $\underline{C_{\mathbb{R}}} := (C_{\mathbb{R}}, \subseteq)$ be the ordered set of all real open intervals where $C_{\mathbb{R}}$ consists of the open intervals

$$]r, s[:= \{x \in \mathbb{R} \mid r < x < s\}$$

with $r \in \mathbb{R} \cup \{-\infty\}$ and $s \in \mathbb{R} \cup \{+\infty\}$. In this ordered set, the convex hull of the set-theoretic union of open intervals is the *supremum* and $]-\infty, +\infty[$ is the greatest element. The $\wedge$-irreducible elements form the *dense chains* $C_1 := \{]-\infty, r[\mid r \in \mathbb{R}\}$ and $C_2 := \{]r, +\infty[\mid r \in \mathbb{R}\}$; there exists an *antiisomorphism* between C_1 and C_2 defined by $]-\infty, r[\mapsto]r, +\infty[$. The pairs $(]-\infty, r[,]r, +\infty[)$ with $r \in \mathbb{R}$ are called the "*cuts*" of $\underline{C_{\mathbb{R}}}$.

The number-theoretic description of a one-dimensional continuum structure gives rise to a more universal order-theoretic definition of linear continuum structures: In general, a *linear continuum structure* is defined as an ordered set $\underline{C} := (C, \leq)$ satisfying the following conditions:

(1) $\underline{C}$ is a $\bigvee$-semilattice with greatest element 1 and no smallest element;
(2) the $\wedge$-irreducible elements of $\underline{C}$ form two disjoint dense chains C_1 and C_2 without greatest and smallest element, where $c_1 \vee c_2 = 1$ for all $c_1 \in C_1$ and $c_2 \in C_2$;
(3) $c_1 \wedge c_2 = d_1 \wedge d_2$ implies $c_1 = d_1$ and $c_2 = d_2$ for all $c_1, d_1 \in C_1$ and $c_2, d_2 \in C_2$;
(4) there exists an antiisomorphism $c^{\dashv} \mapsto c^{\vdash}$ from C_1 onto C_2 such that $C = \{1\} \cup C_1 \cup C_2 \cup \{c^{\dashv} \wedge d \mid c^{\dashv} \in C_1 \text{ and } d \in C_2 \text{ with } c^{\vdash} < d\}$.

The elements of $\underline{C}$ are called (*linear*) *continua*, and the pairs $(c^{\dashv}, c^{\vdash})$ of corresponding elements $c^{\dashv} \in C_1$ and $c^{\vdash} \in C_2$ are called the *cuts* of $\underline{C}$.

The ordered set $\underline{C_{\mathbb{R}}}$ is obviously a linear continuum structure which, according to the Filter-Ideal-Theorem, can be embedded by the mapping ι into the concept lattice $\underline{\mathfrak{B}}(\mathbb{K}(\underline{C_{\mathbb{R}}}))$. This can be illustrated by a linear ordered set $(\check{\mathbb{R}}, \trianglelefteq)$ extending $(\mathbb{R}, \leq)$. $(\check{\mathbb{R}}, \trianglelefteq)$ is defined by

$$\check{\mathbb{R}} := (\mathbb{R} \times \{-1, +1\}) \cup \{(-\infty, +1), (+\infty, -1)\} \text{ and}$$
$$(r, u) \trianglelefteq (s, v) :\iff r < s \text{ or } (r = s \text{ and } u \leq v).$$

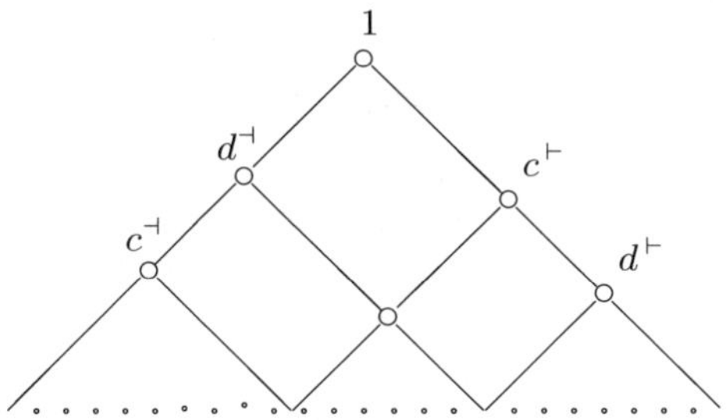

Fig. 3. A linear continuum as an ordered set

The linear ordered set $(\check{\mathbb{R}} \setminus \{(-\infty,+1),(+\infty,-1)\}, \trianglelefteq)$ clearly evolves out of $(\mathbb{R}, \leq)$ by dividing each real number r into the two elements $(r,-1) \lhd (r,+1)$. $\check{\mathbb{R}}$ is bijectively mapped onto the set of all atoms of $\underline{\mathfrak{B}}(\mathbb{K}(\underline{C}_{\mathbb{R}}))$ by the mapping α with

$$\begin{aligned}
&\alpha(-\infty,+1) := \gamma(C_1 \cup \{]-\infty,+\infty[\}),\\
&\alpha(r,-1) := \gamma\{X \in C \mid X \supseteq O \cap]-\infty, r[\text{ for some } O \in C_2 \text{ with }]r,+\infty[\subset O\},\\
&\alpha(r,+1) := \gamma\{Y \in C \mid Y \supseteq O \cap]r,+\infty[\text{ for some } O \in C_1 \text{ with }]-\infty, r[\subset O\},\\
&\alpha(+\infty,-1) := \gamma(C_2 \cup \{]-\infty,+\infty[\}),
\end{aligned}$$

because

$$\begin{aligned}
&C_1 \cup \{]-\infty,+\infty[\},\\
&\{X \in C \mid X \supseteq O \cap]-\infty, r[\text{ for some } O \in C_2 \text{ with }]r,+\infty[\subset O\},\\
&\{Y \in C \mid Y \supseteq O \cap]r,+\infty[\text{ for some } O \in C_1 \text{ with }]-\infty, r[\subset O\},\\
&C_2 \cup \{]-\infty,+\infty[\},
\end{aligned}$$

are exactly the maximal filters of the ordered set $\underline{C}_{\mathbb{R}}$. Simplifying conventions are $-\infty := \alpha(-\infty,+1)$, $r- := \alpha(r,-1)$, $r+ := \alpha(r,+1)$, and $+\infty := \alpha(+\infty,-1)$. The linear order of $(\check{\mathbb{R}}, \leq)$ is transferred onto the set of all atoms of $\underline{\mathfrak{B}}(\mathbb{K}(\underline{C}_{\mathbb{R}}))$ by

$$\alpha(r,u) \trianglelefteq \alpha(s,v) :\Longleftrightarrow (r,u) \trianglelefteq (s,v);$$

according to this order $\trianglelefteq$, $-\infty$ is the smallest atom, $+\infty$ is the greatest atom, and $r- \lhd r+ \lhd s- \lhd s+$ if $r < s$ in $\mathbb{R}$. The continua of the real linear continuum structure $\underline{C}_{\mathbb{R}}$ are represented in the concept lattice by the formal concepts $\iota(]r,s[)$, respectively. Since $\iota(]r,s[) = (r+) \vee (s-)$ the atoms below $\iota(]r,s[)$ are exactly the atoms $\mathfrak{a}$ with $r+ \trianglelefteq \mathfrak{a} \trianglelefteq s-$; therefore it is meaningful to say that the point concepts $r+$ and $s-$ are the limits of the continuum concept $\iota(]r,s[)$. The cuts of the real linear continuum structure are represented in the concept lattice by the pairs $(r-,r+)$; in this conceptual connection $r-$ and $r+$ are standing for the two irreducible subpoints of the reducible point described by the real number r which is represented by the formal concept $(r-) \vee (r+)$.

For a linear continuum structure $\underline{C}$ in general, such structure analysis can be performed using the next theorem (cf. [Wi07b]) in which the following notions are presumed: In the ordered set $\underline{C}$, $F_1 := C_1 \cup \{1\}$ and $F_2 := C_2 \cup \{1\}$ are the 'extreme' ideal-maximal filters and $I_1 := \{x \in C \mid x \leq c \text{ for some } c \in C_1\}$ and

$I_2 := \{x \in C \mid x \leq c \text{ for some } c \in C_2\}$ are the 'extreme' filter-maximal ideals. The cuts $(c^{\dashv}, c^{\vdash})$ of $\underline{C}$ supply the other I-maximal filters of $\underline{C}$ by

$$F_{c^{\dashv}} := \{x \in C \mid x \geq c^{\dashv} \wedge d \text{ for some } d \in C_2 \text{ with } c^{\vdash} < d\},$$
$$F_{c^{\vdash}} := \{y \in C \mid y \geq c^{\vdash} \wedge d \text{ for some } d \in C_1 \text{ with } c^{\dashv} < d\},$$

and the other F-maximal ideals of $\underline{C}$ by

$$I_{(c^{\dashv})} := \{x \in C \mid x < c^{\dashv}\} \text{ and } I_{(c^{\vdash}]} := \{y \in C \mid y \leq c^{\vdash}\},$$
$$I_{(c^{\dashv}]} := \{x \in C \mid x \leq c^{\dashv}\} \text{ and } I_{(c^{\vdash})} := \{y \in C \mid y < c^{\vdash}\}.$$

Theorem 2. *In the concept lattice of the formal context* $\mathbb{K}(\underline{C}) := (\mathfrak{F}(\underline{C}), \mathfrak{I}(\underline{C}), \Delta)$ *of a linear continuum structure* $\underline{C}$,

(1) $\iota(1)(= \gamma F_1 \vee \gamma F_2)$ *is the greatest element of* $\mathbb{K}(\underline{C})$,
(2) $\gamma\mathfrak{F}_0(\underline{C})$ *is the set of all atoms and is the disjoint union of the sets* $A_1 := \{\gamma F_1\} \cup \{\gamma F_{c^{\vdash}} \mid c^{\vdash} \in C_2\}$ *and* $A_2 := \{\gamma F_2\} \cup \{\gamma F_{c^{\dashv}} \mid c^{\dashv} \in C_1\}$,
(3) $\mu\mathfrak{I}_0(\underline{C})(= \{\gamma F_1 \vee \gamma F \mid F \in A_1\} \cup \{\gamma F_2 \vee \gamma F \mid F \in A_2\})$ *is the set of all* $\wedge$*-irreducible elements and is the disjoint union of the convex chains* $[\gamma F_1, \iota(1)[$ *and* $[\gamma F_2, \iota(1)[$,
(4) *for each cut* $(c^{\dashv}, c^{\vdash})$, *we have* $\gamma F_1 \vee \gamma F_{c^{\dashv}} = \mu I_{(c^{\dashv}]}$ *and* $\gamma F_2 \vee \gamma F_{c^{\vdash}} = \mu I_{(c^{\vdash}]}$, $\mu I_{(c^{\dashv}]}$ *is a lower neighbour of* $\mu I_{(c^{\dashv}]} \vee \gamma F_{c^{\vdash}}$ *and an upper neighbour of* $\mu I_{(c^{\dashv})}$, $\mu I_{(c^{\vdash}]}$ *is a lower neighbour of* $\mu I_{(c^{\vdash}]} \vee \gamma F_{c^{\dashv}}$ *and an upper neighbour of* $\mu I_{(c^{\vdash})}$,
(5) *for* $x = d^{\dashv} \wedge c^{\vdash}$, *we have* $\iota(x) := (\{F \in \mathfrak{F}(\underline{C}) \mid x \in F\}, \{I \in \mathfrak{I}(\underline{C}) \mid x \in I\}) = \gamma F_{c^{\vdash}} \vee \gamma F_{d^{\dashv}} = \mu I_{(d^{\dashv}]} \wedge \mu I_{(c^{\vdash}]}$.

The schema of a concept lattice diagram of a general linear continuum structure shown in Fig. 4 stands for the attempt to make phenomena and their related structures conceptually understandable. In the case of linear continua, this has led to an explanation of the distinction between the phenomenological nature of linear continua (represented by ordered sets) and the conceptual nature of points (represented by formal concepts). How the two levels of explanations can work together, this can be demonstrated by the following example: The duration of a journey by train may be imagined as a linear continuum, but if one asks about the departure time and the arrival time of that journey then one asks about time points being the limits of the journey continuum.

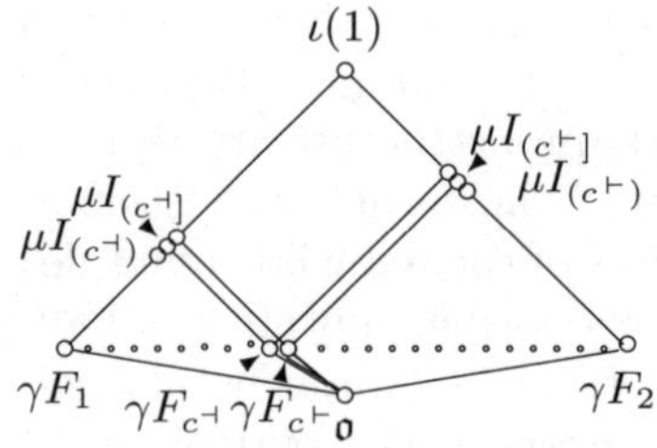

Fig. 4. The schema of a concept lattice diagram of a linear continuum structure

4 Measurement of Conceptual Knowledge Structures

In the preceding sections, the discussed Conceptual Knowledge Structures are all *qualitative* in nature. But, it is also valuable to use Formal Concept Analysis for examining Conceptual Knowledge Structures which are *quantitative* in nature. This has already been done by extending Formal Concept Analysis to the so-called *Algebraic Concept Analysis* in which the formal contexts are also algebraically structured (see [VW94]). More precisely, an *algebraic context* $(\mathbf{A}, B, J)$ is a formal context in which $\mathbf{A} := (A, F)$ is an algebra with A as underlying set and F as family of operations on A having the property that E is an extent of (A, B, J) if and only if E is the underlying set of a subalgebra of $(\mathbf{A}, B, J)$. Dually, the triple $(A, \mathbf{B}, J)$ is said to be a *dual algebraic context* if $(\mathbf{B}, A, J^{-1})$ is an algebraic context for the algebra $\mathbf{B} := (B, G)$. Finally, $(\mathbf{A}, \mathbf{B}, J)$ is called a *bialgebraic context* if $(\mathbf{A}, B, J)$ is an algebraic context and if $(A, \mathbf{B}, J)$ is a dual algebraic context.

Prototypes of bialgebraic contexts are the contexts $(\mathbf{V}, \mathbf{V}^*, \perp_r)$ where $\mathbf{V}$ is a finite-dimensional vector space over a field $\mathbf{K}$, $\mathbf{V}^*$ is its dual space (i.e., the vector space of all linear maps from $\mathbf{V}$ into $\mathbf{K}$ with pointwise operations), and $\perp_r$ is defined by $v \perp_r \varphi :\Longleftrightarrow \varphi(v) = r \in \mathbf{K}$ for all $v \in \mathbf{V}$ and $\varphi \in \mathbf{V}^*$. In case $r = 0$, the extents of the bialgebraic context are exactly the linear subspaces of $\mathbf{V}$ and the intents are exactly the linear subspaces of $\mathbf{V}^*$. In case $r \neq 0$, the extents of the bialgebraic context are exactly V and the affine subspaces of $\mathbf{V}$ not containing 0 and the intents are exactly V^* and the affine subspaces of $\mathbf{V}^*$ not containing 0; to understand $(\mathbf{V}, \mathbf{V}^*, \perp_r)$ also in this case as bialgebraic context, $\mathbf{V}$ resp. $\mathbf{V}^*$ have to be considered as partial algebras which have the linear combinations $r_1x_1 + r_2x_2 + \cdots + r_kx_k$ as partial operations applied only to those k-tuples $(v_1, v_2, \ldots, v_k)$ resp. $(\varphi_1, \varphi_2, \ldots, \varphi_k)$ with

$$0 \notin v_1 + < v_2 - v_1, \ldots, v_k - v_1 > \text{ resp. } \varphi_0 \notin \varphi_1 + < \varphi_2 - \varphi_1, \ldots, \varphi_k - \varphi_1 >$$

where $\varphi_0(x) := 0 \in \mathbf{K}$ for all $x \in \mathbf{V}$.

Theorem 3. (Inversion of Affine Subspaces)[Wl91] *The derivation operators of the bialgebraic context $(\mathbf{V}, \mathbf{V}^*, \perp_r)$ with $r \neq 0$ yield mutually inverse antiisomorhisms between the lattices consisting of the total space of V resp. V^* and of all affine subspaces not containing zero; in particular, if $\mathbf{V} = \mathbb{R}^n = \mathbf{V}^*$, $v \perp_r w :\Leftrightarrow v \cdot w = r$, and $r > 0$, the derivation operators yield the well-known inversion in the hypersphere of radius $\sqrt{r}$ and center 0 in the euclidian space $\mathbb{R}^n$.*

The preceding theorem shows how methods of Formal Concept Analysis can be used to clear up the inversion of affine subspaces conceptually. The appearing connection between Geometry and Formal Concept Analysis shall be made more explicit by explaining the case of dimension two: the *inversion in a circle*, the construction of which is sketched in Fig.5 (cf. [Wi05], p.14ff.):

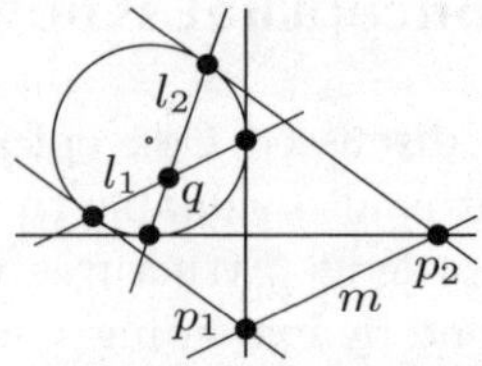

Fig. 5. Inversion in a circle

Geometrically, a point outside the circle is mapped by the inversion onto the line through the two points of contact of the two tangents through the outside point, respectively; for example, p_1 is mapped to l_1 and p_2 is mapped to l_2. By the inverse construction, each line which meets the circle in two points, but does not meet the center of the circle, is mapped to the intersection point of the tangents through the two common points of line and circle; for example, l_1 is mapped to p_1 and l_2 is mapped to p_2. Points in the circle (except the center) and lines outside the circle interchange by the inversion as, for example, the point q and the line m in Fig.5, and a point on the circle interchanges with the tangent through that point. Using the set $\mathbb{R}$ of all *real numbers*, an analogous algebraic representation of the (graphic) plane by $\mathbb{R}^2$ yields a very economic conceptualization of the inversion in a circle: For the circle with radius $\sqrt{r}$, this conceptualization is based on the formal context $(\mathbb{R}^2, \mathbb{R}^2, \perp_r)$ with $(a,b) \perp_r (c,d) :\Leftrightarrow a \cdot c + b \cdot d = r$. For each point $(u,v) \in \mathbb{R}^2$, the derivation $\{(u,v)\}^{\perp_r}$ is a line (and conversely). It follows that the derivations of the formal context $(\mathbb{R}^2, \mathbb{R}^2, \perp_r)$ represent the inversions in the circle with center $(0,0)$ and radius $\sqrt{r}$.

Further research on Algebraic Concept Analysis has been performed via suitably chosen formal contexts for *ordered vector spaces* [Wl99], *finite abelian groups* [Vo95],[Gr99], *modules* [KV96], and *algebraic geometry* [Be99],[Be05],[Be07]. General investigations of *bialgebraic contexts* in which both context sets carry an algebraic structure are presented in [Vo94].

That algebra serves humans with a language for operational descriptions becomes particularly apparent in geometry. Since Graeco-Roman times, algebra plays a significant role in *geometric measurement* which is aiming at the support of human thought and action by making realities graphic, intelligible, and workable (cf. [WW03]). Geometric measurement is to a large extent prototypical for mathematically representing realities which is indicated by the comprehensive theory of representational measurement presented in the three volumes "Foundations of measurement" [KLST71].

The relational structures of measurement theory are idealized models of empirical relationships. They are usually derived from empirical data wherefore data tables are the most important interfaces for abstracting realities to a mathematical theory of measurement. Hence a suitable *mathematization of data tables* - as developed in Formal Concept Analysis [GW99] and presented in the following definition - may substantially contribute to representational measurement theory.

A (*complete*) *many-valued context* has been defined as a set structure (G, M, W, I) where G, M, and W are sets and $I \subseteq G \times M \times W$ such that

$(g, m, v) \in I$ and $(g, m, w) \in I$ imply $v = w$; the elements of G, M, and W are called *objects*, *attributes*, and *attribute values*, respectively. An *ordinal context* has been defined as a set structure $(G, M, (W_m, \geq_m)_{m \in M}, I)$ for which $\geq_m$ is a (partial) order on W_m for all $m \in M$ and $(G, M, \dot{\bigcup}_{m \in M} W_m, I)$ is a many-valued context with $(g, m, w) \in I \Rightarrow w \in W_m$.

If ordinal contexts are considered as relational structures for geometric measurement, a basic task is to investigate representations of ordinal contexts by real vector spaces. An elementary type of such representations is given by the following definition [Wl00]: A finite ordinal context $(G, M, (W_m, \geq_m)_{m \in M}, I)$ with $M = \{m_0, m_1, \ldots, m_n\}$ is *linearly representable* in an n-dimensional real vector space if there are mappings $f_i : W_i \to \mathbb{R}$ $(i = 1, \ldots, n)$ which satisfy, for $s = 1, \ldots, n$ and $g, h \in G$, the following implications:

$$\begin{array}{rcl} m_s(g) >_{m_s} m_s(h) & \Rightarrow & f_s(m_s(g)) > f_s(m_s(h)), \\ m_0(g) >_{m_0} m_0(h) & \Rightarrow & \sum_{i=1}^n f_i(m_i(g)) > \sum_{i=1}^n f_i(m_i(h)), \\ m_0(g) = m_0(h) & \Rightarrow & \sum_{i=1}^n f_i(m_i(g)) = \sum_{i=1}^n f_i(m_i(h)). \end{array}$$

By using *Scott's representation theorem* of finite ordinal structures [Sc64], the following characterization of finite ordinal contexts which are linearly representable in real vector spaces has been proven [Wl00]:

Theorem 4. *A finite ordinal context $(G, M, (W_m, \geq_m)_{m \in M}, I)$ with $M = \{m_0, m_1, \ldots, m_n\}$ is linearly representable in an n-dimensional real vector space if and only if the following condition holds for every $k \in \mathbb{N}$, for all sequences $g_1, \ldots, g_k \in G$, $h_1, \ldots, h_k \in G$ with $m_0(g_j) \geq_{m_0} m_0(h_j)$ $(j = 1, \ldots, k)$, and for all permutations $\sigma_1, \ldots, \sigma_n$ on $\{1, \ldots, k\}$:*

If $m_i(h_{\sigma_i(j)}) \geq_{m_i} m_i(g_j)$ for $j = 1, \ldots, k$ and $i = 1, \ldots, n$, then $m_0(g_j) = m_0(h_j)$ and $m_i(g_j) = m_i(h_{\sigma_i(j)})$ for $j = 1, \ldots, k$ and $i = 1, \ldots, n$.

Although Theorem 4 is a quite elegant representation theorem, it almost never can be substantially applied in practice. A main reason is that, as a rule, data in practice tend to be not densely connected enough. Usually, the most present relationships in data are of ordinal nature. Therefore it is advisable to start with a development of geometric measurement based on ordinal data. In the first step of mathematization ordinal data are modelled by ordinal contexts as defined above. For preparing the second step we define the "*ordinal space*" of an ordinal context:

Let $\mathbb{K} := (G, M, (W_m, \geq_m)_{m \in M}, I)$ be an ordinal context. For each attribute-value-pair $(m, w) \in M \times W_m$ we define the object set $(m, w)^{\geq_m} := \{g \in G \mid m(g) \geq_m w\}$ and $(m, w)^{\leq_m} := \{g \in G \mid m(g) \leq_m w\}$; if $m(g) = w$, we also write $[g)_m$ for $(m, w)^{\geq_m}$ and $(g]_m$ for $(m, w)^{\leq_m}$. The set structure $\underline{\Gamma}(\mathbb{K}) := (G, \mathfrak{H}(\mathbb{K}))$ with $\mathfrak{H}(\mathbb{K}) := \{[g)_m \mid g \in G, m \in M\} \cup \{(g]_m \mid g \in G, m \in M\}$ is called the *ordinal space* of the ordinal context $\mathbb{K}$ and the subsets in $\mathfrak{H}(\mathbb{K})$ are said to be the *generating subspaces* of $\underline{\Gamma}(\mathbb{K})$.

The third step in the development of geometric measurement changes from the synthetic to the *analytic view* where the generating subspaces are desrcibed by linear inequalities. In general, finite-dimensional ordered vector spaces yield the analytic description of ordinal contexts and their ordinal spaces as follows (cf. [Wl99]):

Let V be a finite-dimensional vector space over a (partially) ordered field $(K, \leq)$ and let V^* be its dual space. Then the many-valued context

$$\mathbb{K}(V) := (V, V^*, (K, \leq), E) \text{ with } (v, \varphi, k) \in E :\Longleftrightarrow \varphi(v) = k$$

is called the *ordered bilinear context* of V. This context may also be considered as an ordinal context, the ordinal space of which is given by $(V, \{\{v \in V \mid \varphi(v) \leq k\} \mid \varphi \in V^* \text{ and } k \in K\})$.

As Theorem 4 has already shown, representations of ordinal contexts into ordered vector spaces (over the reals) require strong assumptions. Therefore, to cover a wider spectrum of empirical situations, we offer a more general approach of *geometric measurement by weaker algebraic structures.* For this, further notions concerning ordinal spaces are needed (cf. [WW03]): Let $\mathbb{K} := (G, M, (W_m, \geq_m)_{m \in M}, I)$ be an ordinal context. A *subspace* of the ordinal space $\underline{\Gamma}(\mathbb{K})$ is an arbitrary intersection of generating subspaces of $\underline{\Gamma}(\mathbb{K})$; in particular, G itself is a subspace. For each $m \in M$, the equivalence relation Θ_m on G is defined by its equivalence classes $[g]_m := \{h \in G \mid m(g) = m(h)\}$ $(g \in G)$, which are all subspaces because $[g]_m = [g)_m \cap (g]_m$; furthermore, $\underline{\Delta} := \bigcap_{m \in M} \Theta_m$. $\mathbb{K}$ and $\underline{\Gamma}(\mathbb{K})$ are called *clarified* if $\underline{\Delta}$ is the identity on G.

The basic axioms for our approach are given by the following *antiordinal dependences* between many-valued attributes $m_0, m_1, \ldots, m_n$ in an ordinal context $\mathbb{K} := (G, M, (W_m, \geq_m)_{m \in M}, I)$ $(i = 0, 1, \ldots, n)$:

$$(A_i) \qquad \forall g, h \in G\, (\forall j \in \{0, 1, \ldots, n\} \setminus \{i\} : (g]_{m_j} \subseteq (h]_{m_j}) \\ \Longrightarrow (h]_{m_i} \subseteq (g]_{m_i}.$$

In addition, for the subspaces $[g]_{ij} := \bigcap^{n}_{\substack{k=0 \\ i \neq k \neq j}} [g]_{m_k}$ $(i, j = 0, 1, \ldots, n$ with $i \neq j)$, we consider the following *solvability conditions*:

$$(P_{ij}) \qquad \forall g, h \in G : [g]_{ij} \cap [h]_{m_i} \neq \emptyset.$$

Since the solvability conditions are essential for coordinatizing ordinal spaces, the following embedding theorem (cf. [WW93], [WW96]) yields the important step from empirical models to synthetic geometrical models:

Theorem 5. (**Embedding Theorem**) *The ordinal space $\underline{\Gamma}(\mathbb{K})$ of a clarified ordinal context $\mathbb{K}$ with the attribute set $M = \{m_0, m_1, \ldots, m_n\}$ satisfying (A_i) $(i = 0, 1, \ldots, n)$ can always be embedded into the ordinal space $\underline{\Gamma}(\hat{\mathbb{K}})$ of a clarified ordinal context $\hat{\mathbb{K}}$ with the same attribute set so that $\underline{\Gamma}(\hat{\mathbb{K}})$ satisfies (A_i) and (P_{ij}) $(i, j = 0, 1, \ldots, n$ with $i \neq j)$.*

As a by-product of the Embedding Theorem, we obtain that the solvability conditions cannot be rejected by finite data in the class of ordinal spaces with $n+1$ generating subspaces satisfying the antiordinal dependency axioms. Hence the

(P_{ij}) are only technical conditions which can be added without destroying connections to the empirical models. The resulting spaces can now be coordinatized by ordered n-loops defined as follows (cf. [WW96]):

An *ordered n-loop* is an ordered algebra $\underline{L} := (L, f, 0, \leq)$ for which f is an order-preserving n-ary operation on the ordered set $(L, \leq)$ uniquely solvable in each of its components always respecting the order and with 0 as neutral element. For each ordered n-loop $\underline{L}$, there is a corresponding clarified ordinal context $\mathbb{K}_{\underline{L}} := (L^n, \{m_0, m_1, \ldots, m_n\}, (L, \leq_i)_{i \in \{1,\ldots,n\}}, I_{\underline{L}})$ with $m_0 := f$, $\leq_0 := \geq$ and, for $i = 1, \ldots, n$, $m_i(x_1, \ldots, x_n) := x_i$ and $\leq_i := \leq$. The ordinal space of $\mathbb{K}_{\underline{L}}$ is also denoted by $\underline{\Gamma}(L, f, \leq)$ to emphasize the coordinate structure $(L, f, 0, \leq)$. On the set L^n, the following quasiorders are considered:

$$(x_1, \ldots, x_n) \lesssim_i (y_1, \ldots, y_n) :\Longleftrightarrow x_i \leq y_i \quad \text{for } i = 1, \ldots, n,$$
$$(x_1, \ldots, x_n) \lesssim_0 (y_1, \ldots, y_n) :\Longleftrightarrow f(y_1, \ldots, y_n) \leq f(x_1, \ldots, x_n).$$

Theorem 6. (**Coordinatization Theorem**) *The ordinal space $\underline{\Gamma}(\hat{\mathbb{K}})$ of a clarified ordinal context $\hat{\mathbb{K}}$ with the attribute set $M = \{m_0, m_1, \ldots, m_n\}$ satisfying (A_i) and (P_{ij}) $(i, j = 0, 1, \ldots, n$ with $i \neq j)$ is isomorphic to the ordinal space $\underline{\Gamma}(L, f, \leq)$ of a suitable ordered n-loop $(L, f, 0, \leq)$.*

The Emdedding Theorem and the Coordinatization Theorem together yield a representation of empirical models into algebraic models. Those representations can be explicitly described which give rise to a general representation theorem clarifying the basic role of the antiordinal dependency axioms (cf. [WW95]): Let $\mathbb{K}$ be a clarified ordinal context with the object set G and the attribute set $M = \{m_0, m_1, \ldots, m_n\}$. A *representation map* of its ordinal space $\underline{\Gamma}(\mathbb{K}) := (G, \{[g)_m \mid g \in G, m \in M\} \cup \{(g]_m \mid g \in G, m \in M\})$ into the ordinal space $\underline{\Gamma}(L, f, \leq)$ of an ordered n-loop $(L, f, 0, \leq)$ is defined to be an (injective) mapping $\lambda : G \to L^n$ with $(g]_{m_i} \subseteq (h]_{m_i} \iff \lambda(g) \lesssim_i \lambda(h)$ for all $g, h \in G$ and $i \in \{0, 1, \ldots, n\}$.

Theorem 7. (**General Representation Theorem**) *For the ordinal space $\underline{\Gamma}(\mathbb{K})$ of a clarified ordinal context $\mathbb{K}$ with the attribute set $M = \{m_0, m_1, \ldots, m_n\}$, there exists a representation map from $\underline{\Gamma}(\mathbb{K})$ into the ordinal space $\underline{\Gamma}(L, f, \leq)$ of a suitable ordered n-loop $(L, f, 0, \leq)$ if and only if $\underline{\Gamma}(\mathbb{K})$ satisfies the antiordinal dependency axioms (A_i) for $i = 0, 1, \ldots, n$.*

The question remains how unique are those representation maps? For answering this question we need the following definition (cf. [WW95]): Let $\underline{L} := (L, f, 0, \leq)$ and $\underline{M} := (M, g, 0, \leq)$ be ordered n-loops. $(\iota_0, \iota_1, \ldots, \iota_n)$ is called a *partial isotopy* from $\underline{L}$ into $\underline{M}$ if, for $i = 0, 1, \ldots, n$, ι_i is an isomorphism from $(dom(\iota_i), \leq)$ onto $(im(\iota_i), \leq)$ with $dom(\iota_i) \subseteq L$ and $im(\iota_i) \subseteq M$ such that $\iota_0 f(x_1, \ldots, x_n) = g(\iota_1 x_1, \ldots, \iota_n x_n)$ for all $(f(x_1, \ldots, x_n), x_1, \ldots, x_n) \in dom(\iota_0) \times dom(\iota_1) \times \cdots \times dom(\iota_n)$. For an n-tuple $(\iota_1, \ldots, \iota_n)$ of (partial) maps we define $(\iota_1 \times \cdots \times \iota_n) : dom(\iota_1) \times \cdots \times dom(\iota_n) \longrightarrow im(\iota_1) \times \cdots \times im(\iota_n)$ by $(\iota_1 \times \cdots \times \iota_n)(x_1, \ldots, x_n) := (\iota_1 x_1, \ldots, \iota_n x_n)$.

Theorem 8. (**General Uniqueness Theorem**) *Let $\mathbb{K}$ be a clarified ordinal context with the object set G and the attribute set $M = \{m_0, m_1, \ldots, m_n\}$ and let $\underline{\Gamma}(\mathbb{K})$ be its ordinal space; further, let $\underline{L} := (L, f, 0, \leq)$ and $\underline{M} := (M, g, 0, \leq)$ be ordered n-loops, let λ be a representation map from $\underline{\Gamma}(\mathbb{K})$ into $\underline{\Gamma}(L, f, \leq)$, and let μ be a mapping from G into M^n. Then μ is a representation map from $\underline{\Gamma}(\mathbb{K})$ into $\underline{\Gamma}(M, g, \leq)$ if and only if there exists a partial isotopy $(\iota_0, \iota_1, \ldots, \iota_n)$ from $\underline{L}$ into $\underline{M}$ with $im(f \circ \lambda) \subseteq dom(\iota_0)$, $im(\lambda) \subseteq dom(\iota_1 \times \cdots \times \iota_n)$ and $\mu = (\iota_1 \times \cdots \times \iota_n) \circ \lambda$.*

The Coordinatization Theorem can be specialized to characterize those ordinal spaces which have a representation by ordered Abelian groups and ordered vector spaces (over the reals), respectively (see [Wl95],[WW96]). Of course, as richer the algebraic structures are as better will be the mathematical support for analyzing the empirical models. Therefore, further representation and uniqueness theorems for relevant types of empirical models are desirable, in particular, to grasp even richer dependency structures (cf. [Wl96], [Wl97]).

Although the embeddings of ordinal contexts into bilinear contexts over the reals are not finitely axiomatizable in first order logic [Wl00], the General Representation Theorem can still be used for concrete data which shall finally be demonstrated by the example context presented in Fig.6 (cf. [Wi92]). The data

Receptor	Violet 430	Blue 458	Blue 485	Blue-Green 498
1	147	153	89	57
2	153	154	110	75
3	145	152	125	100
4	99	101	122	140
5	46	85	103	127
6	73	78	85	121
7	14	2	46	52
8	44	65	77	73
9	87	59	58	52
10	60	27	23	24
11	0	0	40	39

Fig. 6. Colour absorption of 11 receptors in a goldfish retina

context in this figure describes the amounts of absorption of four colour stimuli by eleven receptors in a goldfish retina (cf. [SF68]). The data context is viewed as an ordinal context $\mathbb{K}_{col}$ whose integer values are ordered in the natural way. For representing such an ordinal context in a real vector space, by the General Representation Theorem, we have to determine the antiordinal attribute dependencies of the ordinal context (cf. [WW96]). In [GW86], it is shown that the ordinal dependencies of an ordinal context $\mathbb{K} := (G, M, (W_m, \leq_m)_{m \in M}, I)$ are exactly the attribute implications of the formal context $\mathbb{K}_o := (G^2, M, I_o)$ with

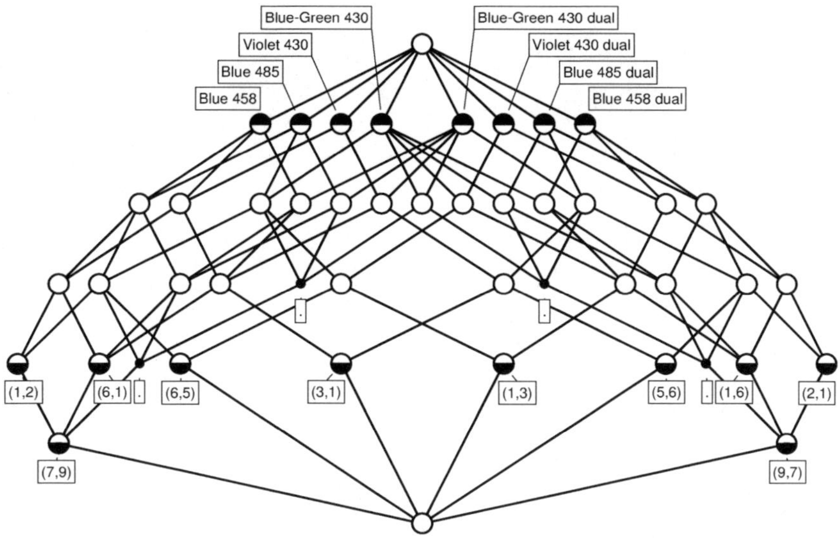

Fig. 7. Concept lattice of $\tilde{\mathbb{K}}_o$

$(g,h)I_o m :\iff m(g) \leq_m m(h)$. Since an ordinal dependency $m_1,\ldots,m_n \xrightarrow{o} m$ is defined by

$$(m_i(g) \leq_{m_i} m_i(h) \text{ for all } i = 1,\ldots,n) \Longrightarrow m(g) \leq_m m(h),$$

for checking the conditions (A_i), it helps to extend $\mathbb{K}$ to the ordinal context $\tilde{\mathbb{K}} := (G, M\dot{\cup}M^d, (W_n, \leq_n)_{n\in M\dot{\cup}M^d}, I\dot{\cup}I^d)$ where $M^d := \{m^d \mid m \in M\}$, $W_{m^d} := W_m$, $v \leq_{m^d} w :\iff w \leq_m v$, and $I^d := \{(g, m^d, w) \mid (g,m,w) \in I\}$. Then we obtain the dependencies described by the (A_i) as implications of $\tilde{\mathbb{K}}_o$ which can be read from the line diagram of the concept lattice of $\tilde{\mathbb{K}}_o$. For our example, this line diagram is depicted in Fig.7 (cf. [WW96] and [Wi04], p.490).

It shows that the conditions (A_i) are satisfied by the ordinal contexts corresponding to the attribute sets

$\{Violet\,430,\ Blue\,458\,dual,\ Blue - Green\,498\}$ and
$\{Violet\,430,\ Blue\,485\,dual,\ Blue - Green\,498\}$.

Therefore, we have two ordered-2-loop-representations which can even be simultanously represented in the Euclidean plane as shown in Fig.8 (cf. [Wi92]). An interesting outcome is that, in the figure, the four colours are "ordered" according to the colour circle.

5 Further Relationships

The main source of further relationships between Formal Concept Analysis and abstract Lattice Theory is the monograph *"Formal Concept Analysis: Mathematical Foundations"* [GW99]. Here we primarily want to point out relationships concerning structural properties, connections, methods, and principles.

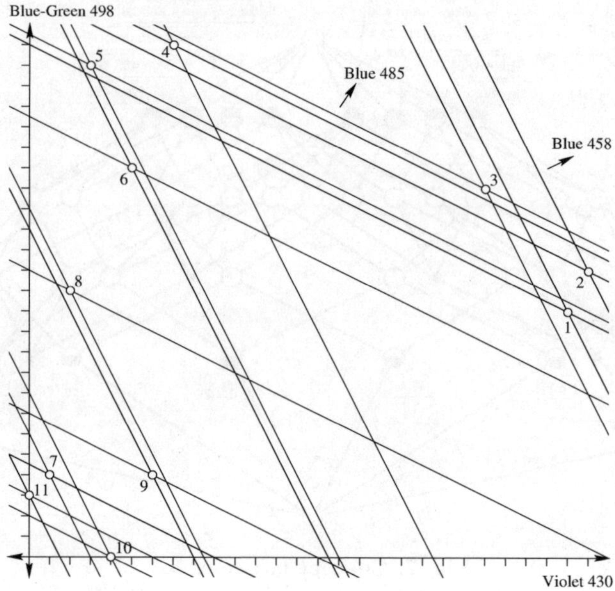

Fig. 8. Representation of the data of Fig.6 in the Euclidean plane

One of the most basic principles is the *duality principle for lattices*; it is based on the observation that the inverse relation $\geq$ of an order relation $\leq$ is an order relation again. A representation of the inverse relation $\geq$, also called the *dual relation* of $\leq$, can be obtained by turning a line diagram of $\leq$ upside-down. The duality principle for lattices says that the identity ι on a lattice $(L, \leq)$ is always an anti-isomorphism from $(L, \leq)$ onto its dual $(L, \leq)^d := (L, \geq)$, i.e., $\iota(x \wedge y) = \iota(x) \vee \iota(y)$ and $\iota(x \vee y) = \iota(x) \wedge \iota(y)$. The duality principle for lattices gives rise to the *duality principle for concept lattices* which can be fomulated as follows: For a formal context (G, M, I), the derived formal context (M, G, I^{-1}) is a formal context for which the mapping $(B, A) \mapsto (A, B)$ is an anti-isomorphism from $\mathfrak{B}(M, G, I^{-1})$ onto $\mathfrak{B}(G, M, I)$ (cf. [GW99], p.22). In applications, an elementary use of the duality principle is based on viewing the formal objects as the formal attributes and vice versa. By the duality principle, we obtain the dual version of Theorem 1 yielding the construction of $\bigwedge$-*Embedding into Direct Products of Concept Lattices.*

One of the most used methods of Formal Concept Analysis in practice is *"Conceptual Scaling"* [GW89] which is derived from the lattice-theoretic construction of *subdirect products of* $\bigvee$-*semilattices.* Conceptual Scaling has been invented for turning (in the plain case) a (complete) many-valued context (G, M, W, I) into a formal context (G, N, J) with $N := \bigcup_{m \in M} \{m\} \times M_m$ and $g \in J(m, n) \Leftrightarrow m(g) = w$ and $wI_m n$, where M_m and I_m are taken from a meaningfully chosen scale context $\mathbb{S}_m := (G_m, M_m, I_m)$ with $m \in M$ and $m(G) \subseteq G_m$. How this is further explained and exemplified by a many-valued context about drive concepts for motor cars can be found in [GW99], p.37ff.

Basic for Formal Concept Analysis in practice is that the formal extents and the formal intents form *closure systems*, respectively. In general, closure systems are complete lattices with respect to set-incusion. For applications, the closure system of all intents of a formal context (G, M, I) is of great interest because it consists of all those sets of attributes which are *closed under attribute implications* of the form $A \to B$ with $A, B \subseteq M$. Formally, a subset T of attributes is closed if and only if, for all attribute implications $A \to B$, $A \subseteq T$ implies $B \subseteq T$ (cf. [GW99], p.69ff.). For many applications it is desirable to know principally all attribute implications of the considered data context. This problem has been solved by introducing the so-called *pseudo-intents* which were used as premises of a canonical basis, the so-called Duquenne-Guigues-Basis [GD86], while the intent closures of the pseudo-intents became the corresponding conclusions. The idea to introduce the pseudo-intents grew out of a lattice-theoretic analysis of the closure system of all intents as $\bigwedge$-subsemilattice within the Boolean lattice of all subsets of the underlying attribute set. How powerful the application of the Duquenne-Guigues-Basis is can already be seen by the *attribute exploration* of nine properties of binary relations (used for ordinal measurement) as elaborated in [GW99], p.85ff.

The rich structure theory of lattices gave rise to an also rich development of *construction and decomposision methods for Formal Concept Analysis*. A systematic examination of different parts and quotients of concept lattices has been performed which led to different types of decompositions such as subdirect decompositions, Atlas-decompositions, substitution decompositions, and tensorial decompositions, but also to several types of constructions such as subdirect product constructions, gluings, local doublings, and tensorial constructions (cf. [GW99], chap. 3 - 5). How such structural tools may be activated shall be sketched by the *method of paired comparisons* which is frequently used to analyse dominance between objects. Here we restrict to paired comparison data consisting of preference judgments on a given set A of alternatives, which can be represented by a finite formal context (A, A, I) with either $(a_1, a_2) \in I$ or $(a_2, a_1) \in I$ for all pairs of alternatives $a_1 \neq a_2$; $(a_1, a_2) \in I$ means that alternative a_2 is preferred to alternative a_1 (cf. [LW88]). For conceptually analysing the represented data, substitution decompositions of the context (A, A, I) and its concept lattice are useful. Two of such decompositions of (A, A, I) lead to the same number of indecomposable subcontexts which are pairwise isomorphic. The line diagrams of the concept lattices of those subcontexts may be used to draw a well readable line diagram of the concept lattice of (A, A, I) by a computer which makes the conceptual structure of the paired comparision data graphically transparent (cf. [LW87], [LW88]).

Finally, it shall be emphasized that it is important for applications to identify meaningful *properties* of concept lattices. Lattice Theory serves Formal Concept Analysis with such properties. In [GW99], chap. 6, a number of interesting properties are discussed, namely *distributivity*, *modularity* and *semimodularity*, *semidistributivity* and *local distributivity*. A different type of properties is given by notions of *dimension.* For formal contexts (G, M, I), the notion of Ferrers dimension is of particular interest. A subset F of $G \times M$ is called a *Ferrers relation* of (G, M, I) if the concept lattice $\mathfrak{B}(G, M, I)$ is a chain; F is a Ferrers relation if and only if $(g, m) \in F$ and $(h, n) \in F$ always imply $(g, n) \in F$ or $(h, m) \in F$. The *Ferrers dimension* of a formal context (G, M, I) is the smallest number of Ferrers relations between G and M the intersection of which equals the relation I. For applications, the following *dimension theorem* is useful (cf. [Wi89], p.36ff.): The Ferrers dimension of a formal context (G, M, I) is equal to the *order dimension* of $\mathfrak{B}(G, M, I)$, i.e. the smallest number of chains which admit an order-embedding of $\mathfrak{B}(G, M, I)$ into their direct product. This theorem can be applied, for instance, to the task of establishing well readable line diagrams of concept lattices which ar essential for reaching valuble interpretations of empirical data represented by formal contexts (cf. [Wi04], p.469ff.).

References

[BM70] Barbut, M., Monjardet, B.: Ordre et Classification. Algébre et Combinatoire. Collection Hachette Université. Paris, Librairie Hachette (1970)

[BH05] Becker, P., Hereth Correia, J.: The ToscanaJ Suite for implementing conceptual information systems. In: [GSW 2005], pp. 324–348 (2005)

[Be99] Becker, T.: Formal Concept Analysis and Algebraic Geometry. Dissertation, TU Darmstadt. Shaker, Aachen (1999)

[Be05] Becker, T.: Features of interaction between Formal Concept Analysis and Algebraic Geometry. In: [GSW 2005], pp. 49–80 (2005)

[Be07] Becker, T.: Polynomial embeddings and representations. In: [KS 2007], pp. 281–303 (2007)

[Bi40] Birkhoff, G.: Lattice theory, 1st edn. American Mathematical Society, Providende (1940)

[Bu03] Burmeister, P.: Formal Concept Analysis with ConImp: Introduction to the basic features. TU Darmstadt (2003), `http://www.mathematik.tu-darmstadt.de/~ Burmeister`

[CH68] Chomsky, N., Halle, M.: The sound pattern of English. Harper & Row, New York (1968)

[DP02] Davey, B.A., Priestley, H.A.: Introduction to lattices and order, 2nd edn. Cambridge University Press, Cambridge (2002)

[DEW04] Denecke, K., Erné, M., Wismath, S.L. (eds.): Galois connections and applications. Kluwer, Dordrecht (2004)

[EW07] Eklund, P., Wille, R.: Semantology as basis for Conceptual Knowledge Processing. In: Kuznetsov, S.O., Schmidt, S. (eds.) ICFCA 2007. LNCS (LNAI), vol. 4390, pp. 18–38. Springer, Heidelberg (2007)

[GSW05] Ganter, B., Stumme, G., Wille, R. (eds.): Formal Concept Analysis: foundations and applications. State-of-the-Art Survey. LNCS (LNAI), vol. 3626. Springer, Heidelberg (2005)

[GW86] Ganter, B., Wille, R.: Implikationen und Abhängigkeiten zwischen Merkmalen. In: Degens, P.O., Hermes, H.-J., Opitz, O. (eds.) Die Klassifikation und ihr Umfeld, pp. 171–185. Indeks, Frankfurt (1986)

[GW89] Ganter, B., Wille, R.: Conceptual scaling. In: Roberts, F. (ed.) Applications of combinatorics and graph theory in the biological and social sciences, pp. 139–167. Springer, New York (1989)

[GW99] Ganter, B., Wille, R.: Formal Concept Analysis: mathematical foundations. Springer, Heidelberg (1999)

[GW06] Wille, R., Gehring, P.: Semantology: Basic Methods for Knowledge Representations. In: Schärfe, H., Hitzler, P., Øhrstrøm, P. (eds.) ICCS 2006. LNCS (LNAI), vol. 4068, pp. 215–228. Springer, Heidelberg (2006)

[GD86] Guigues, J.-L., Duquenne, V.: Familles minimales d'implications informatives resultant d'un tableau de données binaires. Math. Sci. Humaines 95, 5–18 (1986)

[Gr99] Großkopf, A.: Conceptual structures of finite abelian groups. Dissertation, TU Darmstadt. Shaker, Aachen (1999)

[Ha92] Hartung, G.: A topological representation of lattices. Algebra Universalis 29, 273–299 (1992)

[He74] von Hentig, H.: Magier oder Magister? Über die Einheit der Wissenschaft im Verständigungsprozess. Suhrkamp, Frankfurt (1974)

[Ka88] Kant, I.: Logic. Dover, Mineola (1988)

[KV96] Kearnes, K.A., Vogt, F.: Bialgebraic contexts from dualities. Australian Math. Society (Series A) 60, 389–404 (1996)

[KSVW94] Kollewe, W., Skorsky, M., Vogt, F., Wille, R.: TOSCANA - ein Werkzeug zur begrifflichen Analyse und Erkundung von Daten. In: [WZ 1994], pp. 267–288 (1994)

[KLST71] Krantz, D., Luce, D., Suppes, P., Tversky, A.: Foundation of measurement, vol. 1, vol. 2, vol. 3. Academic Press, London (1971) (1989) (1990)

[KS07] Kuznetsov, S.O., Schmidt, S. (eds.): ICFCA 2007. LNCS (LNAI), vol. 4390. Springer, Heidelberg (2007)

[LW87] Luksch, P., Wille, R.: Substitution decomposition of concept lattices. In: Contributions to General Algebra, vol. 5, pp. 213–220. Hölder-Pichler-Tempsky, Wien (1987)

[LW88] Luksch, P., Wille, R.: Formal concept analysis of paired comparisons. In: Bock, H.H. (ed.) Classification and related methods of data analysis, pp. 567–576. North-Holland, Amsterdam (1988)

[Pi70] Piaget, J.: Genetic Epistomology. Columbia University Press, New York (1970)

[SVW93] Scheich, P., Skorsky, M., Vogt, F., Wachter, C., Wille, R.: Conceptual data systems. In: Opitz, O., Lausen, B., Klar, R. (eds.) Information and classification. Concepts, methods and applications, pp. 72–84. Springer, Heidelberg (1993)

[SF68] Schiffmann, H., Falkenberg, Ph.: The organization of stimuli and sensory neurons. Physiology and Behavior 3, 197–201 (1968)

[Sc64] Scott, D.: Measurement structures and linear inequalities. Journal of Mathematical Psychologie 1, 233–244 (1964)

[SW86] Stahl, J., Wille, R.: Preconcepts and set representations of concepts. In: Gaul, W., Schader, M. (eds.) Classification as a tool of research, North-Holland, Amsterdam (1986)

[SW00] Stumme, G., Wille, R. (Hrsg.): Begriffliche Wissensverarbeitung: Methoden und Anwendungen. Springer, Heidelberg (2000)

[Ur78] Urquhart, A.: A topological representation theory for lattices. Algebra Universalis 8, 45–58 (1978)

[Vo94] Vogt, F.: Bialgebraic contexts. Dissertation, TU Darmstadt. Shaker, Aachen (1994)

[Vo95] Vogt, F.: Subgroup lattices of finite Abelian groups. In: Baker, K.A., Wille, R. (eds.) Lattice theory and its applications, pp. 241–259. Heldermann Verlag, Lemgo (1995)

[VWW91] Vogt, F., Wachter, C., Wille, R.: Data analysis based on a conceptional file. In: Bock, H.H., Ihm, P. (eds.) Classification, data analysis, and knowledge organisation, pp. 131–142. Springer, Heidelberg (1991)

[VW94] Vogt, F., Wille, R.: Ideas of Algebraic Concept Analysis. In: Bock, H.-H., Lenski, W., Richter, M.M. (eds.) Information systems and data analysis, pp. 193–205. Springer, Heidelberg (1994)

[VW95] Vogt, F., Wille, R.: TOSCANA – A graphical tool for analyzing and exploring data. In: Tamassia, R., Tollis, I(Y.) G. (eds.) GD 1994. LNCS, vol. 894, pp. 226–233. Springer, Heidelberg (1995)

[We72] Weizsäcker, C.F.: Möglichkeit und Bewegung. Eine Notiz zur aristotelischen Physik. In: C. F. Weizsäcker: Die Einheit der Natur. 3. Aufl. Hanser, München, 428-440 (1972)

[Wi82] Wille, R.: Restructuring lattice theory: an approach based on hierarchies of concepts. In: Rival, I. (ed.) Ordered Sets, pp. 445–470. Reidel, Dordrecht-Boston (1982)

[Wi84] Wille, R.: Liniendiagramme hierarchischer Begriffssysteme. In: Bock, H.H.(ed.) (Hrsg.): Anwendungen der Klassifikation: Datenanalyse und numerische Klassifikation. pp. 32–51 Indeks-Verlag, Frankfurt (1984); Übersetzung ins Englische: Line diagrams of hierachical concept systems. International Classification 11, 77–86 (1984)

[Wi89] Wille, R.: Lattices in data analysis: How to draw them with a computer. In: Rival, I. (ed.) Algorithms and order, pp. 33–58. Kluwer, Dordrecht (1989)

[Wi92] Wille, R.: Concept lattices and conceptual knowledge systems. Computers & Mathematics with Applications 23, 493–515 (1992)

[Wi04] Wille, R.: Dyadic mathematics - abstractions of logical thought. In: Denecke, K., Erné, M., Wismath, S.L. (eds.) Galois Connections and Applications, pp. 453–498. Kluwer, Dordrecht (2004)

[Wi05] Wille, R.: Formal Concept Analysis as mathematical theory of concepts and concept hierarchies. In: [GSW 2005], pp. 1–33 (2005)

[Wi06] Wille, R.: Methods of Conceptual Knowledge Processing. In: Missaoui, R., Schmidt, J. (eds.) Formal Concept Analysis. LNCS (LNAI), vol. 3874, pp. 1–29. Springer, Heidelberg (2006)

[Wi07a] Wille, R.: The basic theorem on labelled line diagrams of finite concept lattices. In: [KS 2007], pp. 303–312 (2007)

[Wi07b] Wille, R.: Formal Concept Analysis of one-dimensional continuum structures. Algebra Universalis (to appear)

[WW93] Wille, R., Wille, U.: On the controversy over Huntington's equations. When are such equations meaningful? Mathematical Social Sciences 25, 173–180 (1993)

[WW95] Wille, R., Wille, U.: Uniqueness of coordinatizations of ordinal structures. Contributions to General Algebra 9, 321–324 (1995)

[WW96] Wille, R., Wille, U.: Coordinatization of ordinal structures. Order 13, 281–284 (1996)

[WW03] Wille, R., Wille, U.: Restructuring general geometry: measurement and visualization of spatial structures. In: Contributions to General Algebra 14, pp. 189–203. Johannes Heyn Verlag, Klagenfurt (2003)

[WZ94] Wille, R., Zickwolff, M. (eds.): Begriffliche Wissensverarbeitung: Grundfragen und Aufgaben. B.I.-Wissenschaftsverlag, Mannheim (1994)

[Wl91] Wille, U.: Eine Axiomatisierung bilinearer Kontexte. Mitteilungen des Mathematischen Seminars Gießen 200, 71–112 (1991)

[Wl95] Wille, U.: Geometric representation of ordinal contexts. Shaker Verlag, Aachen (1996)

[Wl96] Wille, U.: Representation of ordinal contexts by ordered n-quasigroups. European Journal of Combinatorics 17, 317–333

[Wl97] Wille, U.: The role of synthetic geometry in representational measurement theory. Journal of Mathematical Psychology 41, 71–78 (1997)

[Wl99] Wille, U.: Characterization of ordered bilinear contexts. Journal of Geometry 64, 167–207 (1999)

[Wl00] Wille, U.: Linear measurement models - axiomatizations and axiomatizability. Journal of Mathematical Psychology 44, 617–650 (2000)

Direct Factorization by Similarity of Fuzzy Concept Lattices by Factorization of Input Data*

Radim Belohlavek[1,2], Jan Outrata[2], and Vilem Vychodil[1,2]

[1] Dept. Systems Science and Industrial Engineering
T. J. Watson School of Engineering and Applied Science
Binghamton University–SUNY, PO Box 6000, Binghamton, NY 13902–6000, USA
{rbelohla,vychodil}@binghamton.edu
[2] Dept. Computer Science, Palacky University, Olomouc
Tomkova 40, CZ-779 00 Olomouc, Czech Republic
jan.outrata@upol.cz

Abstract. The paper presents additional results on factorization by similarity of fuzzy concept lattices. A fuzzy concept lattice is a hierarchically ordered collection of clusters extracted from tabular data. The basic idea of factorization by similarity is to have, instead of a possibly large original fuzzy concept lattice, its factor lattice. The factor lattice contains less clusters than the original concept lattice but, at the same time, represents a reasonable approximation of the original concept lattice and provides us with a granular view on the original concept lattice. The factor lattice results by factorization of the original fuzzy concept lattice by a similarity relation. The similarity relation is specified by a user by means of a single parameter, called a similarity threshold. Smaller similarity thresholds lead to smaller factor lattices, i.e. to more comprehensible but less accurate approximations of the original concept lattice. Therefore, factorization by similarity provides a trade-off between comprehensibility and precision. We first recall the notion of factorization. Second, we present a way to compute the factor lattice of a fuzzy concept lattice directly from input data, i.e. without the need to compute the possibly large original concept lattice.

1 Introduction and Motivation

Formal concept analysis (FCA) is a method of exploratory data analysis which aims at extracting a hierarchical structure of clusters from tabular data describing objects and their attributes. The history of FCA goes back to Wille's paper [19], foundations, algorithms, and a survey of applications can be found in [11,12].

The clusters $\langle A, B\rangle$, called formal concepts, consist of a collection A (concept extent) of objects and a collection B (concept intent) of attributes which

* Supported by grant No. 1ET101370417 of GA AV ČR, by grant No. 201/05/0079 of the Czech Science Foundation, and by institutional support, research plan MSM 6198959214.

S. Ben Yahia et al. (Eds.): CLA 2006, LNAI 4923, pp. 68–79, 2008.

are maximal with respect to the property that each object from A has every attribute from B. The extent-intent definition of formal concepts goes back to traditional Port-Royal logic. Alternatively, formal concepts can be thought of as maximal rectangles contained in object-attribute data table. Formal concepts can be partially ordered by a natural subconcept-superconcept relation (narrower clusters are under larger ones). The resulting partially ordered set of concepts forms a complete lattice, called a concept lattice, and can be visualized by a labelled Hasse diagram. In the basic setting, the attributes are binary, i.e. each table entry contains either 0 or 1. FCA was extended to data tables with fuzzy attributes, i.e. tables with entries containing degrees to which a particular attribute applies to a particular object, see e.g. [4,5,18].

A direct user comprehension and interpretation of the partially ordered set of clusters may be difficult due to a possibly large number of clusters extracted from a data table. A way to go is to consider, instead of the whole concept lattice, its suitable factor lattice which can be considered a granular version of the original concept lattice: its elements are classes of clusters and the factor lattice is smaller. A method of factorization by a so-called compatible reflexive and symmetric relation (a tolerance) on the set of clusters was described in [12]. Interpreting the tolerance relation as similarity on clusters/concepts, the elements of the factor lattice are classes of pairwise similar clusters. The specification of the tolerance relation is, however, left to the user. In [2], a method of parameterized factorization of concept lattices computed from data with fuzzy attributes was presented: the tolerance relation is induced by a threshold (parameter of factorization) specified by a user. Using a suitable measure of similarity degree of clusters/concepts (see later), the method does the following. Given a threshold a (e.g. a number from $[0,1]$), the elements of the factor lattice are similarity blocks determined by a, i.e. maximal collections of formal concepts which are pairwise similar to degree at least a. The smaller a, the smaller the factor lattice, i.e. the larger the reduction. For a user, the factor lattice provides a granular view on the original concept lattice, where the granules are the similarity blocks.

In order to compute the factor lattice directly by definition, we have to compute the whole concept lattice (this can be done by an algorithm with a polynomial time delay, see [3]) and then compute all the similarity blocks, i.e. elements of the factor lattice (again, this can be accomplished by an algorithm with polynomial time delay).

In this paper, we present a way to compute the factor lattice directly from data. The resulting algorithm is significantly faster than computing first the whole concept lattice and then computing the similarity blocks. In addition to that, the smaller the similarity threshold, the faster the computation of the factor lattice. This feature corresponds to a rule "the more tolerance to imprecision, the faster the result" which is characteristic for human categorization. The method presented can be seen as an alternative to a method of fast factorization of concept lattices by similarity presented in [6].

The paper is organized as follows. Section 2 presents preliminaries on fuzzy sets and formal concept analysis of data with fuzzy attributes. In Section 3, we present the main results. Examples and experiments demonstrating the speed-up are contained in Section 4. Section 5 presents a summary and an outline of a future research.

2 Preliminaries

2.1 Fuzzy Sets and Fuzzy Logic

In this section, we recall necessary notions from fuzzy sets and fuzzy logic. We refer to [4,14,16] for further details. The concept of a fuzzy set generalizes that of an ordinary set in that an element may belong to a fuzzy set in an intermediate truth degree not necessarily being 0 or 1. As a structure of truth degrees, equipped with operations for logical connectives, we use complete residuated lattices, i.e. structures $\mathbf{L} = \langle L, \wedge, \vee, \otimes, \rightarrow, 0, 1\rangle$, where $\langle L, \wedge, \vee, 0, 1\rangle$ is a complete lattice with 0 and 1 being the least and greatest element of L, respectively; $\langle L, \otimes, 1\rangle$ is a commutative monoid (i.e. $\otimes$ is commutative, associative, and $a \otimes 1 = 1 \otimes a = a$ for each $a \in L$); and $\otimes$ and $\rightarrow$ satisfy so-called adjointness property, i.e. $a \otimes b \leq c$ iff $a \leq b \rightarrow c$ for each $a, b, c \in L$. Elements a of L are called truth degrees, $\otimes$ and $\rightarrow$ are (truth functions of) "fuzzy conjunction" and "fuzzy implication".

The most commonly used set L of truth degrees is the real interval $[0, 1]$; with $a \wedge b = \min(a, b)$, $a \vee b = \max(a, b)$. The three most important pairs of "fuzzy conjunction" and "fuzzy implication" are: Łukasiewicz, with $a \otimes b = \max(a + b - 1, 0)$, $a \rightarrow b = \min(1 - a + b, 1)$; minimum, with $a \otimes b = \min(a, b)$, $a \rightarrow b = 1$ if $a \leq b$ and $= b$ else; and product, with $a \otimes b = a \cdot b$, $a \rightarrow b = 1$ if $a \leq b$ and $= b/a$ else. Often, we need a finite chain $\{a_0 = 0, a_1, \ldots, a_n = 1\}$ ($a_0 < \cdots < a_n$); with corresponding Łukasiewicz ($a_k \otimes a_l = a_{\max(k+l-n,0)}$, $a_k \rightarrow a_l = a_{\min(n-k+l,n)}$) or minimum ($a_k \otimes a_l = a_{\min(k,l)}$, $a_k \rightarrow a_l = a_n$ for $a_k \leq a_l$ and $a_k \rightarrow a_l = a_l$ otherwise) connectives. Note that complete residuated lattices are basic structures of truth degrees used in fuzzy logic, see [13,14]. Residuated lattices cover many particular structures, i.e. sets of truth degrees and fuzzy logical connectives, used in applications of fuzzy logic.

A fuzzy set A in a universe set U is a mapping $A : U \rightarrow L$ with $A(u)$ being interpreted as a degree to which u belongs to A. To make $\mathbf{L}$ explicit, fuzzy sets are also called $\mathbf{L}$-sets. By $\mathbf{L}^U$ or L^U we denote the set of all fuzzy sets in universe U, i.e. $L^U = \{A \mid A \text{ is a mapping of } U \text{ to } L\}$. If $U = \{u_1, \ldots, u_n\}$ then A is denoted by $A = \{ {}^{a_1}/u_1, \ldots, {}^{a_n}/u_n\}$ meaning that $A(u_i)$ equals a_i. For brevity, we omit elements of U whose membership degree is zero. A binary fuzzy relation I between sets X and Y is a fuzzy set in universe $U = X \times Y$, i.e. a mapping $I : X \times Y \rightarrow L$ assigning to each $x \in X$ and $y \in Y$ a degree $I(x, y)$ to which x is related to y.

For $A \in L^U$ and $a \in L$, a set ${}^aA = \{u \in U \mid A(u) \geq a\}$ is called an a-cut of A (the ordinary set of elements from U which belong to A to degree at least a); a fuzzy set $a \rightarrow A$ in U defined by $(a \rightarrow A)(u) = a \rightarrow A(u)$ is called an a-shift of A; a fuzzy set $a \otimes A$ in U defined by $(a \otimes A)(u) = a \otimes A(u)$ is called an a-multiple of A.

Given $A, B \in \mathbf{L}^U$, we define a subsethood degree $S(A,B) = \bigwedge_{u \in U}(A(u) \to B(u))$, which generalizes the classical subsethood relation $\subseteq$. $S(A,B)$ represents a degree to which A is a subset of B. In particular, we write $A \subseteq B$ iff $S(A,B) = 1$ (A is fully contained in B). As a consequence, $A \subseteq B$ iff $A(u) \leq B(u)$ for each $u \in U$.

2.2 Fuzzy Concept Lattices

A data table with fuzzy attributes can be identified with a triplet $\langle X, Y, I\rangle$ where X is a non-empty set of objects (table rows), Y is a non-empty set of attributes (table columns), and I is a (binary) fuzzy relation between X and Y, i.e. $I : X \times Y \to L$. In formal concept analysis, the triplet $\langle X, Y, I\rangle$ is called a formal fuzzy context. For $x \in X$ and $y \in Y$, a degree $I(x,y) \in L$ is interpreted as a degree to which object x has attribute y (table entry corresponding to row x and column y). For $L = \{0,1\}$, formal fuzzy contexts can be identified in an obvious way with ordinary formal contexts.

For fuzzy sets $A \in L^X$ and $B \in L^Y$ we define fuzzy sets $A^{\Uparrow} \in L^Y$ and $B^{\Downarrow} \in L^X$ (denoted also $A^{\Uparrow_I}$ and $B^{\Downarrow_I}$ to make I explicit) by

$$A^{\Uparrow}(y) = \textstyle\bigwedge_{x \in X}(A(x) \to I(x,y)), \tag{1}$$

$$B^{\Downarrow}(x) = \textstyle\bigwedge_{y \in Y}(B(y) \to I(x,y)). \tag{2}$$

Using basic rules of predicate fuzzy logic one can see that $A^{\Uparrow}$ is a fuzzy set of all attributes common to all objects from A, and $B^{\Downarrow}$ is a fuzzy set of all objects sharing all attributes from B. The set

$$\mathcal{B}(X,Y,I) = \{\langle A, B\rangle \mid A^{\Uparrow} = B,\ B^{\Downarrow} = A\}$$

of all fixpoints of $\langle{}^{\Uparrow}, {}^{\Downarrow}\rangle$ is called a fuzzy concept lattice associated to $\langle X, Y, I\rangle$; elements $\langle A, B\rangle \in \mathcal{B}(X,Y,I)$ are called formal concepts of $\langle X, Y, I\rangle$; A and B are called the extent and intent of $\langle A, B\rangle$, respectively. Under a partial order $\leq$ defined on $\mathcal{B}(X,Y,I)$ by

$$\langle A_1, B_1\rangle \leq \langle A_2, B_2\rangle \quad \text{iff} \quad A_1 \subseteq A_2,$$

$\mathcal{B}(X,Y,I)$ happens to be a complete lattice. The following theorem, so-called *main theorem of fuzzy concept lattices*, describes the structure of $\mathcal{B}(X,Y,I)$, see [4] for details.

Theorem 1. *$\mathcal{B}(X,Y,I)$ is under $\leq$ a complete lattice where the infima and suprema are given by*

$$\textstyle\bigwedge_{j \in J}\langle A_j, B_j\rangle = \left\langle \bigcap_{j \in J} A_j, (\bigcup_{j \in J} B_j)^{\Downarrow\Uparrow}\right\rangle, \tag{3}$$

$$\textstyle\bigvee_{j \in J}\langle A_j, B_j\rangle = \left\langle (\bigcup_{j \in J} A_j)^{\Uparrow\Downarrow}, \bigcap_{j \in J} B_j\right\rangle. \tag{4}$$

Moreover, an arbitrary complete lattice $\mathbf{K} = \langle K, \leq\rangle$ is isomorphic to some $\mathcal{B}(X,Y,I)$ iff there are mappings $\gamma : X \times L \to K$, $\mu : Y \times L \to K$ such that

(i) $\gamma(X \times L)$ is $\bigwedge$-dense in K, $\mu(Y \times L)$ is $\bigvee$-dense in K and
(ii) $\gamma(x,a) \leq \mu(y,b)$ iff $a \otimes b \leq I(x,y)$.

3 Factorization of $\mathcal{B}(X, Y, I)$ by Similarity

3.1 The Notion of Factorization of Fuzzy Concept Lattice by Similarity

We need to recall the parameterized method of factorization introduced in [2]. Given $\langle X, Y, I\rangle$, introduce a binary fuzzy relation $\approx_{\mathrm{Ext}}$ on $\mathcal{B}(X, Y, I)$ by

$$(\langle A_1, B_1\rangle \approx_{\mathrm{Ext}} \langle A_2, B_2\rangle) = \textstyle\bigwedge_{x\in X}(A_1(x) \leftrightarrow A_2(x)) \quad (5)$$

for $\langle A_i, B_i\rangle \in \mathcal{B}(X, Y, I)$, $i = 1, 2$. Here, $\leftrightarrow$ is a so-called biresiduum (i.e., a truth function of equivalence connective) defined by

$$a \leftrightarrow b = (a \to b) \wedge (b \to a).$$

$(\langle A_1, B_1\rangle \approx_{\mathrm{Ext}} \langle A_2, B_2\rangle)$, called a degree of similarity of $\langle A_1, B_1\rangle$ and $\langle A_2, B_2\rangle$, is just the truth degree of "for each object x: x is covered by A_1 iff x is covered by A_2". One can also consider a fuzzy relation $\approx_{\mathrm{Int}}$ defined by

$$(\langle A_1, B_1\rangle \approx_{\mathrm{Int}} \langle A_2, B_2\rangle) = \textstyle\bigwedge_{y\in Y}(B_1(y) \leftrightarrow B_2(y)). \quad (6)$$

It can be shown [4] that measuring similarity of formal concepts via intents B_i coincides with measuring similarity via extents A_i, i.e. $\approx_{\mathrm{Ext}}$ coincides with $\approx_{\mathrm{Int}}$, corresponding naturally to the duality of extent/intent view. As a result, we write also just $\approx$ instead of $\approx_{\mathrm{Ext}}$ and $\approx_{\mathrm{Int}}$. Note also that $\approx$ is a fuzzy equivalence relation on $\mathcal{B}(X, Y, I)$.

Given a truth degree $a \in L$ (a similarity threshold specified by a user), consider the thresholded relation ${}^a\!\approx$ on $\mathcal{B}(X, Y, I)$ defined by

$$\langle\langle A_1, B_1\rangle, \langle A_2, B_2\rangle\rangle \in {}^a\!\approx \quad \text{iff} \quad (\langle A_1, B_1\rangle \approx \langle A_2, B_2\rangle) \geq a.$$

That is, ${}^a\!\approx$ is an ordinary relation "being similar to degree at least a" and we thereby call it simply similarity (relation). ${}^a\!\approx$ is a reflexive and symmetric binary relation (i.e., a tolerance relation) on $\mathcal{B}(X, Y, I)$. However, ${}^a\!\approx$ need not be transitive (it is transitive if, for instance, $a \otimes b = a \wedge b$ holds true in $\mathbf{L}$). ${}^a\!\approx$ is said to be compatible if it is preserved under arbitrary suprema and infima in $\mathcal{B}(X, Y, I)$, i.e. if $\langle c_j, c'_j\rangle \in {}^a\!\approx$ for $j \in J$ implies both $\langle \bigwedge_{j\in J} c_j, \bigwedge_{j\in J} c'_j\rangle \in {}^a\!\approx$ and $\langle \bigvee_{j\in J} c_j, \bigvee_{j\in J} c'_j\rangle \in {}^a\!\approx$ for any $c_j, c'_j \in \mathcal{B}(X, Y, I)$, $j \in J$. We call $\approx$ compatible if ${}^a\!\approx$ is compatible for each $a \in L$.

Call a subset B of $\mathcal{B}(X, Y, I)$ an ${}^a\!\approx$-block if it is a maximal subset of $\mathcal{B}(X, Y, I)$ such that any two formal concepts from B are similar to degree at least a, i.e., for any $c_1, c_2 \in B$ we have $\langle c_1, c_2\rangle \in {}^a\!\approx$. Note that the notion of an ${}^a\!\approx$-block generalizes that of an equivalence class: if ${}^a\!\approx$ is an equivalence relation then ${}^a\!\approx$-blocks are exactly the equivalence classes of ${}^a\!\approx$. Denote by $\mathcal{B}(X, Y, I)/{}^a\!\approx$ the collection of all ${}^a\!\approx$-blocks. It follows from the results on tolerances on complete lattices [12] that if ${}^a\!\approx$ is compatible, then ${}^a\!\approx$-blocks are special intervals in the concept lattice $\mathcal{B}(X, Y, I)$. For a formal concept

$\langle A,B\rangle \in \mathcal{B}(X,Y,I)$, denote by $\langle A,B\rangle_a$ and $\langle A,B\rangle^a$ the infimum and the supremum of the set of all formal concepts which are similar to $\langle A,B\rangle$ to degree at least a, that is,

$$\langle A,B\rangle_a = \bigwedge\{\langle A',B'\rangle \mid (\langle A,B\rangle,\langle A',B'\rangle) \in {}^a\!\approx\}, \tag{7}$$
$$\langle A,B\rangle^a = \bigvee\{\langle A',B'\rangle \mid (\langle A,B\rangle,\langle A',B'\rangle) \in {}^a\!\approx\}. \tag{8}$$

Operators $\ldots_a$ and $\ldots^a$ are important in description of ${}^a\!\approx$-blocks [12]:

Lemma 1. *${}^a\!\approx$-blocks are exactly intervals of $\mathcal{B}(X,Y,I)$ of the form $[\langle A,B\rangle_a, (\langle A,B\rangle_a)^a]$, i.e.,*

$$\mathcal{B}(X,Y,I)/{}^a\!\approx \;=\; \{[\langle A,B\rangle_a, (\langle A,B\rangle_a)^a] \mid \langle A,B\rangle \in \mathcal{B}(X,Y,I)\}.$$

Note that an interval with lower bound $\langle A_1,B_1\rangle$ and upper bound $\langle A_2,B_2\rangle$ is the subset $[\langle A_1,B_1\rangle,\langle A_2,B_2\rangle] = \{\langle A,B\rangle \in \mathcal{B}(X,Y,I) \mid \langle A_1,B_1\rangle \leq \langle A,B\rangle \leq \langle A_2,B_2\rangle\}$.

Now, define a partial order $\preceq$ on blocks of $\mathcal{B}(X,Y,I)/{}^a\!\approx$ by

$$[c_1,c_2] \preceq [d_1,d_2] \quad \text{iff} \quad c_1 \leq d_1 \quad (\text{iff} \quad c_2 \leq d_2) \tag{9}$$

for any $[c_1,c_2],[d_1,d_2] \in \mathcal{B}(X,Y,I)/{}^a\!\approx$. Then we have [2]:

Theorem 2. *$\mathcal{B}(X,Y,I)/{}^a\!\approx$ equipped with $\preceq$ is a partially ordered set which is a complete lattice, the so-called factor lattice of $\mathcal{B}(X,Y,I)$ by similarity $\approx$ and threshold a.*

Elements of $\mathcal{B}(X,Y,I)/{}^a\!\approx$ can be seen as similarity-based granules of formal concepts/clusters from $\mathcal{B}(X,Y,I)$. $\mathcal{B}(X,Y,I)/{}^a\!\approx$ thus provides a granular view on the possibly large $\mathcal{B}(X,Y,I)$. For further details and properties of $\mathcal{B}(X,Y,I)/{}^a\!\approx$ we refer to [2].

3.2 Similarity-Based Factorization of Input Data $\langle X, Y, I\rangle$ and Direct Computing of the Factor Lattice $\mathcal{B}(X,Y,I)/{}^a\!\approx$

We now turn our attention to the problem of how to compute the factor lattice. One way is to follow the definition and to split the computation of $\mathcal{B}(X,Y,I)/{}^a\!\approx$ into two steps: (1) compute the possibly large fuzzy concept lattice $\mathcal{B}(X,Y,I)$ and (2) compute the ${}^a\!\approx$-blocks, i.e. the elements of $\mathcal{B}(X,Y,I)/{}^a\!\approx$. Although there are efficient algorithms for both (1) and (2), computing $\mathcal{B}(X,Y,I)/{}^a\!\approx$ this way is time demanding. In what follows, we present a way to obtain $\mathcal{B}(X,Y,I)/{}^a\!\approx$ directly, without the need to compute $\mathcal{B}(X,Y,I)$ first and then to compute the blocks of ${}^a\!\approx$. We need the following lemmas.

Lemma 2 ([6]). *For $\langle A,B\rangle \in \mathcal{B}(X,Y,I)$, we have*

(a) $\langle A,B\rangle_a = \left\langle (a \otimes A)^{\Uparrow\Downarrow}, a \rightarrow B\right\rangle$,
(b) $\langle A,B\rangle^a = \left\langle a \rightarrow A, (a \otimes B)^{\Downarrow\Uparrow}\right\rangle$.

Lemma 3. *If A is an extent then we have $a \to A = (a \to A)^{\Uparrow\Downarrow}$; similarly for an intent B.*

Proof. Follows from Lemma 2, cf. [4].

Remark 1. Thus we have $(\langle A, B\rangle_a)^a = \langle a \to (a \otimes A)^{\Uparrow\Downarrow}, (a \otimes (a \to B))^{\Downarrow\Uparrow}\rangle$.

Let us now introduce the construction of a similarity-based factorization assigning to $\langle X, Y, I\rangle$ a "factorized data" $\langle X, Y, I\rangle/a$. For a formal fuzzy context $\langle X, Y, I\rangle$ and a (user-specified) threshold $a \in L$, introduce a formal fuzzy context $\langle X, Y, I\rangle/a$ by

$$\langle X, Y, I\rangle/a \;:=\; \langle X, Y, a \to I\rangle.$$

$\langle X, Y, I\rangle/a$ will be called the factorized context of $\langle X, Y, I\rangle$ by threshold a. That is, $\langle X, Y, I\rangle/a$ has the same objects and attributes as $\langle X, Y, I\rangle$, and the incidence relation of $\langle X, Y, I\rangle/a$ is $a \to I$. Since

$$(a \to I)(x, y) \;=\; a \to I(x, y),$$

computing $\langle X, Y, I\rangle/a$ from $\langle X, Y, I\rangle$ is easy. Note that objects and attributes are more similar in $\langle X, Y, I\rangle/a$ than in the original context $\langle X, Y, I\rangle$. Indeed, for any $x_1, x_2 \in X$ and $y_1, y_2 \in Y$ one can easily verify that

$$I(x_1, y_1) \leftrightarrow I(x_2, y_2) \le (a \to I)(x_1, y_1) \leftrightarrow (a \to I)(x_2, y_2)$$

which intuitively says that in the factorized context, the table entries are more similar (closer) than in the original one.

A way to obtain the factor lattice $\mathcal{B}(X, Y, I)/^a\!\approx$ directly from input data $\langle X, Y, I\rangle$ is based on the next theorem.

Theorem 3. *For a formal fuzzy context $\langle X, Y, I\rangle$ and a threshold $a \in L$ we have*

$$\mathcal{B}(X, Y, I)/^a\!\approx \;\cong\; \mathcal{B}(X, Y, a \to I).$$

In words, $\mathcal{B}(X, Y, I)/^a\!\approx$ is isomorphic to $\mathcal{B}(X, Y, a \to I)$. Moreover, under the isomorphism, $[\langle A_1, B_1\rangle, \langle A_2, B_2\rangle] \in \mathcal{B}(X, Y, I)/^a\!\approx$ corresponds to $\langle A_2, B_1\rangle \in \mathcal{B}(X, Y, a \to I)$.

Proof. Let $^\Uparrow$ and $^\Downarrow$ denote the operators (1) and (2) induced by I and $^{\Uparrow_a}$ and $^{\Downarrow_a}$ denote the operators induced by $a \to I$, that is, for $A \in L^X$ and $B \in L^Y$ we have

$$A^{\Uparrow_a}(y) = \bigwedge_{x \in X} A(x) \to (a \to I)(x, y),$$

$$B^{\Downarrow_a}(y) = \bigwedge_{y \in Y} B(y) \to (a \to I)(x, y).$$

Take any $A \in L^X$. Then we have

$$A^{\Uparrow_a}(y) = \bigwedge_{x\in X} A(x) \to (a \to I(x,y)) =$$
$$= \bigwedge_{x\in X} a \to (A(x) \to I(x,y)) =$$
$$= a \to \bigwedge_{x\in X} (A(x) \to I(x,y)) = a \to A^{\Uparrow}(x),$$

and

$$A^{\Uparrow_a \Downarrow_a}(x) = \bigwedge_{y\in Y} A^{\Uparrow_a}(y) \to (a \to I(x,y)) =$$
$$= \bigwedge_{y\in Y} a \to (A^{\Uparrow_a}(y) \to I(x,y)) = a \to \bigwedge_{y\in Y} (A^{\Uparrow_a}(y) \to I(x,y)) =$$
$$= a \to \bigwedge_{y\in Y} ([\bigwedge_{x\in X} a \to (A(x) \to I(x,y))] \to I(x,y)) =$$
$$= a \to \bigwedge_{y\in Y} ([\bigwedge_{x\in X} (a \otimes A(x)) \to I(x,y)] \to I(x,y)) =$$
$$= a \to \bigwedge_{y\in Y} ((a \otimes A)^{\Uparrow}(x) \to I(x,y)) = a \to (a \otimes A)^{\Uparrow\Downarrow}(x),$$

i.e.

$$A^{\Uparrow_a} = a \to A^{\Uparrow} \quad \text{and} \quad A^{\Uparrow_a \Downarrow_a} = a \to (a \otimes A)^{\Uparrow\Downarrow}. \tag{10}$$

Now, let $[\langle A_1, B_1\rangle, \langle A_2, B_2\rangle] \in \mathcal{B}(X,Y,I)/^a\!\approx$. By Lemmas 1, 2 and 3, there is $\langle A, B\rangle \in \mathcal{B}(X,Y,I)$ such that $\langle A_1, B_1\rangle = \langle A, B\rangle_a = \langle (a \otimes A)^{\Uparrow\Downarrow}, a \to B\rangle$ and $\langle A_2, B_2\rangle = (\langle A, B\rangle_a)^a = \langle a \to (a \otimes A)^{\Uparrow\Downarrow}, (a \otimes (a \to B))^{\Downarrow\Uparrow}\rangle$. Since $\langle A, B\rangle = \langle A, A^{\Uparrow}\rangle$, (10) yields

$$A_2 = a \to (a \otimes A)^{\Uparrow\Downarrow} = A^{\Uparrow_a \Downarrow_a}$$

and

$$B_1 = a \to B = a \to A^{\Uparrow} = A^{\Uparrow_a}.$$

This shows $\langle A_2, B_1\rangle \in \mathcal{B}(X,Y,a \to I)$.

Conversely, if $\langle A_2, B_1\rangle \in \mathcal{B}(X,Y,a \to I)$ then using (10), $B_1 = A_2^{\Uparrow_a} = a \to A_2^{\Uparrow}$ and $A_2 = A_2^{\Uparrow_a \Downarrow_a} = a \to (a \otimes A_2)^{\Uparrow\Downarrow}$. By Lemma 1 and Lemma 2, $[\langle B_1^{\Downarrow}, B_1\rangle, \langle A_2, A_2^{\Uparrow}\rangle] \in \mathcal{B}(X,Y,I)/^a\!\approx$. The proof is complete.

Remark 2. (1) The blocks of $\mathcal{B}(X,Y,I)/^a\!\approx$ can be reconstructed from the formal concepts of $\mathcal{B}(X,Y,a \to I)$:
If $\langle A, B\rangle \in \mathcal{B}(X,Y,a \to I)$ then $[\langle B^{\Downarrow}, B\rangle, \langle A, A^{\Uparrow}\rangle] \in \mathcal{B}(X,Y,I)/^a\!\approx$.

(2) Computing $\mathcal{B}(X, Y, a \to I)$ means computing of the ordinary fuzzy concept lattice. This can be done by an algorithm of polynomial time delay complexity, see [3].

This shows a way to obtain $\mathcal{B}(X, Y, I)/^{a}\!\approx$ without computing first the whole $\mathcal{B}(X, Y, I)$ and then computing the factorization. Note that in [6], we showed an alternative way to speed up the computation of $\mathcal{B}(X, Y, I)/^{a}\!\approx$ by showing that suprema of blocks of $\mathcal{B}(X, Y, I)/^{a}\!\approx$ are fixed points of a certain fuzzy closure operator. Compared to that, the present approach shows that the blocks of $\mathcal{B}(X, Y, I)/^{a}\!\approx$ can be interpreted as formal concepts in a "factorized context" $\langle X, Y, I\rangle/a$.

4 Examples and Experiments

In this section we demonstrate the effect of reduction of size of a fuzzy concept lattice by factorization by similarity, and the speed-up achieved by our algorithm based on Theorem 3. By reduction of size of a fuzzy concept lattice given by a data table $\langle X, Y, I\rangle$ with fuzzy attributes and a user-specified threshold a, we mean the ratio

$$\frac{|\mathcal{B}(X,Y,I)/^{a}\!\approx|}{|\mathcal{B}(X,Y,I)|}$$

of the number $|\mathcal{B}(X,Y,I)/^{a}\!\approx|$ of elements of $\mathcal{B}(X,Y,I)/^{a}\!\approx$, i.e. the number of elements of the factor lattice, to the number $|\mathcal{B}(X,Y,I)|$ of elements of $\mathcal{B}(X,Y,I)$,

Table 1. Data table with fuzzy attributes

	1	2	3	4	5	6	7
1 Czech	0.4	0.4	0.6	0.2	0.2	0.4	0.2
2 Hungary	0.4	1.0	0.4	0.0	0.0	0.4	0.2
3 Poland	0.2	1.0	1.0	0.0	0.0	0.0	0.0
4 Slovakia	0.2	0.6	1.0	0.0	0.2	0.2	0.2
5 Austria	1.0	0.0	0.2	0.2	0.2	1.0	1.0
6 France	1.0	0.0	0.6	0.4	0.4	0.6	0.6
7 Italy	1.0	0.2	0.6	0.0	0.2	0.6	0.4
8 Germany	1.0	0.0	0.6	0.2	0.2	1.0	0.6
9 UK	1.0	0.2	0.4	0.0	0.2	0.6	0.6
10 Japan	1.0	0.0	0.4	0.2	0.2	0.4	0.2
11 Canada	1.0	0.2	0.4	1.0	1.0	1.0	1.0
12 USA	1.0	0.2	0.4	1.0	1.0	0.2	0.4

attributes: 1 – High Gross Domestic Product per capita (USD), 2 – High Consumer Price Index (1995=100) , 3 – High Unemployment Rate (percent - ILO), 4 – High production of electricity per capita (kWh), 5 – High energy consumption per capita (GJ), 6 – High export per capita (USD), 7 – High import per capita (USD)

Table 2. Łukasiewicz fuzzy logical connectives, $\mathcal{B}(X,Y,I)$ of data from Tab. 1: $|\mathcal{B}(X,Y,I)| = 774$, time for computing $\mathcal{B}(X,Y,I) = 2292$ ms; table entries for thresholds $a = 0.2, 0.4, 0.6, 0.8$.

	0.2	0.4	0.6	0.8
size $\lvert\mathcal{B}(X,Y,I)/^{a}\!\approx\rvert$	8	57	193	423
size reduction	0.010	0.073	0.249	0.546
naive algorithm (ms)	8995	9463	8573	9646
our algorithm (ms)	23	214	383	1517
speed-up	391.09	44.22	22.38	6.36

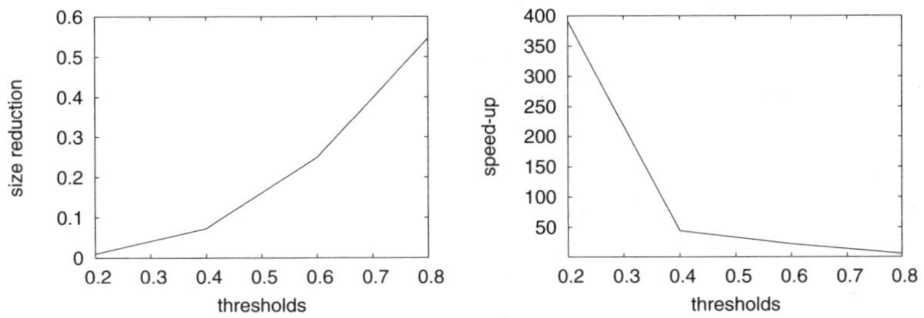

Fig. 1. Size reduction and speed-up from Tab. 2

Table 3. Minimum-based fuzzy logical connectives, $\mathcal{B}(X,Y,I)$ of data from Tab. 1: $|\mathcal{B}(X,Y,I)| = 304$, time for computing $\mathcal{B}(X,Y,I) = 341$ ms; table entries for thresholds $a = 0.2, 0.4, 0.6, 0.8$.

	0.2	0.4	0.6	0.8
size $\lvert\mathcal{B}(X,Y,I)/^{a}\!\approx\rvert$	8	64	194	304
size reduction	0.026	0.210	0.638	1.000
naive algorithm (ms)	1830	1634	3787	4440
our algorithm (ms)	23	106	431	1568
speed-up	79.57	15.42	8.79	2.83

i.e. the number of elements of the original lattice. By a speed-up we mean the ratio of the time for computing the factor lattice $\mathcal{B}(X,Y,I)/^{a}\!\approx$ by a naive algorithm to the time for computing $\mathcal{B}(X,Y,I)/^{a}\!\approx$ by our algorithm. By "our algorithm" we mean the algorithm computing $\mathcal{B}(X,Y,I)/^{a}\!\approx$ directly by reduction to the computation of $\mathcal{B}(\langle X,Y,I\rangle/a)$, described in subsection 3.2. By "naive algorithm" we mean computing $\mathcal{B}(X,Y,I)/^{a}\!\approx$ by first generating $\mathcal{B}(X,Y,I)$ (by a polynomial time-delay algorithm from [3]) and subsequently generating the $^{a}\!\approx$-blocks by producing $[\langle A,B\rangle_a, (\langle A,B\rangle_a)^a]$.

Consider the data table depicted in Tab. 1. The data table contains countries (objects from X) and some of their economic characteristics (attributes from Y). The values of the characteristics are scaled to interval $[0,1]$ so that the characteristics can be considered as fuzzy attributes.

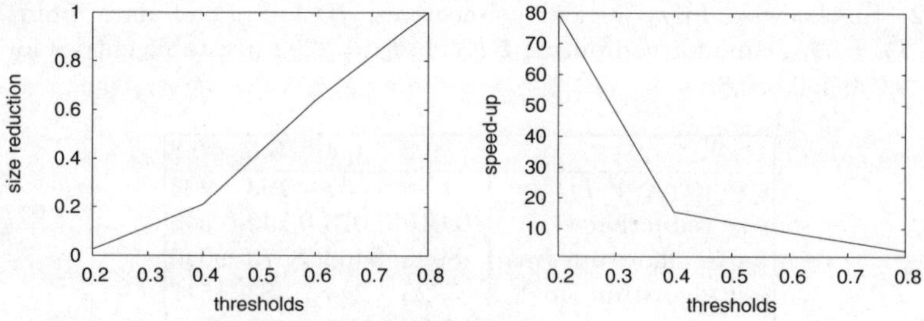

Fig. 2. Size reduction and speed-up from Tab. 3

Tab. 2 summarizes the results when using Łukasiewicz fuzzy logical operations and threshold values $a = 0.2, 0.4, 0.6, 0.8$. The whole concept lattice $\mathcal{B}(X,Y,I)$ contains 774 formal concepts, computing $\mathcal{B}(X,Y,I)$ using the polynomial time delay algorithm from [3] takes 2292ms.

The example demonstrates that smaller thresholds lead to both larger size reduction and speed-up. Furthermore, we can see that the time needed for computing the factor lattice $\mathcal{B}(X,Y,I)/{}^a\!\approx$ is smaller than time for computing the original concept lattice $\mathcal{B}(X,Y,I)$.

Note also that since computing $\mathcal{B}(X,Y,I)$ takes 2292 ms, most of the time consumed by the naive algorithm is spent on factorization. For instance, for $a = 0.2$, 8995 ms is consumed in total of which 2292 ms is spent on computing $\mathcal{B}(X,Y,I)$ and $6703 = 8995 - 2292$ ms is spent on factorization, i.e. on computing $\mathcal{B}(X,Y,I)/{}^a\!\approx$ from $\mathcal{B}(X,Y,I)$.

Fig. 1 contains graphs depicting reduction $|\mathcal{B}(X,Y,I)/{}^a\!\approx| / |\mathcal{B}(X,Y,I)|$ and speed-up from Tab. 2.

Tab. 3 and Fig. 2 show the same characteristics when using the minimum-based fuzzy logical operations (instead of Łukasiewicz fuzzy logical operations).

5 Conclusions and Future Research

We presented an additional method of factorization of fuzzy concept lattices. A factor lattice represents an approximate version of the original fuzzy concept lattice. The size of the factor lattice is controlled by a user-specified threshold. The factor lattice can be computed directly from input data, without first computing the possibly large original fuzzy concept lattice.

Our future research will focus on factorization of further types of fuzzy concept lattices. In particular, [7] presents a method of fast factorization of fuzzy concept lattices with hedges, see [8], which can be seen as a generalization of the method from [6]. Fuzzy concept lattices with hedges serve as a common platform for some of the types of fuzzy concept lattices, see [9], and also [10,17]. An immediate problem is whether and to what extent the results presented in this paper can be accommodated for the setting of fuzzy concept lattices with hedges.

References

1. Belohlavek, R.: Fuzzy Galois connections. Math. Logic Quarterly 45(4), 497–504 (1999)
2. Belohlavek, R.: Similarity relations in concept lattices. J. Logic and Computation 10(6), 823–845 (2000)
3. Belohlavek, R.: Algorithms for fuzzy concept lattices. In: Proc. RASC 2002, Nottingham, UK, December 12–13, 2002, pp. 200–205 (2002)
4. Belohlavek, R.: Fuzzy Relational Systems: Foundations and Principles. Kluwer, Academic/Plenum Publishers, New York (2002)
5. Belohlavek, R.: Concept lattices and order in fuzzy logic. Ann. Pure Appl. Logic 128, 277–298 (2004)
6. Belohlavek, R., Dvořák, J., Outrata, J.: Fast factorization by similarity in formal concept analysis of data with fuzzy attributes. Journal of Computer and System Sciences 73(6), 1012–1022 (2007)
7. Belohlavek, R., Outrata, J., Vychodil, V.: On factorization by similarity of fuzzy concepts with hedges. Int. J. of Foundations of Computer Science (to appear)
8. Belohlavek, R., Vychodil, V.: Reducing the size of fuzzy concept lattices by hedges. In: FUZZ-IEEE 2005, The IEEE International Conference on Fuzzy Systems, Reno (Nevada, USA), May 22–25, 2005, pp. 663–668 (2005)
9. Belohlavek, R., Vychodil, V.: What is a fuzzy concept lattice? In: Proc. CLA 2005, 3rd Int. Conference on Concept Lattices and Their Applications, Olomouc, Czech Republic, September 7–9, 2005, pp. 34–45., http://ceur-ws.org/Vol-162/
10. Ben Yahia, S., Jaoua, A.: Discovering knowledge from fuzzy concept lattice. In: Kandel, A., Last, M., Bunke, H. (eds.) Data Mining and Computational Intelligence, pp. 167–190. Physica-Verlag, Heidelberg New York (2001)
11. Carpineto, C., Romano, G.: Concept Data Analysis. Theory and Applications. J. Wiley, Chichester (2004)
12. Ganter, B., Wille, R.: Formal Concept Analysis. Mathematical Foundations. Springer, Berlin (1999)
13. Goguen, J.A.: The logic of inexact concepts. Synthese 18(9), 325–373 (1968)
14. Hájek, P.: Metamathematics of Fuzzy Logic. Kluwer, Dordrecht (1998)
15. Hájek, P.: On very true. Fuzzy Sets and Systems 124, 329–333 (2001)
16. Klir, G.J., Yuan, B.: Fuzzy Sets and Fuzzy Logic. Theory and Applications. Prentice-Hall, Englewood Cliffs (1995)
17. Krajči, S.: Cluster based efficient generation of fuzzy concepts. Neural Network World 5, 521–530 (2003)
18. Pollandt, S.: Fuzzy Begriffe. Springer, Berlin/Heidelberg (1997)
19. Wille, R.: Restructuring lattice theory: an approach based on hierarchies of concepts. In: Rival, I. (ed.) Ordered Sets, Reidel, Dordrecht, Boston, pp. 445–470 (1982)

Succinct System of Minimal Generators: A Thorough Study, Limitations and New Definitions

T. Hamrouni[1,2], S. Ben Yahia[1], and E. Mephu Nguifo[2]

[1] Department of Computer Science, Faculty of Sciences of Tunis, Tunis, Tunisia
{tarek.hamrouni,sadok.benyahia}@fst.rnu.tn
[2] CRIL-CNRS, IUT de Lens, Lens, France
{hamrouni,mephu}@cril.univ-artois.fr

Abstract. Minimal generators (MGs) are the minimal ones (*w.r.t.* the number of items) among equivalent itemsets sharing a common set of objects, while their associated closed itemset (CI) is the largest one. The pairs - composed by MGs and their associated CI - divide the itemset lattice into distinct equivalence classes. Such pairs were at the origin of various works related to generic association rule bases, concise representations of *frequent* itemsets, arbitrary Boolean expressions, etc. Furthermore, the MG set presents some important properties like the order ideal. The latter helped some level-wise bottom-up and even slightly modified depth-first algorithms to efficiently extract interesting knowledge. Nevertheless, the inherent absence of a unique MG associated to a given CI motivates an in-depth study of the possibility of discovering a kind of redundancy within the MG set. This study was started by Dong *et al.* who introduced the succinct system of minimal generators (SSMG) as an attempt to eliminate the redundancy within this set. In this paper, we give a thorough study of the SSMG as formerly defined by Dong *et al*. Then, we show that the latter suffers from some drawbacks. After that, we introduce new definitions allowing to overcome the limitations of their work. Finally, an experimental evaluation shows that the SSMG makes it possible to eliminate without information loss an important number of *redundant* MGs.

1 Introduction

One efficient way to characterize the itemset lattice is to divide it into different equivalence classes [1]. The minimal elements (*w.r.t.* the number of items) in each equivalence class are called *minimal generators* (MGs) [2] (also referred to as **0**-free itemsets [3] and key itemsets [4]) and the largest element is called a *closed itemset* (CI) [5]. The set of *frequent* CIs is among the first concise representations of the whole set of *frequent* itemsets that were introduced in the literature. This set has been extensively studied and tens of algorithms were proposed to efficiently extract it [6,7]. In the contrary, and despite the important role played by the MGs, they have been paid little attention. Indeed, the MG set is, in general, extracted as a means to achieve *frequent* itemset computations [1,8], *frequent* CI computations [4,5,7], the Iceberg lattice construction [9], etc. The use of the MGs was mainly motivated by their small sizes (they are hence the first elements to be reached in each equivalence class) and by the fact that the MG set fulfills the order

S. Ben Yahia et al. (Eds.): CLA 2006, LNAI 4923, pp. 80–95, 2008.

ideal property which clearly increased the efficiency of both level-wise bottom-up algorithms [4,5,9] and even slightly modified depth-first ones [7]. Nevertheless, some work has been done on the semantic advantages offered by the use of MGs. These works are mainly related to generic association rule bases [2,10,11,12,13], concise representations of *frequent* itemsets [10,11,14,15,16], arbitrary Boolean expressions [17], etc.

The inherent absence of a unique MG associated to a given CI motivates an in-depth study to try to discover a kind of redundancy within the MGs associated to a given CI. This study was started thanks to Dong *et al.* who recently note that some MGs associated to a given CI can be derived from other ones [18]. Indeed, they consider the set of MGs by distinguishing two distinct categories: *succinct* MGs and *redundant* ones. Thus, Dong *et al.* introduce the succinct system of minimal generators (SSMG) as a concise representation of the MG set. They state that *redundant* MGs can be pruned out from the MG set since they can straightforwardly be inferred, without loss of information, using the information gleaned from *succinct* ones [18].

In this paper, we give a thorough study of the SSMG as formerly defined by Dong *et al.* [18]. Then, we show that the *succinct* MGs, as defined in [18], proves *not* to be an *exact* representation of the MG set (*i.e.*, no loss of information *w.r.t. redundant* MGs) in contrary to authors' claims. Furthermore, we also show that the different SSMGs associated to an extraction context do not necessarily share the same size, in contrary to what was stated in [18]. After that, we introduce new definitions allowing to overcome the limitations of the work of Dong *et al.* Indeed, our definitions allow, on the one hand, the SSMG to act as an *exact* representation and, on the other hand, the different SSMGs associated to an extraction context to have the same size. Finally, carried out experiments show that the SSMG makes it possible to eliminate without loss of information an important number of *redundant* MGs and, hence, to almost reach the ideal case: *only one succinct* MG per equivalence class.

The organization of the paper is as follows: Section 2 recalls some preliminary notions that will be used in the remainder of the paper. Section 3 presents a detailed formal study of the SSMG as formerly defined by Dong *et al.* [18], sketches its limitations, and gives new definitions allowing to go beyond the drawbacks of their work. Section 4 is dedicated to the related work. In Section 5, several experiments illustrate the utility of the SSMG towards the redundancy removal within the MG set. Finally, Section 6 concludes this paper and points out our future work.

2 Preliminary Notions

In this section, we present some notions that will be used in the following.

Definition 1. (EXTRACTION CONTEXT) *An extraction context is a triplet* $\mathcal{K} = (\mathcal{O}, \mathcal{I}, \mathcal{R})$*, where* $\mathcal{O}$ *represents a finite set of objects,* $\mathcal{I}$ *is a finite set of items and* $\mathcal{R}$ *is a binary (incidence) relation (i.e.,* $\mathcal{R} \subseteq \mathcal{O} \times \mathcal{I}$*). Each couple* $(o, i) \in \mathcal{R}$ *indicates that the object* $o \in \mathcal{O}$ *has the item* $i \in \mathcal{I}$*.*

Example 1. Consider the extraction context in Table 1 where $\mathcal{O} = \{1, 2, 3, 4\}$ and $\mathcal{I} = \{a, b, c, d, e, f, g\}$. The couple $(2, d) \in \mathcal{R}$ since it is crossed in the matrix.

Table 1. An extraction context $\mathcal{K}$

	a	b	c	d	e	f	g
1			×	×	×	×	×
2	×	×	×	×	×		
3	×	×	×			×	×
4	×	×	×	×	×		×

For arbitrary sets $I \subseteq \mathcal{I}$ and $O \subseteq \mathcal{O}$, the following derivation operators are defined [19]: $I' = \{o \in \mathcal{O} \mid \forall\, i \in I, (o, i) \in \mathcal{R}\}$, and, $O' = \{i \in \mathcal{I} \mid \forall\, o \in O, (o, i) \in \mathcal{R}\}$. The composite operators $''$ define closures on $(2^{\mathcal{I}}, \subseteq)$ and $(2^{\mathcal{O}}, \subseteq)$. A pair (I, O), of mutually corresponding subsets, *i.e.*, $I = O'$ and $O = I'$, is called a (formal) concept [19], where I is the intent and O is the extent (*e.g.*, (*cde*, 124)[1] is a concept from Table 1). Once applied, the corresponding operator $''$ induces an equivalence relation on the power set of items $2^{\mathcal{I}}$ partitioning it into distinct subsets called *equivalence classes* [1], which will further be denoted γ*-equivalence classes*. In each class, all itemsets appear in the same set of objects and, hence, have the same closure. The largest element (*w.r.t.* set inclusion) is called a *closed itemset* (CI) – the intent part of a formal concept – while the minimal incomparable ones are called *minimal generators* (MGs). These notions are defined as follows:

Definition 2. (CLOSED ITEMSET)*[5] An itemset $f \subseteq \mathcal{I}$ is said to be closed if and only if $f'' = f$.*

Example 2. Given the extraction context depicted by Table 1, the itemset "*cdeg*" is a closed itemset since it is the maximal set of items common to the set of objects $\{1, 4\}$. The itemset "*cdg*" is not a closed itemset since all objects containing the itemset "*cdg*" also contain the item "*e*".

Definition 3. (MINIMAL GENERATOR)*[2] An itemset $g \subseteq \mathcal{I}$ is said to be a minimal generator of a closed itemset f if and only if $g'' = f$ and $\nexists\, g_1 \subset g$ s.t. $g_1'' = f$.*

The set MG_f of the MGs associated to an CI f is hence $\text{MG}_f = \{g \subseteq \mathcal{I} \mid g'' = f \wedge \nexists\, g_1 \subset g \text{ s.t. } g_1'' = f\}$.

Example 3. Consider the CI "*cdeg*" described by the previous example. "*cdeg*" has "*dg*" as an MG. Indeed, $(dg)'' = cdeg$ and the closure of every subset of "*dg*" is different from "*cdeg*". Indeed, $(\emptyset)'' = c$, $(d)'' = cde$ and $(g)'' = cg$. The CI "*cdeg*" has also another MG which is "*eg*". Hence, $\text{MG}_{cdeg} = \{dg, eg\}$. "*cdeg*" is then the largest element of its γ-equivalence class, whereas "*dg*" and "*eg*" are the minimal incomparable ones. All these itemsets share the set of objects $\{1, 4\}$.

Since in practice, we are mainly interested in itemsets that occur at least in a given number of objects, we introduce the notion of support and frequency.

Definition 4. (SUPPORT AND FREQUENCY) *The support of an itemset $I \subseteq \mathcal{I}$, denoted by Supp(I), is equal to the number of objects in $\mathcal{K}$ that have all items from I. While*

[1] We use a separator-free form for the sets, *e.g.*, the set *cde* stands for $\{c, d, e\}$.

the frequency of I in $\mathcal{K}$ is equal to $\frac{Supp(I)}{|\mathcal{O}|}$. I is said to be frequent in $\mathcal{K}$ if Supp(I) is greater than or equal to a minimum support threshold, denoted minsupp.

In the remainder, we will mainly use the support of itemsets instead of their frequency.

Example 4. Consider the itemset "*cde*" of the extraction context depicted by Table 1. The objects 1, 2 and 4 contain the itemset "*cde*". Hence, Supp(*cde*) = **3**. If *minsupp* = **2**, then "*cde*" is *frequent* in $\mathcal{K}$ since Supp(*cde*) = $\mathbf{3 \geq 2}$.

3 Succinct System of Minimal Generators

In this section, and as a first step, we study the main structural properties of the succinct system of minimal generators (SSMG) as formerly defined by Dong *et al.* [18][2]. As a second step, we highlight some drawbacks of their work. Finally, we propose new definitions allowing to overcome these limitations.

3.1 A Thorough Study

Recently, Dong *et al.* note the existence of a certain form of intra-redundancy within the set of the minimal generators (MGs) associated to a given closed itemset (CI), *i.e.*, that one can derive some MGs from the others. They, hence, presented a study [18] in which they split the set of MGs associated to a given CI into three distinct subsets. The formalization of these subsets, introduced in Definition 6, requires that we adopt a total order relation among itemsets defined as follows:

Definition 5. (TOTAL ORDER RELATION) *Let $\preceq$ be a total order relation among item literals, i.e., $\forall\ i_1, i_2 \in \mathcal{I}$, we have either $i_1 \preceq i_2$ or $i_2 \preceq i_1$. This relation is extended to also cope with itemsets of different sizes by first considering their cardinality. This is done as follows: Let X and Y be two itemsets and let Card(X) and Card(Y) be their respective cardinalities. We then have:*

- *If Card(X) < Card(Y), then $X \prec Y$.*
- *If Card(X) = Card(Y), then X and Y are compared using their lexicographic order. Hence, $X \prec Y$ if and only if $X \preceq Y$ and $X \neq Y$.*

Example 5. Consider the alphabetic order on items as the basis for the total order relation $\preceq$ on itemsets[3]:

- Since Card(*d*) < Card(*be*), then $d \prec be$.
- Since Card(*abd*) = Card(*abe*), then *abd* and *abe* are compared using their lexicographic order. We then have $abd \prec abe$ since $abd \preceq abe$ and $abd \neq abe$.

[2] Please note that we mainly refer to the SSMG_MINER algorithm proposed by the authors [18]. In fact, the concrete examples related to SSMG_MINER are the only source of precise information about several aspects of the target structure.

[3] In the remainder of the paper, we will only mention the criterion used to order items (*e.g.*, alphabetic order, ascending/descending support order, etc). The latter is then extended to be the total order relation on itemsets, as shown in Definition 5.

Please note that the cardinality factor preserves the spirit of MGs as the smallest itemset in a γ-equivalence class will necessarily be an MG. Three categories of MGs emerge [18], which we formalize as follows:

Definition 6. (MINIMAL GENERATORS' CATEGORIES) *The set* MG_f, *of the* MG*s associated to an* CI f, *can be portioned into three distinct subsets as follows:*

1. $\text{MG}\texttt{rep}_f = \{g \in \text{MG}_f \mid \nexists\, g_1 \in \text{MG}_f \text{ s.t. } g_1 \prec g\}$ *contains the smallest* MG, *given a total order relation* $\preceq$, *which constitutes the* ***representative*** MG *of* f.
2. $\text{MG}\texttt{can}_f = \{g \in \text{MG}_f \mid (g \notin \text{MG}\texttt{rep}_f) \wedge (\forall\, g_1 \subset g,\ g_1 \in \text{MG}\texttt{rep}_{f_1} \text{ where } f_1 = g_1'')\}$ *contains the* ***canonical*** MG*s of* f. *A canonical* MG *is not the smallest one in* MG_f *and, hence, is not the representative* MG *of* f. *Nevertheless, all its subsets are the representative* MG*s of their respective closures.*
3. $\text{MG}\texttt{red}_f = \{g \in \text{MG}_f \mid \exists\, g_1 \subset g \text{ s.t. } g_1 \notin \text{MG}\texttt{rep}_{f_1} \text{ where } f_1 = g_1''\}$ *contains the* ***redundant*** MG*s of* f.

An MG is said to be *succinct* if it is either a *representative* or a *canonical* one [18]. The set $\text{MG}\texttt{suc}_f$, composed by the *succinct* MGs associated to the CI f, is then equal to the union of $\text{MG}\texttt{rep}_f$ and $\text{MG}\texttt{can}_f$: $\text{MG}\texttt{suc}_f = \text{MG}\texttt{rep}_f \bigcup \text{MG}\texttt{can}_f$. Hence, $\text{MG}\texttt{red}_f = \text{MG}_f \backslash \text{MG}\texttt{suc}_f$.

Example 6. Let us consider the extraction context $\mathcal{K}$ depicted by Table 1. The total order relation $\preceq$ is set to the alphabetic order. Table 2 shows, for each CI, the following information: its MGs, its *succinct* MGs and its support. In the fourth column, the *representative* MG is marked with bold letters. The others are hence *canonical* ones. Note that for **11** CIs, there are **23** MGs, from which only **13** are *succinct* ones (**11** are *representative* MGs and only **2** are *canonical* ones). The MG "*ad*" is a *representative* one, since it is the smallest MG, *w.r.t.* $\preceq$, among those of "*abcde*". Indeed, $ad \preceq ae$, $ad \preceq bd$ and $ad \preceq be$. The MG "*e*" is not the *representative* of its CI "*cde*", since $d \preceq e$. Nevertheless, its unique subset (*i.e.*, "$\emptyset$") is the *representative* MG of its CI "*c*". Hence, "*e*" is a *canonical* MG. Finally, the MG "*bdg*" is a *redundant* one, since at least one of its subsets is not a *representative* MG ("*bg*", for example).

The definition of the SSMG is as follows [18]:

Definition 7. (SUCCINCT SYSTEM OF MINIMAL GENERATORS) *Given a total order relation* $\preceq$, *a succinct system of minimal generators* (SSMG) *consists of, for each* CI f, *the set* $\text{MG}\texttt{rep}_f$ *containing its representative* MG *and, if there is any, the set* $\text{MG}\texttt{can}_f$ *containing its canonical* MG*s.*

It is important to mention that, for a given extraction context, the SSMG is not unique since it closely depends on the choice of the total order relation $\preceq$ (*e.g.*, the alphabetic order, the ascending/descending support order, etc.).

In the remainder, the set of *representative* (*resp. canonical*, *redundant*, *succinct* and all) MGs extracted from a context $\mathcal{K}$ will be denoted $\mathcal{MG}\texttt{rep}_{\mathcal{K}}$ (*resp.* $\mathcal{MG}\texttt{can}_{\mathcal{K}}$, $\mathcal{MG}\texttt{red}_{\mathcal{K}}$, $\mathcal{MG}\texttt{suc}_{\mathcal{K}}$ and $\mathcal{MG}_{\mathcal{K}}$). The set of CIs extracted from $\mathcal{K}$ will be denoted $\mathcal{CI}_{\mathcal{K}}$. The letter $\mathcal{F}$ will be added to each notation if the respective set is restricted to its *frequent* elements.

Table 2. The CIs extracted from $\mathcal{K}$ and for each one, the corresponding MGs, *succinct* MGs and support

#	CI	MGs	Succinct MGs	Support
1	*c*	$\emptyset$	$\emptyset$	4
2	*abc*	*a, b*	***a**, b*	3
3	*cde*	*d, e*	***d**, e*	3
4	*cg*	*g*	***g***	3
5	*cfg*	*f*	***f***	2
6	*abcde*	*ad, ae, bd, be*	***ad***	2
7	*abcg*	*ag, bg*	***ag***	2
8	*abcfg*	*af, bf*	***af***	1
9	*cdeg*	*dg, eg*	***dg***	2
10	*cdefg*	*df, ef*	***df***	1
11	*abcdeg*	*adg, aeg, bdg, beg*	***adg***	1

Proposition 1. *The total order relation $\preceq$ ensures the uniqueness of the representative* MG *associated to a given* CI. *Hence, the cardinality of the set of representative* MGs *is equal to that of* CIs *(i.e., Card($\mathcal{MG}\texttt{rep}_{\mathcal{K}}$) = Card($\mathcal{CI}_{\mathcal{K}}$)).*

The proof is trivial. Indeed, there is only one representative MG per γ-equivalence class (see Definition 6).

Remark 1. The respective sizes of both sets $\mathcal{MG}\texttt{can}_{\mathcal{K}}$ and $\mathcal{MG}\texttt{red}_{\mathcal{K}}$ are closely related to the nature of the extraction context, *i.e.*, whether the objects are highly/weakly correlated. Nevertheless, if the set $\mathcal{MG}\texttt{can}_{\mathcal{K}}$ is empty, then the set $\mathcal{MG}\texttt{red}_{\mathcal{K}}$ is also empty (the reverse is not always true).

To show that the set $\mathcal{MG}\texttt{suc}_{\mathcal{K}}$ is an order ideal (or down-set) in $(2^{\mathcal{I}}, \subseteq)$ [19], we have to prove that all subsets of a *representative* MG are also *representative* ones. This is done thanks to Proposition 3 whose the proof requires Lemma 1 and Proposition 2.

Lemma 1. *[19] Let X and Y be two itemsets. If $X'' = Y''$, then $\forall$ $Z \subseteq \mathcal{I}$, $(X \bigcup Z)''$ $= (Y \bigcup Z)''$.*

In our context, with $X \bigcup Y$, we will indicate the ordered set of items, *w.r.t.* the total order relation $\preceq$, contained in X or in Y.

Proposition 2. *Let X, Y, Z be three itemsets s.t. $X \bigcap Z = \emptyset$ and $Y \bigcap Z = \emptyset$. If $X \preceq Y$, then $(X \bigcup Z) \preceq (Y \bigcup Z)$.*

Proposition 3. *All subsets of a representative* MG *are also representative ones.*

Proof. Let g be a representative MG *and f its closure. Suppose, we have $g_1 \subset g$ and $g_1 \notin$* MG$\texttt{rep}_{f_1}$ *where $f_1 = g_1''$. Let g_2 be the representative* MG *of f_1. Consequently, $g_2 \prec g_1$. Since, $g_1'' = g_2''$, then, according to Lemma 1, we have $(g_1 \bigcup (g \setminus g_1))'' = (g_2 \bigcup (g \setminus g_1))''$ and, hence, $g'' = (g_2 \bigcup (g \setminus g_1))''$. Let g_3 be equal to $(g_2 \bigcup (g \setminus g_1))$.*

According to the second case in Definition 5 and to Proposition 2, we have $g_3 \prec g$ *since* $g_2 \prec g_1$, $g_2 \bigcap (g \setminus g_1) = \emptyset$ *and* $g_1 \bigcap (g \setminus g_1) = \emptyset$. *If* g_3 *is an* MG, *then* g *can not be a representative* MG *what is in contradiction with the initial assumption that* g *is a representative* MG. *If* g_3 *is not an* MG, *then it exists an* MG g_4 *s.t.* $g_4 \subset g_3$ *and* $g_4'' = g_3''$. *Since* $Card(g_4) < Card(g_3)$, *then* $g_4 \prec g_3$ *(according to the first case in Definition 5) and, hence,* $g_4 \prec g$. *This result is also in contradiction with the starting assumption. Thus, we can conclude that each subset of* g *is necessarily a representative* MG. ♦

Hence, according to Proposition 3, if f is an CI, then MGsuc_f=$\text{MGrep}_f \bigcup \text{MGcan}_f$ = $\{g \in \text{MG}_f \mid \forall\, g_1 \subset g, g_1 \in \text{MGrep}_{f_1}$ where $f_1 = g_1''\}$.

Thanks to Proposition 4, given below with its proof, we show that the *succinctness* of MGs is an anti-monotone constraint. Hence, the set $\mathcal{MG}\text{suc}_{\mathcal{K}}$ is an order ideal in $(2^{\mathcal{I}}, \subseteq)$.

Proposition 4. (ANTI-MONOTONE CONSTRAINT) *Let* g *be an itemset.* g *fulfills the following two properties:*

1. *If* $g \in \mathcal{MG}\text{suc}_{\mathcal{K}}$, *then* $\forall\, g_1$ *s.t.* $g_1 \subset g$, $g_1 \in \mathcal{MG}\text{suc}_{\mathcal{K}}$.
2. *If* $g \notin \mathcal{MG}\text{suc}_{\mathcal{K}}$, *then* $\forall\, g_1$ *s.t.* $g_1 \supset g$, $g_1 \notin \mathcal{MG}\text{suc}_{\mathcal{K}}$.

Proof.
1. $g \in \mathcal{MG}\text{suc}_{\mathcal{K}} \Longrightarrow \forall\, g_1$ *s.t.* $g_1 \subset g$, $g_1 \in \text{MGrep}_{f_1}$ *where* $f_1 = g_1''$ *(according to Definition 6.)* $\Longrightarrow \forall\, g_1$ *s.t.* $g_1 \subset g$, $g_1 \in \text{MGsuc}_{f_1}$ *(since* $\text{MGrep}_{f_1} \subseteq \text{MGsuc}_{f_1}$.*)* $\Longrightarrow \forall\, g_1$ *s.t.* $g_1 \subset g$, $g_1 \in \mathcal{MG}\text{suc}_{\mathcal{K}}$ *(since* $\text{MGsuc}_{f_1} \subseteq \mathcal{MG}\text{suc}_{\mathcal{K}}$.*).*
2. $g \notin \mathcal{MG}\text{suc}_{\mathcal{K}} \Longrightarrow \forall\, g_1$ *s.t.* $g \subset g_1$, $g_1 \in \text{MGred}_{f_1}$ *where* $f_1 = g_1''$ *(indeed,* g_1 *has at least a non-representative subset, namely* g*, since the latter is not a succinct* MG *and, hence, is not a representative one.)* $\Longrightarrow \forall\, g_1$ *s.t.* $g \subset g_1$, $g_1 \notin \text{MGsuc}_{f_1}$ *(according to Definition 6,* g_1 *can not be both redundant and succinct at the same time.)* $\Longrightarrow \forall\, g_1$ *s.t.* $g \subset g_1$, $g_1 \notin \mathcal{MG}\text{suc}_{\mathcal{K}}$ *(we have* $g_1 \notin \text{MGsuc}_{f_1}$. *In addition,* $g_1 \notin (\mathcal{MG}\text{suc}_{\mathcal{K}} \setminus \text{MGsuc}_{f_1})$ *since the closure of* g_1 *is unique and is equal to* f_1.*).* ♦

Since the frequency constraint is also anti-monotone, it is easy to show that the set $\mathcal{FMG}\text{suc}_{\mathcal{K}}$, of the *succinct frequent* MGs extracted from the context $\mathcal{K}$, is also an order ideal in $(2^{\mathcal{I}}, \subseteq)$. This interesting property allowed us to propose an efficient algorithm to extract the SSMG according to the definition of Dong *et al.* (see [20] for more details).

3.2 Limitations of the Work of Dong *et al.*

Starting form Definition 7, the main facts that can be pointed out from the work of Dong *et al.* can be unraveled by the following claims [18]:

Claim 1: The cardinality of an SSMG is insensitive to the considered total order relation $\preceq$, *i.e.*, whatever the total order relation, the number of *canonical* MGs is the same. Recall that the number of *representative* ones is exactly equal to that of CIs, as stated by Proposition 1.

Claim 2: A SSMG is an *exact representation* of the MG set, *i.e.*, if g is a *redundant* MG, then g can be inferred from the SSMG without loss of information. To do so, for each γ-equivalence class, Dong *et al.* propose to infer its *redundant* MGs by replacing the subsets (one or more) of its *succinct* MGs by *non-representative* MGs having, respectively, the same closures as those of the replaced subsets [18]. For example, the *redundant* MG "*bdg*", extracted from the context sketched by Table 1, can be inferred from the *succinct* MG "*adg*" by replacing its subset "*ad*" by "*bd*" (both MGs "*ad*" and "*bd*" have the same closure).

In what follows, we show that, according to the current definition of the SSMG, the cardinality of the latter closely depends on the selected total order relation (contrary to the statement of **Claim 1**). Furthermore, we give an example where the SSMG presents a loss of information (contrary to the statement of **Claim 2**).

As mentioned in the previous subsection, Dong *et al.* claimed that the shift of the total order relation $\preceq$ does not affect the size of the associated SSMG [18]. Such a claim seems to hold when confronted to the extraction context depicted by Table 1. Indeed, for different total order relations (*e.g.*, the alphabetic order, the ascending/descending support order, etc.), we obtain the same number of *succinct* minimal generators (MGs). It is the same for the running example used in the proper paper of Dong *et al.* (see [18]). However, if we consider the extraction context sketched by Table 3 (Left), we find that their claim is erroneous. Indeed, as shown by Table 3 (Right), the total number of *succinct* MGs is equal to **23** if the alphabetic order is of use. Whereas, it is equal to **22** in the case of the ascending support order, and **25** in the case of the descending support order. Hence, the number of the *succinct* MGs closely depends on the chosen total order relation. The difference occurs within the γ-equivalence class number **11** (shown with bold letters). The other γ-equivalence classes do not contain any *redundant* MGs and, hence, are not of interest in our explanations.

Furthermore, if we adopt the ascending support order as a total order relation $\preceq$, then we find that, given the *succinct* MGs, it is not possible to infer *all redundant* ones. Indeed, from the *succinct* MGs "*ea*" and "*acd*", only both *redundant* MGs "*adf*" and "*cdf*" can be inferred by replacing the subset "*ac*" of "*acd*" by the MGs having its closure, *i.e.*, "*af*" and "*cf*". Hence, for example, the *redundant* MG "*edf*" will be missed if we need to infer *all* MGs.

Even if the first "bug" (*i.e.*, that related to the size of the different SSMGs associated to a given extraction context) can be regarded as not having a dramatic consequence, fixing the second one is of paramount importance, since the need for *exact* compact representation is always conditioned by the ability to discover *all redundant* information without looking back at the extraction context. Hence, aiming to ensure the completeness of the derivation of *redundant* MGs, we introduce, in the next section, new definitions allowing to go beyond the limitations of the work proposed by Dong *et al.*

3.3 Succinct System of Minimal Generators: New Definitions

The set MG_f of the MGs associated to a given closed itemset (CI) f can be divided into different equivalence classes thanks to a substitution process. To avoid confusion with the γ-equivalence classes induced by the closure operator $''$, the substitution-based ones will be denoted *σ-equivalence classes*. The substitution process uses an operator denoted *Subst*. This substitution operator is a partial one allowing to substitute a subset

Table 3. (**Left**) An extraction context $\mathcal{K}'$. (**Right**) The CIs extracted from $\mathcal{K}'$ and for each one, the corresponding MGs for different total order relations (the *succinct* MGs, according to the definition of Dong *et al.*, are indicated with bold letters).

	a	*b*	*c*	*d*	*e*	*f*
1	×	×				
2	×		×			×
3	×			×		
4		×	×	×	×	
5	×	×	×	×	×	×
6			×	×		
7		×				×
8				×		×
9		×			×	×

#	*alphabetic* order		*ascending support* order		*descending support* order	
	CI	MGs	CI	MGs	CI	MGs
1	∅	∅	∅	∅	∅	∅
2	*a*	***a***	*a*	***a***	*a*	***a***
3	*b*	***b***	*b*	***b***	*b*	***b***
4	*c*	***c***	*c*	***c***	*c*	***c***
5	*d*	***d***	*d*	***d***	*d*	***d***
6	*be*	***e***	*eb*	***e***	*be*	***e***
7	*f*	***f***	*f*	***f***	*f*	***f***
8	*ab*	***ab***	*ab*	***ab***	*ba*	***ba***
9	*acf*	***ac***, ***af***, ***cf***	*acf*	***ac***, ***af***, ***cf***	*fac*	***fa***, ***fc***, ***ac***
10	*ad*	***ad***	*ad*	***ad***	*da*	***da***
11	*abcdef*	***ae***, ***abc***, *abd*, *abf*, ***acd***, *adf*, *bcf*, *bdf*, *cdf*, *cef*, *def*	*eacbdf*	***ea***, *ecf*, *edf*, *acb*, ***acd***, *abd*, *abf*, *adf*, *cbf*, *cdf*, *bdf*	*bdface*	***ae***, ***bdf***, ***bda***, ***bfa***, *bfc*, *bac*, ***dfa***, *dfc*, *dfe*, *dac*, *fce*
12	*bcde*	***bc***, ***bd***, ***ce***, ***de***	*ecbd*	***ec***, ***ed***, ***cb***, ***bd***	*bdce*	***bd***, ***bc***, ***de***, ***ce***
13	*bf*	***bf***	*bf*	***bf***	*bf*	***bf***
14	*cd*	***cd***	*cd*	***cd***	*dc*	***dc***
15	*df*	***df***	*df*	***df***	*df*	***df***
16	*bef*	***ef***	*ebf*	***ef***	*bfe*	***fe***

of an itemset X, say Y, by another itemset Z belonging to the same γ-equivalence class of Y (*i.e.*, $Y'' = Z''$). This operator is then defined as follows:

Definition 8. (SUBSTITUTION OPERATOR) *Let X, Y and Z be three itemsets s.t. $Y \subset X$ and $Y'' = Z''$. The substitution operator Subst, w.r.t. X, Y and Z, is defined as follows: $Subst(X, Y, Z) = (X \backslash Y) \bigcup Z$.*

To prove that X and $Subst(X, Y, Z)$ have the same closure, we need the following lemma.

Lemma 2. *[19] Let X and Y be two itemsets. X and Y verify the following property: $(X \bigcup Y)'' = (X'' \bigcup Y'')''$.*

Proposition 5. *X and $Subst(X, Y, Z)$ belong to the same γ-equivalence class.*

Proof. *Let W be the result of $Subst(X, Y, Z)$, i.e., $W = (X\backslash Y) \bigcup Z$. We will show that X and W have the same closure.*
Using Lemma 2, we have: $X'' = ((X\backslash Y) \bigcup Y)'' = ((X\backslash Y)'' \bigcup Y'')''$. Since $Y'' = Z''$, then $X'' = ((X\backslash Y)'' \bigcup Y'')'' = ((X\backslash Y)'' \bigcup Z'')'' = ((X\backslash Y) \bigcup Z)'' = W''$. Hence, $X'' = W''$. Thus, we can conclude that X and W necessarily belong to the same γ-equivalence class. ♦

For each γ-equivalence class $\mathcal{C}$ (or equivalently, for each CI f), the substitution operator induces an equivalence relation on the set MG_f of the MGs of f portioning it into distinct σ-equivalence classes. The definition of a σ-equivalence class requires that we redefine the notion of *redundant* MG under the substitution process point of view. Indeed, according to the definition given by Dong *et al.* (see Definition 6), *redundant* MGs are blindly pruned according to purely syntactic properties, only consisting in checking the order of their subsets *w.r.t* $\preceq$, in their respective γ-equivalence classes. Hence, we propose to incorporate a semantic part based on the actual concept of redundancy.

Definition 9. (MINIMAL GENERATORS' REDUNDANCY) *Let g and g_1 be two* MG*s belonging to the same γ-equivalence class.*

- *g is said to be a* ***direct redundant*** *(resp. derivable) with respect to (resp. from) g_1, denoted $g_1 \vdash g$, if* Subst$(g_1, g_2, g_3) = g$ *where $g_2 \subset g_1$ and $g_3 \in \mathcal{MG}_{\mathcal{K}}$ s.t. $g_3'' = g_2''$.*
- *g is said to be a* ***transitive redundant*** *with respect to g_1, denoted $g_1 \vdash^{+} g$, if it exists a sequence of n* MG*s ($n \geq 2$), gen_1, gen_2, $\ldots$, gen_n, s.t. $gen_i \vdash gen_{i+1}$ ($i \in$ [1..(n-1)]) where $gen_1 = g_1$ and $gen_n = g$.*

Proposition 6. *The substitution relations $\vdash$ and $\vdash^{+}$ have the following properties:*

- *The substitution relation $\vdash$ is reflexive, symmetric but not necessarily transitive.*
- *The substitution relation $\vdash^{+}$ is reflexive, symmetric and transitive.*

The formal definition of a σ-equivalence class is then as follows:

Definition 10. (σ-EQUIVALENCE CLASS) *The operator $\vdash^{+}$ induces an equivalence relation on the set* MG$_f$*, of the* MG*s associated to an* CI *f, portioning it into distinct subsets called σ-equivalence classes. If $g \in$* MG$_f$*, then the σ-equivalence class of g, denoted by [g], is the subset of* MG$_f$ *consisting of all elements that are transitively redundant w.r.t. g. In other words, we have: [g] = $\{g_1 \in \mathrm{MG}_f \mid g \vdash^{+} g_1\}$.*

The smallest MG *in each σ-equivalence class, w.r.t. the total order relation $\preceq$, will be considered as its* ***succinct*** MG*. While, the other* MG*s will be qualified as* ***redundant*** MG*s.*

The following pseudo-code offers a straightforward way to extract the different σ-equivalence classes associated to an CI f. A σ-equivalence class will be denoted σ-*Equiv_Class*.

Function 1: σ-EQUIVALENCE CLASSES MINER

Input: The set MG$_f$ of the MGs associated to f.
Output: The σ-equivalence classes associated to f.

Begin
  $\mathcal{S}$ = MG$_f$;
  $i = 0$;
  While ($\mathcal{S} \neq \emptyset$) **do**
    $i = i + 1$;
    $g_s = \min_{\preceq}(\mathcal{S})$; /*$g_s$ is the smallest MG in $\mathcal{S}$ *w.r.t.* $\preceq$.*/
    σ-*Equiv_Class*$_i$ = $\{g_s\} \bigcup \{g \in \mathcal{S} \mid g_s \vdash^{+} g\}$;
    $\mathcal{S} = \mathcal{S} \backslash \sigma$-*Equiv_Class*$_i$;
  return $\bigcup_{j=1}^{j \leq i} \sigma$-*Equiv_Class*$_j$;
End

Example 7. Let us consider the extraction context depicted by Table 3, the ascending support order as a total order relation $\preceq$ and the γ-equivalence class having for CI "*eacbdf*". Using Function 1, the MGs associated to "*eacbdf*" are divided as follows:

1. First, $\mathcal{S}$ = MG$_{eacbdf}$ = $\{ea, ecf, edf, acb, acd, abd, abf, adf, cbf, cdf, bdf\}$ and i = **1**. "*ea*" is the *smallest* MG in $\mathcal{S}$. Hence, σ-*Equiv_Class*$_1$ = $\{ea\} \bigcup \{g \in \mathcal{S} \mid ea \vdash^+ g\}$. However, none MG can be deduced from "*ea*". Thus, σ-*Equiv_Class*$_1$ = $\{ea\}$.
2. Second, $\mathcal{S} = \mathcal{S} \backslash \sigma$-*Equiv_Class*$_1$ = $\{ea, ecf, edf, acb, acd, abd, abf, adf, cbf, cdf, bdf\} \backslash \{ea\}$ = $\{ecf, edf, acb, acd, abd, abf, adf, cbf, cdf, bdf\}$ and i = **2**. "*ecf*" is the *smallest* one in $\mathcal{S}$. Hence, σ-*Equiv_Class*$_2$ = $\{ecf\} \bigcup \{g \in \mathcal{S} \mid ecf \vdash^+ g\}$ = $\{ecf\}$ $\bigcup \{edf, acb, abd, abf, cbf, bdf\}$. Indeed, *Subst*$(ecf, ec, ed)$ = *edf* $\in$ MG$_{eacbdf}$ ($ecf \vdash edf$ and, hence, $ecf \vdash^+ edf$), *Subst*(ecf, ec, cb) = *cbf* $\in$ MG$_{eacbdf}$ ($ecf \vdash cbf$ and, hence, $ecf \vdash^+ cbf$), *Subst*(cbf, cf, ac) = *acb* $\in$ MG$_{eacbdf}$ ($ecf \vdash^+ acb$ since $ecf \vdash cbf$ and then, $cbf \vdash acb$), etc.
3. Finally, $\mathcal{S} = \mathcal{S} \backslash \sigma$-*Equiv_Class*$_2$ = $\{ecf, edf, acb, acd, abd, abf, adf, cbf, cdf, bdf\}$ $\backslash \{ecf, edf, acb, abd, abf, cbf, bdf\}$ = $\{acd, adf, cdf\}$ and i = **3**. "*acd*" is the *smallest* MG in $\mathcal{S}$. Hence, σ-*Equiv_Class*$_3$ = $\{acd\} \bigcup \{g \in \mathcal{S} \mid acd \vdash^+ g\}$ = $\{acd\} \bigcup \{adf,$ $cdf\}$ since *Subst*(acd, ac, af) = *adf* ($acd \vdash adf$ and, hence, $acd \vdash^+ adf$) and *Subst*(acd, ac, cf) = *cdf* ($acd \vdash cdf$ and, hence, $acd \vdash^+ cdf$).

In conclusion, MG$_{eacbdf}$ is divided into three σ-equivalence classes as follows (*succinct* MGs are marked with bold letters): MG$_{eacbdf}$ = $\{$***ea***$\} \bigcup \{$***ecf***, *edf, acb, abd, abf, cbf, bdf*$\} \bigcup \{$***acd***, *adf, cdf*$\}$. Note that "***ecf***" was not considered as a succinct *MG* according to the initial definition that was introduced by Dong *et al.* since its subset "*cf*" is not the *representative MG* of its CI "*acf*". Hence, *all* MGs belonging to σ-*Equiv_Class*$_2$ can not be inferred according to their definition, contrary to ours.

Example 8. For the same context, if we consider the descending support order as a total order relation $\preceq$, then we will note that the SSMG, as formerly defined by Dong *et al.*, can even contain redundancy in comparison to our definition. Indeed, thanks to the substitution operator *Subst*, MG$_{bdface}$ is divided as follows: MG$_{bdface}$ = $\{$***ae***$\} \bigcup \{$***bdf***, *bda, bfa, bfc, bac, dfe, fce*$\} \bigcup \{$***dfa***, *dfc, dac*$\}$. The storage of the MGs "*bda*" and "*bfa*" is then redundant and useless since they can simply be inferred starting from the *succinct* MG "*bdf*" ($bdf \vdash^+ bda$ and $bdf \vdash^+ bfa$). Indeed, *Subst*(bdf, bd, bc) = bfc, *Subst*(bfc, fc, fa) = `bfa`, *Subst*(bfa, fa, ac) = *bac* and finally *Subst*(bac, bc, bd) = `bda`.

Proposition 7. *The different* σ*-equivalence classes associated to a given* CI f *fulfill the following properties:*

- $\bigcup_{i=1}^{i \leq Card(\mathrm{MG}sucf)}$ σ-*Equiv_Class*$_i$ = MG$_f$.
- $\forall$ *i, j* $\in$ *[1... Card*(MG$\mathtt{suc}_f$)*] s.t.* $i \neq j$, σ-*Equiv_Class*$_i \bigcap \sigma$-*Equiv_Class*$_j$ = $\emptyset$.

Using the new definitions of both *succinct* and *redundant* MGs (*cf.* Definition 9 and Definition 10), we can now define the succinct system of minimal generators (SSMG) in its new form as follows:

Definition 11. (SUCCINCT SYSTEM OF MINIMAL GENERATORS: NEW DEFINITION) *A succinct system of minimal generators* (SSMG) *is a system where only succinct* MG*s are retained among all* MG*s associated to each* CI.

Thanks to the new consideration of the concept of redundancy within MGs, Proposition 8 and Proposition 9 make it possible to correct the claims of Dong *et al.* [18].

Proposition 8. *Whatever the total order relation $\preceq$, the substitution operator Subst maintains unchanged the elements belonging to each σ-equivalence class.*

Proof. Let $\preceq_1$ and $\preceq_2$ be two different total order relations. Let f be an CI *and* MG_f *be the set of its associated* MG*s. Using $\preceq_1$,* MG_f *will be divided into σ-equivalence classes. Let σ-Equiv_Class$_{\preceq_1}$ be one of them and g_{s_1} be its succinct* MG *(i.e., the smallest one in σ-Equiv_Class$_{\preceq_1}$ w.r.t. $\preceq_1$). σ-Equiv_Class$_{\preceq_1}$ can be represented by a tree, denoted $T_{\preceq_1}$. The root of $T_{\preceq_1}$ contains the succinct* MG *g_{s_1}. In this tree, a node N, which represents an* MG *g, points to a node N_1, which represents an* MG *g_1, if $g \vdash g_1$. Hence, from whatever node in $T_{\preceq_1}$, we can access the remaining nodes as follows: we move downward from the node N to the node N_1 using the relation $g \vdash g_1$ and conversely, from N_1 to N using the dual relation $g_1 \vdash g$. Indeed, if* Subst$(g, g_2, g_3) = g_1$ *where $g_2 \subset g$ and $g_3 \in \mathcal{MG}_\mathcal{K}$ s.t. $g_3'' = g_2''$, then also* Subst$(g_1, g_3, g_2) = g$ *since the operator $\vdash$ is symmetric (cf. Proposition 6).*

Now, consider the set σ-Equiv_Class$_{\preceq_1}$ ordered w.r.t. the second total order relation $\preceq_2$. The obtained new set will be denoted σ-Equiv_Class$_{\preceq_2}$ and its associated succinct MG *will be denoted g_{s_2}. Hence, if we transform the tree $T_{\preceq_1}$ in a new one, denoted $T_{\preceq_2}$ and rooted in g_{s_2}, then we are able to reach all remaining* MG*s contained in σ-Equiv_Class$_{\preceq_2}$ thanks to the substitution application as explained above. Thus, the change of the total order relation does not affect the content of the σ-Equiv_Class$_{\preceq_1}$ since it does not involve the deletion of any node in $T_{\preceq_1}$.*

Furthermore, this change does not augment σ-Equiv_Class$_{\preceq_2}$ by any another redundant MG*. Indeed, suppose that an* MG *denoted g_{new}, not already belonging to σ-Equiv_Class$_{\preceq_1}$, will be added to σ-Equiv_Class$_{\preceq_2}$ once we shift the total order relation from $\preceq_1$ to $\preceq_2$ (i.e., $g_{s_2} \vdash^+ g_{new}$ but $g_{s_1} \nvdash^+ g_{new}$). Since, $g_{s_1} \vdash^+ g_{s_2}$ ($g_{s_2} \in$ σ-Equiv_Class$_{\preceq_1}$) and $g_{s_2} \vdash^+ g_{new}$, then $g_{s_1} \vdash^+ g_{new}$. Indeed, the relation $\vdash^+$ is transitive (cf. Proposition 6). Hence, g_{new} should belong to σ-Equiv_Class$_{\preceq_1}$ (according to Definition 10) what is in contradiction with the starting assumption ($g_1 \nvdash^+ g_{new}$). Thus, $g_2 \nvdash^+ g_{new}$.*

Therefore, we can conclude that the elements belonging to σ-Equiv_Class$_{\preceq_2}$ are exactly the same than those contained in σ-Equiv_Class$_{\preceq_1}$, ordered w.r.t. $\preceq_2$ instead of $\preceq_1$. ♦

Example 9. If we scrutinize both Example 7 and Example 8, we note that *σ-Equiv_Class$_1$*, *σ-Equiv_Class$_2$* and *σ-Equiv_Class$_3$* are exactly the same for both examples. However, they are sorted according to the ascending support order and to the descending support order, respectively.

According to Proposition 8, the number of *succinct* MGs associated to each CI f (*i.e.*, Card(MGsuc_f)) is then equal to the number of σ-equivalence classes induced by the substitution operator, independently of the chosen total order relation. Hence, the cardinality of the set $\mathcal{MG}\text{suc}_\mathcal{K}$, containing the *succinct* MGs that can be extracted from the context $\mathcal{K}$, remains unchanged even if we change the total order relation. In other words, the different SSMGs associated to an extraction context have the same size.

Proposition 9. *The* SSMG *as newly defined ensures the inference of each redundant* MG *g.*

Proof. Since g is a redundant MG*, then g is not the smallest one in its σ-equivalence class. Hence, according to the definition of a σ-equivalence class (see Definition 10), it necessarily exists a succinct* MG g_s *belonging to the* SSMG *whose a substitution process certainly leads to g* ($g_s \vdash^+ g$) *since the number of* MGs *belonging to each σ-equivalence class is finite.* ♦

Proposition 9 states that the new SSMG is an *exact* representation of the MG set.

4 Related Work

In this part, we will mainly concentrate on the concept of **clone items** [21,22], since it is closely related to our work. Clone items can be roughly considered as a *restriction* of the SSMG to γ-equivalence classes where two or more *items* have the same closure, *i.e.*, to MGs of size *one* (like the couple (*a*, *b*) and the couple (*d*, *e*) of the extraction context depicted by Table 1). The authors [21,22] show that, for a couple like (*a*, *b*), items *a* and *b* present symmetries, which can be seen as redundant information since for *all* association rules containing *a* in the premise there exists the same association rules where "*a*" is replaced by "*b*" [22]. Thus, they propose to ignore *all* rules containing "*b*" but not "*a*" without loss of information [22]. This reduction process was applied to the Guigues-Duquenne basis [23] of exact implications. Association rules of this basis present implications between pseudo-closed itemsets [23] in the premise part, and, closed itemsets in the conclusion part. Note that clone items when applied to pseudo-closed itemsets are called *P-clone items* [21].

5 Experimental Study

In order to evaluate the utility of our approach, we conducted series of experiments on four benchmark datasets, frequently used by the data mining community[4]. Characteristics of these datasets are summarized by Table 4. Hereafter, we use a logarithmically scaled ordinate axis for all curves.

Figure 1 shows the effect of the succinct system of minimal generators (SSMG) by comparing the number of the *succinct frequent* minimal generators (MGs) *vs.* that of *all frequent* MGs. For both the PUMSB and the MUSHROOM datasets, a large part of the *frequent* MGs proves to be *redundant*. Indeed, for PUMSB (*resp.* MUSHROOM), in average **52.27%** (*resp.* **50.50%**) of the *frequent* MGs are *redundant*, and the maximum rate of redundancy reaches **64.06%** (*resp.* **53.28%**) for a *minsupp* value equal to **65%** (*resp.* **0.20%**). It is important to mention that for the PUMSB dataset, the redundancy is caused by the fact that there are some couples of items having the same closure (like "*a*" and "*b*" of the extraction context sketched by Table 1). Hence, using only an item, instead of both items forming each couple, was sufficient to eliminate all redundancy, which is not the case for MUSHROOM. Noteworthily, in average, the

[4] These benchmark datasets are downloadable from: *http://fimi.cs.helsinki.fi/data.*

Table 4. Dataset characteristics

Dataset	Number of items	Number of objects	Average object size	*minsupp* interval (%)
PUMSB	7, 117	49, 046	74.00	90 - 60
MUSHROOM	119	8, 124	23.00	1 - 0.01
CONNECT	129	67, 557	43.00	90 - 50
T40I10D100K	1, 000	100, 000	39.61	10 - 1

number of *succinct* (*resp. all*) *frequent* MGs per γ-equivalence class, is equal to **1.00** (*resp.* **2.24**) for the PUMSB dataset, while it is equal to **1.06** (*resp.* **2.13**) for the MUSHROOM dataset. Such statistics explain why the curve representing the number of *frequent* closed itemsets (CIs) is almost overlapped with that depicting the number of *succinct frequent* MGs.

For the CONNECT dataset and although it is widely tagged to be a "dense" one, each *frequent* CI extracted from this dataset has only a unique *frequent* MG and, hence, there are no *redundant* ones. It is the same for the "sparse" T40I10D100K dataset. Hence, it is worth noting that the reduction ratio from the number of *all frequent* MGs to that of *succinct* ones can be considered as a new measure for an improved dataset classification, as mentioned by Dong *et al.* [18].

Obtained results prove that the SSMG allows to almost reach the ideal case: a *unique succinct* MG per γ-equivalence class.

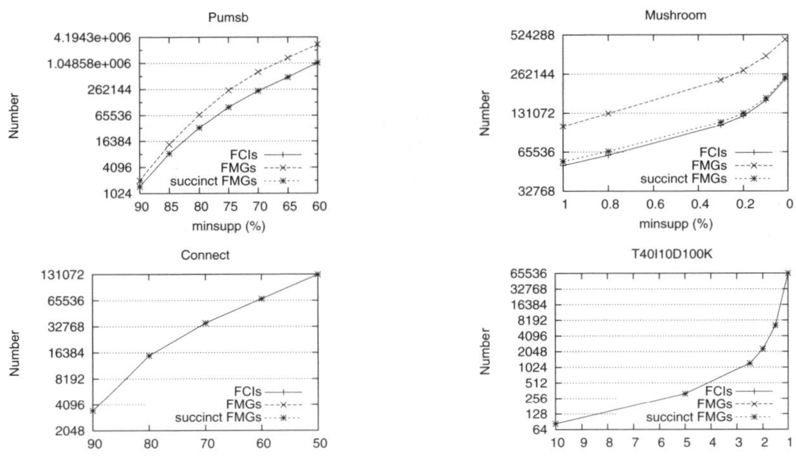

Fig. 1. The number of *frequent* CIs (denoted FCIs), of *frequent* MGs (denoted FMGs) and of *succinct frequent* MGs (denoted succinct FMGs)

6 Conclusion and Future Work

In this paper, we studied the principal properties of the succinct system of minimal generators (SSMG) as formerly defined by Dong *et al.* Once the limitations of the current definition pointed out, we introduced a new one aiming to make of the SSMG an *exact* representation of the minimal generator (MG) set, on the one hand, and, on the other hand, its size independent from the adopted total order relation. After that, we discussed the main related work. Finally, an experimental study confirmed that the application of the SSMG makes it possible to get, in average, almost as many closed itemsets as *succinct* MGs, thanks to the elimination of an important number of *redundant* ones. It is important to mention that our approach can easily be applied when negative items are considered, as well as within the disjunctive search space where itemsets are characterized by their disjunctive support, instead of the conjunctive one [24].

As part of future work, we plan to use the SSMG in an in-depth structural analysis of dataset characteristics. In this setting, we propose to set up a sparseness measure based on the compactness rate offered by the SSMG. Such measure can be used to increase the extraction efficiency by helping to choose the most suitable algorithm according to the data under treatment. The extension of the SSMG to the framework of generic association rules is also an interesting issue. As a first attempt, the work we proposed in [25] gave very encouraging results. Furthermore, we think that the application of the SSMG to some real-life domains like biological applications will be of an added value for users.

Acknowledgements. We are grateful to the anonymous reviewers for their useful remarks and suggestions. This work is partly supported by the French-Tunisian project PAI CMCU 05G1412.

References

1. Bastide, Y., Taouil, R., Pasquier, N., Stumme, G., Lakhal, L.: Mining frequent patterns with counting inference. ACM-SIGKDD Explorations 2(2), 66–75 (2000)
2. Bastide, Y., Pasquier, N., Taouil, R., Stumme, G., Lakhal, L.: Mining Minimal Non-redundant Association Rules Using Frequent Closed Itemsets. In: Palamidessi, C., Moniz Pereira, L., Lloyd, J.W., Dahl, V., Furbach, U., Kerber, M., Lau, K.-K., Sagiv, Y., Stuckey, P.J. (eds.) CL 2000. LNCS (LNAI), vol. 1861, pp. 972–986. Springer, Heidelberg (2000)
3. Boulicaut, J.F., Bykowski, A., Rigotti, C.: Free-sets: A condensed representation of Boolean data for the approximation of frequency queries. Data Mining and Knowledge Discovery (DMKD) 7(1), 5–22 (2003)
4. Stumme, G., Taouil, R., Bastide, Y., Pasquier, N., Lakhal, L.: Computing Iceberg concept lattices with TITANIC. Journal on Knowledge and Data Engineering (KDE) 2(42), 189–222 (2002)
5. Pasquier, N., Bastide, Y., Taouil, R., Stumme, G., Lakhal, L.: Generating a condensed representation for association rules. Journal of Intelligent Information Systems 24(1), 25–60 (2005)
6. Ben Yahia, S., Hamrouni, T., Mephu Nguifo, E.: Frequent closed itemset based algorithms: A thorough structural and analytical survey. ACM-SIGKDD Explorations 8(1), 93–104 (2006)
7. Li, H., Li, J., Wong, L., Feng, M., Tan, Y.: Relative risk and odds ratio: A data mining perspective. In: Proceedings of the 24th ACM SIGMOD-SIGACT-SIGART symposium on Principles Of Database Systems (PODS 2005), Baltimore, Maryland, USA, June 13–15, 2005, pp. 368–377 (2005)
8. Lucchese, C., Orlando, S., Palmerini, P., Perego, R., Silvestri, F.: kDCI: A multi-strategy algorithm for mining frequent sets. In: Proceedings of the IEEE ICDM Workshop on Frequent Itemset Mining Implementations (FIMI 2003), CEUR Workshop Proceedings, Melbourne, Florida, USA. CEUR Workshop Proceedings, vol. 90 (November 19, 2003)

9. Hamrouni, T., Ben Yahia, S., Slimani, Y.: Prince: An algorithm for generating rule bases without closure computations. In: Tjoa, A.M., Trujillo, J. (eds.) DaWaK 2005. LNCS, vol. 3589, pp. 346–355. Springer, Heidelberg (2005)
10. Kryszkiewicz, M.: Concise representation of frequent patterns and association rules. In: Habilitation dissertation, Institute of Computer Science, Warsaw University of Technology, Warsaw, Poland (August 2002)
11. Ceglar, A., Roddick, J.F.: Association mining. ACM Computing Surveys 38(2) (July 2006)
12. Zaki, M.J.: Mining non-redundant association rules. Data Mining and Knowledge Discovery (DMKD) 9(3), 223–248 (2004)
13. Gasmi, G., Ben Yahia, S., Mephu Nguifo, E., Slimani, Y.: $\mathcal{IGB}$: A new informative generic base of association rules. In: Ho, T.-B., Cheung, D., Liu, H. (eds.) PAKDD 2005. LNCS (LNAI), vol. 3518, pp. 81–90. Springer, Heidelberg (2005)
14. Boulicaut, J.-F., Rigotti, C., Calders, T.: A Survey on Condensed Representations for Frequent Sets. In: Boulicaut, J.-F., De Raedt, L., Mannila, H. (eds.) Constraint-Based Mining and Inductive Databases. LNCS (LNAI), vol. 3848, pp. 64–80. Springer, Heidelberg (2005)
15. Liu, G., Li, J., Wong, L., Hsu, W.: Positive borders or negative borders: How to make lossless generator based representations concise. In: Jonker, W., Petković, M. (eds.) SDM 2006. LNCS, vol. 4165, pp. 469–473. Springer, Heidelberg (2006)
16. De Raedt, L., Ramon, J.: Condensed representations for inductive logic programming. In: Proceedings of the Ninth International Conference on Principles of Knowledge Representation and Reasoning (KR 2004), Whistler, Canada, June 2–5, 2004, pp. 438–446 (2004)
17. Zhao, L., Zaki, M.J., Ramakrishnan, N.: BLOSOM: A framework for mining arbitrary Boolean expressions. In: Proceedings of the 12th International Conference on Knowledge Discovery and Data Mining (KDD 2006), Philadelphia, PA, USA, August 20–23, 2006, pp. 827–832 (2006)
18. Dong, G., Jiang, C., Pei, J., Li, J., Wong, L.: Mining succinct systems of minimal generators of formal concepts. In: Zhou, L.-z., Ooi, B.-C., Meng, X. (eds.) DASFAA 2005. LNCS, vol. 3453, pp. 175–187. Springer, Heidelberg (2005)
19. Ganter, B., Wille, R.: Formal Concept Analysis. Springer, Heidelberg (1999)
20. Hamrouni, T., Ben Yahia, S., Mephu Nguifo, E.: Redundancy-free generic bases of association rules. In: Proceedings of the 8th French Conference on Machine Learning (CAp 2006), Presses Universitaires de Grenoble, Trégastel, France, May 22–24, 2006, pp. 363–378 (2006)
21. Gély, A., Medina, R., Nourine, L., Renaud, Y.: Uncovering and reducing hidden combinatorics in Guigues-Duquenne bases. In: Ganter, B., Godin, R. (eds.) ICFCA 2005. LNCS (LNAI), vol. 3403, pp. 235–248. Springer, Heidelberg (2005)
22. Medina, R., Nourine, L., Raynaud, O.: Interactive Association Rules Discovery. In: Missaoui, R., Schmidt, J. (eds.) Formal Concept Analysis. LNCS (LNAI), vol. 3874, pp. 177–190. Springer, Heidelberg (2006)
23. Guigues, J.L., Duquenne, V.: Familles minimales d'implications informatives résultant d'un tableau de données binaires. Mathématiques et Sciences Humaines 24(95), 5–18 (1986)
24. Hamrouni, T., Ben Yahia, S., Mephu Nguifo, E.: A new exact concise representation based on disjunctive closure. In: Proceedings of the 2nd Jordanian International Conference on Computer Science and Engineering (JICCSE 2006), Al-Balqa, Jordan, December 5–7, 2006, pp. 361–373 (2006)
25. Hamrouni, T., Ben Yahia, S., Mephu Nguifo, E.: Generic association rule bases: Are they so succinct? In: Ben Yahia, S., Mephu Nguifo, E., Behlohlavek, R. (eds.) CLA 2006. LNCS (LNAI), vol. 4923, pp. 198–213. Springer, Heidelberg (2008) (this volume)

Closure Systems of Equivalence Relations and Their Labeled Class Geometries

Tim B. Kaiser

Darmstadt University of Technology, 64287 Darmstadt, Germany
tkaiser@mathematik.tu-darmstadt.de

Abstract. The notion of an affine ordered set is specialized to that of a complete affine ordered set, which can be linked to attribute-complete many-valued contexts and is categorically equivalent to the notion of a closed system of equivalence relations (SER). This specialization step enables us to give conditions under which the complete affine ordered set can be interpreted as the set of congruence classes labeled with the congruence relation they stem from yielding a coordinatization theorem for affine ordered sets.

1 Introduction

In [KS04] the notion of affine ordered sets is introduced to provide an order-theoretic, geometric counterpart of (simple) many-valued contexts. Here we specialize the notion of an affine ordered set to that of a complete affine ordered set, which is categorically equivalent to attribute-complete many-valued contexts and to closed systems of equivalence relations (SER). This specialization step enables us to add an algebraic aspect, that is, to give conditions under which the complete affine ordered set can be interpreted as the set of congruence classes of an algebra labeled with the congruence relation they stem from. This approach can be seen in the tradition of coordinatization theorems in geometry where a prominent example is the coordinatization of affine planes via the Theorem of Desargues.

In Section 2 we introduce the notions of attribute-complete many-valued contexts and closed SERs and insinuate the correspondence between the two. The order-theoretic geometric counterpart is introduced as *complete affine ordered set* in Section 3, and Section 4 shows the categorical equivalence between complete affine ordered sets and closed SERs. The second part of the paper, consisting of Section 5, deals with the question how to coordinatize closed SERs and complete affine ordered sets.

2 Attribute-Complete Many-Valued Contexts and Closed Systems of Equivalence Relations

Data tables can be formalized as many-valued contexts as it is common in Formal Concept Analysis [GW99]. Many-valued contexts are also known as *Chu Spaces* [Pr95] or *Knowledge Representation Systems* [Pa91].

S. Ben Yahia et al. (Eds.): CLA 2006, LNAI 4923, pp. 96–106, 2008.

Definition 1 (many-valued context). *A* (complete) many-valued context *is a structure* $\mathbb{K} := (G, M, W, I)$, *where* G *is a set of objects,* M *is a set of attributes,* W *is a set of values and* $I \subseteq G \times M \times W$ *is a ternary relation, where for every* $(g, m) \in G \times M$ *there exists a unique* $w \in W$ *with* $(g, m, w) \in I$*; in the following* m *will be considered as a function from* G *to* W *via* $m(g) := w$.

We call an attribute $m \in M$ an *id attribute* if for any two objects $g_1, g_2 \in G$ the values of g_1 and g_2 regarding to m are different (i.e. $m(g_1) \neq m(g_2)$). The following definition provides a notion of dependency between attributes of a many-valued context.

Definition 2 (functional dependency). *If* M_1 *and* M_2 *are sets of attributes of a many-valued context* (G, M, W, I)*, we call* M_2 functionally dependent *on* M_1 *(in symbols:* $M_1 \to M_2$*) if for all pairs of objects* $g, h \in G$

$$\forall m_1 \in M_1 : m_1(g) = m_1(h) \Rightarrow \forall m_2 \in M_2 : m_2(g) = m_2(h).$$

If $M_1 \to M_2$ *and* $M_2 \to M_1$*, the sets of attributes,* M_1 *and* M_2*, are called* functionally equivalent*, denoted by* $M_1 \leftrightarrow M_2$.

For a map $f : A \to B$ the *kernel* of f is defined as the equivalence relation $\ker(f) := \{(a, b) \in A^2 \mid f(a) = f(b)\}$. It is easily seen that $M_1 \to M_2$ holds if and only if $\bigcap_{m_1 \in M_1} \ker(m_1) \subseteq \bigcap_{m_2 \in M_2} \ker(m_2)$. Accordingly, m_1 and m_2 are functionally equivalent if and only if $\bigcap_{m_1 \in M_1} \ker(m_1) = \bigcap_{m_2 \in M_2} \ker(m_2)$. Many-valued contexts where any two functionally equivalent attributes are equal will be called *simple*.

Definition 3 (attribute-complete many-valued context). *A many-valued context* $\mathbb{K} := (G, M, W, I)$ *is called* attribute-complete *if it is simple, has an id attribute, and*

$$\forall N \subseteq M \, \exists m \in M : N \leftrightarrow \{m\}.$$

Following the main scheme from [KS04], we assign a system of equivalence relations to attribute-complete many-valued contexts in order to describe them geometrically and order-theoretically in a later step. We recall the basic definitions for systems of equivalence relations from [KS04]. We denote the *identity relation* on the set X by $\Delta_X := \{(x, x) \mid x \in X\}$.

Definition 4 (system of equivalence relations). *We call* $\mathbb{E} := (D, E)$ *a* system of equivalence relations (SER)*, if* D *is a set and* E *is a set of equivalence relations on* D*. If* $d \in D$ *and* $\theta \in E$*, we denote the equivalence class of* d *by* $[d]\theta := \{d' \in D \mid d'\theta d\}$*. If* $\Delta_D \in E$ *we will also call* (D, E) *a* SER with identity relation.

Every attribute $m \in M$ induces a partition on the object set via the equivalence classes of $\ker(m)$. So we can regard a simple many-valued context as a set of partitions induced by its attributes. Every block of a partition corresponds to the set of objects with a certain value with respect to a certain attribute. The following definition captures attribute-complete many-valued contexts.

Definition 5 (closed SER). *Let (G, E) be a SER with identity relation. Then we call (G, E) a* closed SER *if E is meet-closed which means that E forms a* closure system of equivalence relations.

To every given closed SER $\mathbb{E} := (D, E)$ we can assign a simple many-valued context $\mathbf{K}(\mathbb{E}) := (D, E, W, I)$, where $W := \{[d]\theta \mid d \in D, \theta \in E\}$ and $I := \{(d, \theta, w) \in D \times E \times W \mid w = [d]\theta\}$. Obviously, $\mathbf{K}(\mathbb{E})$ is attribute-complete. On the other hand we can assign, as described above, a closed SER to every attribute-complete many-valued context. We define $\mathbf{E}(\mathbb{K}) := (G, \{\ker(m) \mid m \in M\})$. We observe that, for every attribute-complete many-valued context $\mathbb{K}$, we have $\mathbf{K}(\mathbf{E}(\mathbb{K})) \simeq \mathbb{K}$ and for every closed SER $\mathbb{E}$ we have $\mathbf{E}(\mathbf{K}(\mathbb{E})) = \mathbb{E}$.

If we have such a closed system of equivalence relations we can assign the lattice of its *labeled equivalence classes* to it. This structure, called *complete affine ordered set*, is axiomatized in the next chapter.

3 Complete Affine Ordered Sets

In [KS04] the labeled equivalence classes of a system of equivalence relations containing the identity relation are characterized order-theoretically using the notion of an *affine ordered set*. We recall this basic definition which we will specialize in the following yielding a corresponding notion to a closed SER.

Definition 6 (affine ordered set). *We call a triple $\mathbb{A} := (Q, \leq, \parallel)$ an* affine ordered set, *if $(Q, \leq)$ is a partially ordered set, $\parallel$ is a equivalence relation on Q, and the axioms* (A1) - (A4) *hold. Let $A(Q) := \mathrm{Min}(Q, \leq)$ denote the set of all minimal elements in $(Q, \leq)$ and $A(x) := \{a \in A(Q) \mid a \leq x\}$.*

(A1) $\forall x \in Q : A(x) \neq \emptyset$
(A2) $\forall x \in Q \, \forall a \in A(Q) \, \exists! t \in Q : a \leq t \parallel x$
(A3) $\forall x, y, x', y' \in Q : x' \parallel x \leq y \parallel y' \;\;\&\;\; A(x') \cap A(y') \neq \emptyset \Rightarrow x' \leq y'$
(A4) $\forall x, y \in Q \, \exists x', y' \in Q : x \not\leq y \;\;\&\;\; A(x) \subseteq A(y)$
$\Rightarrow x' \parallel x \;\;\&\;\; y' \parallel y \;\;\&\;\; A(x') \cap A(y') \neq \emptyset \;\;\&\;\; A(x') \not\subseteq A(y')$.

The elements of $A(Q)$ are called points *and, in general, elements of Q are called* subspaces. *We say that a subspace x is* contained *in a subspace y if $x \leq y$.*

For a point a and a subspace x we denote by $\pi(a|x)$ the subspace which contains a and is parallel to x. Axiom (A2) guarantees that there is exactly one such subspace. For every $x \in Q$ we observe that $\theta(x) := \{(a, b) \in A^2 \mid \pi(a|x) = \pi(b|x)\}$ is an equivalence relation on the set of points.

Homomorphisms between ordered sets with parallelism are defined as follows:

Definition 7 (homomorphism for ordered sets with parallelism). *For ordered sets with parallelism $\mathbb{A} = (Q, \leq, \parallel)$ and $\mathbb{A}_0 = (Q_0, \leq_0, \parallel_0)$ we call a mapping $\alpha : Q \to Q_0$ a homomorphism if*

- *α maps points to points,*
- *α is order preserving (i.e. $x \leq y \implies \alpha(x) \leq_0 \alpha(y)$),*
- *α preserves the parallelism (i.e. $x \parallel y \implies \alpha(x) \parallel_0 \alpha(y)$).*

By Hom$(\mathbb{A}, \mathbb{A}_0)$ *we denote the set of all homomorphisms from* $\mathbb{A}$ *to* $\mathbb{A}_0$.

From [KS04] we know that we can assign to any affine ordered set $\mathbb{A}$ a SER with identity relation via $\mathbf{E}(\mathbb{A}) := (\mathrm{Min}(Q, \leq), \{\theta(x) | x \in Q\})$ and to any SER with diagonal $\mathbb{E}$ an affine ordered set via $\mathbf{A}(\mathbb{E}) := (\{([x]\theta, \theta) \,|\, \theta \in E \; x \in D\}, \leq', \|')$ where $\leq'$ is defined by $([x]\theta_1, \theta_1) \leq' ([y]\theta_2, \theta_2) :\iff [x]\theta_1 \subseteq [y]\theta_2 \;\; \& \;\; \theta_1 \subseteq \theta_2$ and $\|'$ is defined by $([x]\theta_1, \theta_1) \|' ([y]\theta_2, \theta_2) :\iff \theta_1 = \theta_2$.

For any ordered set P we denote by $P_\perp$ the order which results by adding a bottom element, also called the *lifting* of P. Given an affine ordered set $\mathbb{A} := (Q, \leq, \|)$, we will add as an axiom that $(Q, \leq)_\perp$ is a complete lattice to assure that the SER $\mathbf{E}(\mathbb{A})$ is closed, i.e. the set $\{\theta(x) \,|\, x \in Q\}$ forms a closure system.

Definition 8 (complete affine ordered set). *We call an affine ordered set* $\mathbb{C} := (Q, \leq, \|)$ complete affine ordered set *if* $(Q, \leq)_\perp$ *forms a complete lattice.*

As an illustration we give an example for a closed SER and its associated affine ordered set.

Example 1. We construct an affine ordered set from the following relations on a set $U := \{a, b, c, d, e\}$:

- $\triangle_U$
- θ_1 defined by the classes $\{a\}$ and $\{b, c, d, e\}$
- θ_2 defined by the classes $\{a, b\}$, $\{c\}$, and $\{d, e\}$
- θ_3 defined by the classes $\{b, c\}$ plus singletons
- θ_4 defined by the classes $\{d, e, \}$ plus singletons
- ∇_U

The lifting of the constructed affine ordered set is a lattice .

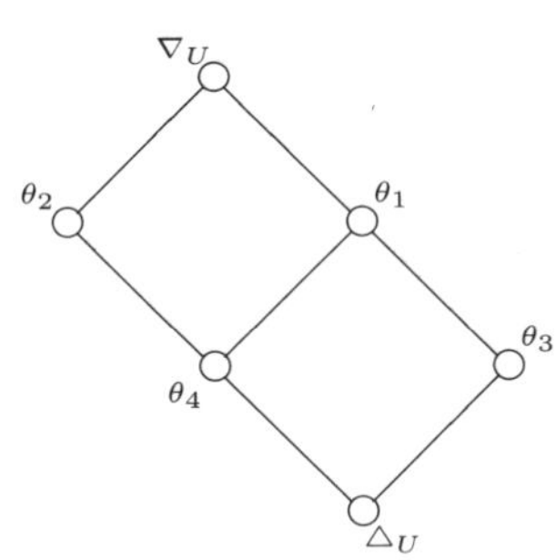

Fig. 1. Set of Equivalence Relations ordered via Set Inclusion

Note that for affine ordered sets we have $x \leq y \iff A(x) \subseteq A(y) \;\; \& \;\; \theta(x) \subseteq \theta(y)$.

Proposition 1. *Let* $\mathbb{A} := (Q, \leq, \|)$ *be an affine ordered set where* $(Q, \leq)_\perp$ *is a lattice and let* $x_i \in Q$ *for* $i \in I$. *Then we have* $A(\bigwedge_I x_i) = \bigcap_I A(x_i)$.

Proof. Let $z := \bigwedge_I x_i$. *We know that* $A(z) \subseteq \bigcap_I A(x_i)$. *Assume that there exists a* $z^* \in \bigcap_I A(x_i)$ *with* $z^* \notin A(z)$. *Assume that* $A(\pi(z^*|z)) \subseteq \bigcap_I A(x_i)$. *This*

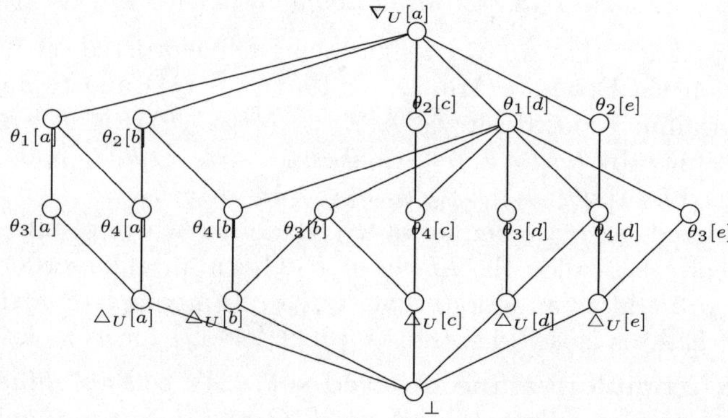

Fig. 2. Lifted Affine Ordered Set

contradicts the assumption that $(Q, \leq)$ *is a lattice, since* $\pi(z^*|z)$ *would be a not comparable to* z *but also a lower bound of the* x_i*. So we have to assume* $A(\pi(z^*|z)) \nsubseteq A(x) \cap A(y)$ *which contradicts* $\theta(z) \subseteq \bigcap_I \theta(x_i)$*.* □

Complete affine ordered sets exhibit a natural connection between parallelism and the meet of the lattice.

Proposition 2. *For a complete affine ordered set* $\mathbb{C} := (Q, \leq, \|)$ *we have*

(P1) $x_i \parallel y_i$ for all $i \in I$ & $\bigcap_{i\in I} A(x_i) \neq \emptyset$ & $\bigcap_{i\in I} A(y_i) \neq \emptyset$
$\Longrightarrow \bigwedge_{i\in I} x_i \parallel \bigwedge_{i\in I} y_i$.

Proof. The premise yields elements $a, b \in A(Q)$ *with* $a \in \bigcap_{i\in I} A(x_i)$ *and* $b \in \bigcap_{i\in I} A(y_i)$*. By* (A3) *we get* $\pi(b|\bigwedge_{i\in I} x_i) \leq \bigwedge_{i\in I} y_i$ *since* $\pi(b|\bigwedge_{i\in I} x_i) \parallel \bigwedge_{i\in I} x_i \leq x_{i_0} \parallel y_{i_0}$ *and* $b \leq \pi(b|\bigwedge_{i\in I} x_i)$ *and* $b \leq y_{i_0}$*. Exchanging the roles of the* x_i *and the* y_i *we dually get* $\pi(a|\bigwedge_{i\in I} y_i) \leq \bigwedge_{i\in I} x_i$*. But now assume* $\pi(b|\bigwedge_{i\in I} x_i) < \bigwedge_{i\in I} y_i$*. This would imply that* $\bigwedge_{i\in I} y_i \nparallel \bigwedge_{i\in I} x_i$ *and therefore we would get* $\pi(a|\bigwedge_{i\in I} y_i) < \bigwedge_{i\in I} x_i$*. But this yields a contradictory configuration as depicted in Figure 3. Therefore we have* $\pi(b|\bigwedge_{i\in I} x_i) = \bigwedge_{i\in I} y_i$ *which completes our proof.* □

4 The Correspondence between Closed SERs and Complete Affine Ordered Sets

In [KS04] it is shown that to any affine ordered set $\mathbb{A} = (Q, \leq, \|)$ the functor $\mathbf{E}$ assigns a SER with identity relation via $\mathbf{E}(\mathbb{A}) := (\mathrm{Min}(Q, \leq), \{\theta(x) | x \in Q\})$ to $\mathbb{A}$. Conversely, for a SER with identity relation $\mathbb{E} = (D, E)$ an affine ordered set $\mathbf{A}(\mathbb{E})$ can be constructed as follows:

- take the labeled equivalence classes $Q := \{([x]\theta, \theta) | x \in D\,,\, \theta \in E\}$ as set of subspaces of the affine ordered set

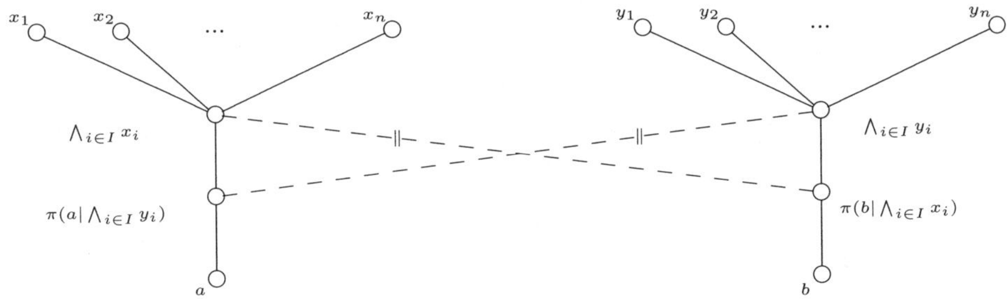

Fig. 3. Contradictory configuration for $I = \{1, ..., n\}$

- define the order $\leq'$ on Q as $([x]\theta_1, \theta_1) \leq' ([y]\theta_2, \theta_2) :\iff [x]\theta_1 \subseteq [y]\theta_2$ & $\theta_1 \subseteq \theta_2$
- define a relation $\|'$ on the set of equivalence classes as $([x]\theta_1, \theta_1) \|' ([y]\theta_2, \theta_2) :\iff \theta_1 = \theta_2$

Theorem 2 in [KS04] includes the assertion that the functors **E** and **A** (extended to the respective homomorphisms) establish a categorical equivalence between SERs with diagonal and affine ordered sets. In the following we will show that these functors also yield a categorical equivalence between the category of closed SERs and the category of complete affine ordered sets.

Theorem 1. *The category of closed SERs and the category of complete affine ordered sets are equivalent.*

Proof. Since we know already that the category of affine ordered sets and the category of SERs with identity are equivalent it remains to show that the functors **E** *and* **A** *move complete affine ordered sets to closed SERs and vice versa. In the following let* $\mathbb{C} := (Q, \leq, \|)$ *be a complete affine ordered set. Firstly, we show that* $\mathbf{E}(\mathbb{C}) = (E, D)$ *is closed. Since we know that it contains the identity we have to show that* D *is a closure system. Let* $R \subseteq Q$*. Then we define* $M_R := \{\theta(x) \mid x \in R\}$*. We want to show that* $\bigcap M_R \in \mathbf{E}(\mathbb{C})$*. For this we construct an equivalence relation* $\theta(z)$ *and prove that* $\theta(z) = \bigcap M_R$*. Let* $a \in A(Q)$ *be an arbitrary but fixed point of* $\mathbb{C}$*. We define* $z := \bigwedge_{x\in R} \pi(a|x)$*. First, we show that* $\theta(z) \subseteq \bigcap M_R$*. For any* $x \in R$ *we have that* $z \leq \pi(a|x)$*. This implies that* $\theta(z) \subseteq \theta(\pi(a|x)) = \theta(x)$*. Second, we show that* $\bigcap M_R \subseteq \theta(z)$*. Let* $(b, c) \in \theta(x)$ *for all* $x \in R$*. Then we have* $\pi(a|x) \| \pi(b|x) = \pi(c|x)$ *for all* $x \in R$*. Since the* $\bigcap_{x\in R} \pi(a|x) \supseteq \{a\}$ *and* $\bigcap_{x\in R} \pi(b|x) \supseteq \{b, c\}$ *we can use* (P1) *to conclude that* $z = \bigwedge_{x\in R} \pi(a|x) \| \bigwedge_{x\in R} \pi(b|x) = \bigwedge_{x\in R} \pi(c|x)$*. But this shows that* $(b, c) \in \theta(z)$*. For the other direction we have to show that* $\mathbf{A}(\mathbb{E})$ *is a complete affine ordered set for a closed SER* $\mathbb{E} := (E, D)$*. We know already that* $\mathbf{A}(\mathbb{E})$ *is an affine ordered set. So it remains to show that* $\mathbf{A}(\mathbb{E})_\perp := (\{([x]\theta, \theta) \mid \theta \in D\}, \leq', \|')_\perp$ *forms a complete lattice. We can consistently interpret* $\perp$ *as* $(\emptyset, \emptyset)$*. Let* $S := \{([x]\theta, \theta) \mid \theta \in E\} \cup \{(\emptyset, \emptyset)\}$ *and let* $R \subseteq S$*. By* π_1 *and* π_2 *we denote the projections on the first and second coordinate. We define* $\bigwedge R := (\bigcap \pi_1(R), \bigcap \pi_2(R))$*. We* *have to show that* $\bigwedge R \in S$. *But since* D *is a closure system* $\bigcap \pi_2(R) \in D$ *having* $\bigcap \pi_1(R)$ *as a class.* □

5 Coordinatization

In this section we give characterizations for the previously studied structures to be coordinatizable, that is, we characterize those structures whose carrier/point set consistently can be seen as the carrier set of an algebra.

5.1 Coordinatization of Closed SERs

At first, we investigate under which conditions a closed SER $\mathbb{E} = (A, D)$ can be coordinatized, that means, under which conditions there exists an algebra $\mathbb{A} := (A, (f)_I)$ with $\mathrm{Con}(\mathbb{A}) = D$.

In the context of a SER $\mathbb{E}$ we define the set of *dilations* $\Delta(\mathbb{E})$ of the SER as all functions mapping points to points and respecting all equivalence relations in $\mathbb{E}$ (a map $\delta \in A^A$ respects an equivalence relation R on A if for all $(a, b) \in R$ we have $(\delta(a), \delta(b)) \in R$), that is, $\Delta(\mathbb{E}) := \{\delta \in A^A \mid \delta \text{ respects all } E \in D\}$. What makes dilations so interesting is the fact that congruence relations can already be characterized by their compatibility with unary polynomial functions. The set of all unary polynomial functions of an algebra $\mathbb{A}$ is denoted by $\Delta(\mathbb{A})$.

Proposition 3 ([Ih93], Theorem 1.4.8). *Let* $\mathbb{A} := (A, (f)_I)$ *be an algebra and* $\theta \in EqA$. *Then* $\theta \in Con(\mathbb{A})$ *if and only if* θ *respects all* $\delta \in \Delta(\mathbb{A})$.

Now it is easy to see that the dilations of a closed SER subsume the unary polynomial functions of a coordinatizing algebra if it exists.

Proposition 4. *Let* $\mathbb{A} := (A, (f)_I)$ *coordinatize* $\mathbb{E} := (A, D)$. *Then* $\Delta(\mathbb{A}) \subseteq \Delta(\mathbb{E})$.

Proof. We get a well known Galois connection between the set EqA of all equivalence relations on a set A and the set of all unary operations $Op_1(A)$ on that same set if we define the relation $I \subseteq EqA \times Op_1(A)$ via $EI\delta \iff \delta$ respects E. Since if $\mathbb{A} := (A, (f)_I)$ coordinatizes $\mathbb{E}$ by Proposition 3 we have $\Delta(\mathbb{A})^I = \mathrm{Con}(\mathbb{A})$, and since $\cdot^{II}$ is a closure operator we get $\Delta(\mathbb{A}) \subseteq \Delta(\mathbb{A})^{II} = \Delta(\mathbb{E})$.

The following proposition gives a constructive view on the principal congruence relations of an algebra which will be useful in the proof of our characterization theorem for closed SERs.

Proposition 5. *Let* $\mathbb{A} := (A, (f)_I)$, *let* $\theta(a, b) \in \mathrm{Con}(\mathbb{A})$ *denote the least congruence relation* θ *with* $(a, b) \in \theta$, *and let* $\Delta(\mathbb{A})$ *denote the set of all unary polynomial functions of* $\mathbb{A}$. *Then* $\theta(a, b)$ *is the reflexive, symmetric, and transitive closure of*

$$\Delta(a, b) := \{(\delta a, \delta b) \mid \delta \in \Delta(\mathbb{A})\}.$$

The next theorem gives a characterization of coordinatizable closed SERs. Note that other characterizations of the set of congruence relations of an algebra are known, e.g. compare [Ih93], p. 56, Theorem 3.4.5. Before presenting our characterization which aims at providing some analogies to Theorem 3.5 in [Wi70] (where a geometry is coordinatized by the (not labeled) congruence classes of an algebra), we need one more definition:

Definition 9. *Let A be a set, let $B \subseteq A$, and let Δ be a set of maps from A to A. Then we define a relation $\equiv \subseteq A \times A$ such that for $a, b \in A$ we have $a \equiv b \bmod (B, \Delta)$ if and only if there exist $\delta_i \in \Delta$ for $i = 0, ..., n$ with $a \in \delta_0(B)$ & $b \in \delta_n(B)$ & $\delta_i(B) \cap \delta_{i+1}(B) \neq \emptyset$ for $i \in \{1, ..., n-1\}$.*

Theorem 2. *Let $\mathbb{E} = (A, D)$ be a closed SER. Then there exists an algebra $\mathbb{A} := (A, (f)_I)$ with $\mathrm{Con}(\mathbb{A}) = D$ if and only if*

(E1) $(c, d) \in \theta(a, b) \iff c \equiv d \bmod (\{a, b\}, \Delta(\mathbb{E}))$
(E2) $[(a, b) \in \theta \Rightarrow \theta_D(a, b) \subseteq \theta] \iff \theta \in D.$

A relation R which fulfills the left hand side of the equivalence (E2) is called one-closed *with respect to the closure operator θ_D. Then condition (E2) can be understood as saying that a one-closed relation is already closed.*

Proof. "$\Rightarrow$": Let $\mathbb{A} := (A, (f)_I)$ be an algebra with $\mathrm{Con}(\mathbb{A}) = D$ and let $(c, d) \in \theta(a, b)$. We will show that the conditions (E1) and (E2) hold. Since $\theta(a, b)$ is the least congruence relation in $\mathrm{Con}(\mathbb{A})$ containing (a, b), we know by Proposition 5 that $\theta(a, b)$ is the reflexive, transitive, and symmetric closure of $\Delta(a, b)$. Therefore there exist mappings $\delta_1, ..., \delta_n \in \Delta(\mathbb{A})$ with $\delta_0(a) = c$, $\delta_n(b) = d$, and $\delta_i \cap \delta_{i+1} \neq \emptyset$ and, using Proposition 4, we have $c \equiv d \bmod (\{a, b\}, \Delta(\mathbb{E}))$, which verifies condition (E1). To verify condition (E2), let θ be one-closed in $\mathrm{Con}(\mathbb{A})$. Now assume $\theta \notin \mathrm{Con}(\mathbb{A})$. Then there exist $(a_j, b_j) \in \theta$ for $j = 1, \ldots, n$ such that for some operation f of $\mathbb{A}$ with arity n we have $(f(a_1, \ldots, a_n), f(b_1, \ldots, b_n)) \notin \theta$. But let us consider the unary polynomial functions $\Gamma_i : A \to A$ for $i = 1, \ldots, n$ where $\Gamma_i(x) := f(b_1, \ldots, b_{i-1}, x, a_{i+1}, \ldots, a_n)$. We get

$$\begin{array}{lll} & \Gamma_1(a_1) & = f(a_1, a_2, a_3, \ldots, a_n) \\ \theta(a_1, b_1) & \Gamma_1(b_1) & = f(b_1, a_2, a_3, \ldots, a_n) \\ = & \Gamma_2(a_2) & = f(b_1, a_2, a_3, \ldots, a_n) \\ \theta(a_2, b_2) & \Gamma_2(b_2) & = f(b_1, b_2, a_3, \ldots, a_n) \\ & \vdots & \\ \theta(a_n, b_n) & \Gamma_n(b_n) & = f(b_1, \ldots, b_{n-1}, b_n). \end{array}$$

Since θ is one-closed, we receive $\theta(a_i, b_i) \subseteq \theta$ for $i = 1, \ldots, n$, and therefore, it follows that $(f(a_1, \ldots, a_n), f(b_1, \ldots, b_n)) \in \theta$, a contradiction.
"$\Leftarrow$": Now let $\mathbb{E} = (A, D)$ be a closed SER satisfying condition (E1) and (E2). We will show that $D = \mathrm{Con}(\mathbb{A})$ for $\mathbb{A} := (A, \Delta(\mathbb{E}))$. By definition of $\Delta(\mathbb{E})$ all relations in D are congruence relations of the constructed algebra, that is, $D \subseteq \mathrm{Con}(\mathbb{A})$. It remains to show that D is "sufficiently large". For this we

deduce that a congruence relation fulfills the left side of the equivalence (E2). Let $\theta \in \mathrm{Con}(\mathbb{A})$ and $(a,b) \in \theta$ and $(c,d) \in \theta_D(a,b)$. We have to show that $(c,d) \in \theta$. Since $\theta_D(a,b) \in D$ by (E1) we get $c \equiv d \bmod (\{a,b\}, \Delta(\mathbb{E}))$ which yields the existence of $\delta_i \in \Delta(\mathbb{E})$ for $i = 0, 1, \ldots, n$ with $a \in \delta_0(\{a,b\})$ & $b \in \delta_n(\{a,b\})$ & $\delta_i(\{a,b\}) \cap \delta_{i+1}(\{a,b\}) \neq \emptyset$ for $i \in \{1, 2, .., n-1\}$. Since θ is a congruence relation $(\delta_i(a), \delta_i(b)) \in \theta$. Transitivity yields $(\delta_0(a), \delta_n(b)) = (c,d) \in \theta$. This completes the proof. □

5.2 Coordinatization of Complete Affine Ordered Sets

In the following we investigate under which conditions a complete affine ordered set $\mathbb{C}$ can be coordinatized, that means, under which conditions there exists an algebra $\mathbb{A}$ such that the elements of the complete affine ordered set can be interpreted as the labeled congruence classes of the algebra, precisely $\mathbb{C} \simeq \mathbf{A}(\mathrm{Con}(\mathbb{A}))$.

In the language of complete affine ordered sets the notion of a dilation reads as:

Definition 10. *Let $\mathbb{C} := (Q, \leq, \|)$ be a complete affine ordered set. Then we call a self map δ on $A(Q)$ a* dilation *if for all $a, b \in A(Q)$ it holds that $\delta(a) \leq \pi(\delta(b)|a \vee b)$. The set of all dilations of a complete affine ordered set is denoted by $\Delta(\mathbb{C})$.*

The dilations of a complete affine ordered set coincide with the dilations of its associated closed SER:

Proposition 6. *Let $\mathbb{C} := (Q, \leq, \|)$ be a complete affine ordered set. Then we have $\Delta(\mathbf{E}(\mathbb{C})) = \Delta(\mathbb{C})$.*

Proof. Let $a, b \in A(Q)$. We have $\delta(a) \leq \pi(\delta(b)|a \vee b)$ if and only if $(\delta(a), \delta(b)) \in \theta(a \vee b)$. □

Using Proposition 4 we see that the dilations of a complete affine ordered set also subsume the unary polynomial functions of a coordinatizing algebra if such an algebra exists.

For a complete affine ordered set $\mathbb{C}$, we call a partition $(C_i)_{i \in I}$ of the points of $\mathbb{C}$ *compatible* if for all $\delta \in \Delta(\mathbb{C})$ and for all $i \in I$ if $a, b \in C_i$ there exists a $j \in I$ such that $\delta(a), \delta(b) \in C_j$. Now we can state the coordinatization theorem for complete affine ordered sets as follows.

Theorem 3 (coordinatization of complete affine ordered sets). *Let $\mathbb{C} := (Q, \leq, \|)$ be a complete affine ordered set. Then $\mathbb{C}$ can be coordinatized if and only if*

(C1) *for any compatible partition $(C_i)_{i \in I}$ of $A(Q)$ there exist $(x_i)_{i \in I}$ with $C_i = A(x_i)$ for $i \in I$ and $x_i \| x_j$ for all $i, j \in I$.*

Proof. "$\Rightarrow$": Let $\mathbb{A} := (A, (f)_I)$ be an algebra that coordinatizes $\mathbb{C}$. To show (C1) let $(C_i)_{i \in I}$ be a compatible partition of $A(Q)$. By supposition $\mathbb{C}$ is isomorphic to $\mathbf{A}(\mathrm{Con}(\mathbb{A}))$. So we know there exists an isomorphism $\epsilon : Q \longrightarrow Q_{\mathbb{A}}$. The points of $\mathbf{A}(\mathrm{Con}(\mathbb{A}))$ are of the form $\{(a, \Delta_A), \mid a \in A\}$ and since we can identify points of $\mathbb{C}$ with points of $\mathbf{A}(\mathrm{Con}(\mathbb{A}))$ via ϵ we can also identify them with the carrier set A of $\mathbb{A}$. So we can recognize $\hat{C} := \bigcup_{i \in I} C_i^2$ as a congruence relation since the dilations of $\mathbb{C}$ subsume the unary polynomial functions of $\mathbb{A}$ and by Proposition 3 this is enough. But since $\hat{C}$ is a congruence relation we know that $(C_i, \hat{C})$ are mutually parallel subspaces of $\mathbf{A}(\mathrm{Con}(\mathbb{A}))$. And since $A(\epsilon^{-1}(C_i, \hat{C})) = \epsilon^{-1}(A(C_i, \hat{C})) = \epsilon^{-1}(\{(c, \Delta_A) \mid c \in C_i\}) = C_i$ we know that the required x_i exist and equal $\epsilon^{-1}(C_i, \hat{C})$.
"$\Leftarrow$": Let (C1) hold for a complete affine ordered set $\mathbb{C}$. In the following we will show that $\mathbb{A}_{\mathbb{C}} := (A(Q), \Delta(\mathbb{C}))$ coordinatizes $\mathbb{C}$. It suffices to prove that $\mathbf{E}(\mathbb{C}) = \mathrm{Con}(\mathbb{A}_{\mathbb{C}})$ since by Theorem 2 in [KS04] we know that $\mathbb{C} \simeq \mathbf{A}(\mathrm{Con}(\mathbb{A}_{\mathbb{C}}))$ is equivalent to $\mathbf{E}(\mathbb{C}) \simeq \mathbf{EA}(\mathrm{Con}(\mathbb{A}_{\mathbb{C}})) \simeq \mathrm{Con}(\mathbb{A}_{\mathbb{C}})$. Let $\theta \in \mathbf{E}(\mathbb{C})$. Obviously, for $(a, b) \in \theta$ we have that $(\delta(a), \delta(b)) \in \theta$ since δ is a dilation. Now assume that $\theta \in \mathrm{Con}(\mathbb{A}_{\mathbb{C}})$. Then $\{[a]\theta \mid a \in A\}$ constitutes a compatible partition of $\mathbb{C}$ since dilations respect the congruence relations of $\mathbb{A}_{\mathbb{C}}$ by construction. But this implies the existence of $(x_i)_{i \in I}$ with $x_i \parallel x_j$ and $A(x_i) = [a]\theta$ for some $a \in A$ and we have $\theta = \theta(x_i) \in \mathbf{E}(\mathbb{C})$ for an arbitrary $i \in I$. □

For a complete affine ordered set we can define a closure operator $\mathfrak{H}$ on the set of points $A(Q)$ via $\mathfrak{H}(P) := A(\bigvee P)$ for $P \subseteq A(Q)$. If a complete affine ordered set can be coordinatized this closure operator coincides with the closure operator assigning to each set of elements of an associated algebra the smallest congruence class they are contained in.

6 Outlook

As a *congruence class geometry* [Wi70] is a closure structure $(A, [\cdot])$ derived from an algebra $\mathbb{A} := (A, f_I)$ where $[\cdot]$ assigns to a set $C \subseteq A$ the smallest congruence class $[C]$ where C is contained in, it would be challenging to investigate the connection between complete affine ordered sets and congruence class geometries. Especially, this could enable utilitizations for data representations, since techniques for using congruence class geometries for data representation were insinuated in [Ka05]. For coordinatizable complete affine ordered sets, the structure $(A, \mathfrak{H})$ – where $\mathfrak{H}$ is defined as in the previous section – is already a congruence class geometry which "sits" in the affine ordered set. A first step could be to study the function which maps a coordinatizable affine ordered set onto the closed sets of a congruence class geometry of the respective algebra, denoted by $\mathrm{Kon}(\mathbb{A})$, via "forgetting the labels"; if we define $\mathrm{Kon}_l(\mathbb{A}) := \{(A, \theta) \in \mathrm{Kon}(\mathbb{A}) \times \mathrm{Con}(\mathbb{A}) \mid A \text{ is a class of } \theta\}$ this map is given by $\varphi : \mathrm{Kon}_l(\mathbb{A}) \to \mathrm{Kon}(\mathbb{A})$ with $\varphi((A, \theta)) := A$. The map φ is $\bigvee$-preserving and therefore residuated (if we attach bottom elements to source and target). Another viable extension of this line of research is constituted by the problem of coordinatizing *projective ordered sets*, the fourth categorical counterpart to simple many-valued contexts as formulated in [KS04].

References

[DP90] Davey, B.A., Priestly, H.A.: Introduction to Lattices and Order. Cambridge University Press, Cambridge (1990)

[GW99] Ganter, B., Wille, R.: Formal Concept Analysis, Mathematical Foundations. Springer, Berlin – Heidelberg – New York (1999)

[Ih93] Ihringer, Th.: Allgemeine Algebra. Teubner, Stuttgart (1993)

[Ka05] Kaiser, T.B.: Representation of Data Contexts and Their Concept Lattices in General Geometric Spaces. In: Dau, F., Mugnier, M.-L., Stumme, G. (eds.) ICCS 2005. LNCS (LNAI), vol. 3596, pp. 195–208. Springer, Heidelberg (2005)

[KS04] Kaiser, T.B., Schmidt, S.E.: Geometry of Data Tables. In: Eklund, P.W. (ed.) ICFCA 2004. LNCS (LNAI), vol. 2961, pp. 222–235. Springer, Heidelberg (2004)

[Pa91] Pawlak, Z.: Rough Sets - Theoretical Aspects of Reasoning about Data. Kluwer Academic Publishers, Dordrecht – Boston – London (1991)

[Pr95] Pratt, V.R.: Chu spaces and their interpretation as concurrent objects. In: van Leeuwen, J. (ed.) Computer Science Today. LNCS, vol. 1000, pp. 392–405. Springer, Heidelberg (1995)

[Wi70] Wille, R.: Kongruenzklassengeometrien. Springer, Berlin – Heidelberg – New York (1970)

Generalizations of Approximable Concept Lattice

Xueyou Chen[1], Qingguo Li[2], Fei Long[2], and Zike Deng[2]

[1] School of Mathematics and Information Science, Shandong University of Technology, Zibo, Shandong 255049, P.R. China
chenxueyou0@yahoo.com.cn

[2] School of Mathematics and Economics, Hunan University, Changsha, Hunan 410012, P.R.China

Abstract. B. Ganter, R. Wille initiated formal concept analysis. Concept lattice is one of the main notions and tools. Some researchers have investigated the fuzzification of the classical crisp concept lattice. One of them was shown by R. Bělohlávek : concept lattice in fuzzy setting. The second one was given by S. Krajči: generalized concept lattice. On the other hand, as a generalization of concept, Zhang, P. Hitzler, Shen defined the notion of approximable concept on a Chu space. In this paper, we introduce two generalizations of approximable concept lattice: approximable concept lattice in the sense of R. Bělohlávek, and generalized approximable concept in the sense of S. Krajči.

Keywords: concept, approximable concept, L-set, generalized concept, Chu space.

1 Introduction

B. Ganter, R. Wille initiated formal concept analysis, which is an order-theoretical analysis of scientific data. Concept lattice is one of the main notions and tools, see [14]. Some researchers have investigated the fuzzification of the classical crisp concept lattice. One is R. Bělohlávek's work ([1]), which considers (L-)fuzzy subsets of objects and (L-)fuzzy subsets of attributes. Another is S. Krajči's work ([20]) which considers fuzzy subsets of attributes and ordinary/classical/crisp subsets of objects. For more details, see [1, 5, 6, 7, 18, 19, 20].

As constructive models of linear logic, Barr and Seely brought Chu space to light in computer science. V. Pratt also investigated Chu space in [21], and Zhang, P. Hitzler, Shen discussed a special form of Chu space in [17, 23, 24].

In theoretical computer science, D. Scott initiated the domain theory which theoretical approximable to the denotational semantics of programming languages at 1960s. At about the same time, in pure mathematics, J. D. Lawsom, K. H. Hofman discovered the theory of algebraic lattices, continuous lattices from the study of the structure of compact semi-lattice, see [15]. From the study of domain theory, Zhang showed that a concept is not an affirmable property([22]),

S. Ben Yahia et al. (Eds.): CLA 2006, LNAI 4923, pp. 107–122, 2008.

see Example 2. As a generalization of concept, in [17, 23, 24], Zhang, P. Hitzler, Shen introduced the notion of approximable concept on a Chu space. They obtained the equivalence between the category of formal contexts with context morphisms, and the category of complete algebraic lattices with Scott continuous functions. For more results, including its applications in data-mining and knowledge discovery, we refer to [23, 24].

In [8], we investigated the relation between approximable concept lattice and formal topology (information base). Thus the connections between the four categories of Domain theory, Formal Concept Analysis, Formal topology, Information System have been constructed in [8, 23, 24].

In this paper, we begin with an overview of algebraic lattices, **L**-sets, which surveys Preliminaries. In Section 3, we introduce Zhang's work. Then in Section 4, we discuss the equivalence between two definitions of approximable concept in fuzzy setting, and prove that all approximable concepts form an algebraic completely lattice of **L**-ordered sets. In the end, we investigate generalized approximable concept, and show that generalized approximable concept lattices represent algebraic lattices.

We generalize R. Bělohlávek, S. Krajči and Zhang's works, also show a connection between Formal Concept Analysis and Domain theory, provide a method in data-mining and knowledge discovery in fuzzy setting.

2 Preliminaries

Let us recall some main notions needed in the paper. i.e., algebraic lattices, **L**-sets. For the other notions, see [3, 15].

2.1 Algebraic Lattices

In [15], the notions of continuous lattice and algebraic lattice were introduced. In the section, we recall some main definitions. For more details, see [15].

Let $(P, \leq, \vee, \wedge, 0, 1)$ be a complete lattice. For $D \subseteq P$, D is called a directed set, $\forall x, y \in D$, if there exists $z \in D$, such that $x \leq z, y \leq z$.

For $x, y \in P$, x is said to be way below y, denoted by $x \ll y$, if for all directed set D with $y \leq \vee D$, there exists $z \in D$, such that $x \leq z$. Let $\Downarrow x = \{y \mid y \ll x\}$. $(P, \leq)$ is called a continuous lattice if for every $x \in P$, we have $x = \vee \Downarrow x$.

$x \in P$ is called a compact element, if $x \ll x$, which is equivalent to: for all directed sets D with $x \leq \vee D$, there exists $z \in D$, satisfying $x \leq z$. Let $K(\ll) = \{x \mid x \text{ is compact }\}$, $K(\ll)$ is not a complete lattice in general.

$(P, \leq)$ is called an algebraic lattice, if for every $x \in P$, there exists a directed set D_x of compact elements, such that $x = \vee D_x$, that is to say,
$x = \vee(\downarrow x \cap K(\ll))$, where $\downarrow x = \{y \mid y \leq x\}$.

In universal algebra, algebraic lattices have become familiar objects as lattices of congruences and lattices of subalgebras of an algebra. Thus they have been extensively studied, and applied in many areas, such that topological theory and domain theory (see [15]). The role of algebraic completely lattice **L**-ordered sets is analogous to the role of algebraic lattices in ordinary relational systems.

2.2 L-Sets

The notion of an L-set was introduced in ([16]), as a generalization of Zadeh's (classical) notion of a fuzzy set. An overview of the theory of L-sets and L-relations (i.e., fuzzy sets and relations in the framework of complete residuated lattices) can be found in [3]. Let us recall some main definitions.

Definition 1. *A residuated lattice is an algebra* $\mathbf{L}= \langle L, \vee, \wedge, \otimes, \rightarrow, 0, 1\rangle$ *such that*

(1) $\langle L, \vee, \wedge, \otimes, \rightarrow, 0, 1\rangle$ *is a lattice with the least element 0 and the greatest element 1.*

(2) $\langle L, \otimes, 1\rangle$ *is a commutative monoid, i.e.,* $\otimes$ *is associative, commutative, and it holds the identity* $a \otimes 1 = a$.

(3) $\otimes, \rightarrow$ *form an adjoint pair, i.e.,*
$x \otimes y \leq z$ *iff* $x \leq y \rightarrow z$ *holds for all* $x, y, z \in L$.

Residuated lattice $\mathbf{L}$ is called complete if $\langle L, \vee, \wedge\rangle$ is a complete lattice. In this paper, we assume that $\mathbf{L}$ is complete.

The following properties of complete residuated lattices will be needed in this paper.

(1) $a \leq b \Rightarrow a \rightarrow c \geq b \rightarrow c$,
(2) $a \leq b \Rightarrow c \rightarrow a \leq c \rightarrow b$,
(3) $a \leq b \Rightarrow a \otimes c \leq b \otimes c$,
(4) $a = 1 \rightarrow a$,
(5) $a \otimes b \leq a \wedge b$,
(6) $a \leq (a \rightarrow b) \rightarrow b$,
(7) $a \otimes (a \rightarrow b) \leq b$,
(8) $a \otimes (b \rightarrow c) \leq b \rightarrow a \otimes c$,
(9) $a \otimes \bigwedge_{i\in I} b_i \leq \bigwedge_{i\in I} (a \otimes b_i)$,
(10) $(\bigvee_{i\in I} a_i) \rightarrow b = \bigwedge_{i\in I} (a_i \rightarrow b)$,
(11) $a \rightarrow \bigwedge_{i\in I} b_i = \bigwedge_{i\in I} (a \rightarrow b_i)$,
(12) $(a \rightarrow b) \otimes (b \rightarrow c) \leq (a \rightarrow c)$.

As discussed in [3], several important algebras are special residuated lattices: Boolean algebras, Heyting algebras, BL-algebras, MV-algebras, Girard monoids and others.

For a universe set X, an $\mathbf{L}$-set in X is a mapping $A : X \rightarrow L$, $A(x)$ indicates that the truth degree of "x belongs to A". We use the symbol L^X to denote the set of all $\mathbf{L}$-sets in X. The concept of an $\mathbf{L}$-relation is defined obviously, and the truth degree to which elements x and y are related by an $\mathbf{L}$-relation I is denoted by $I(x, y)$ or (xIy).

For $a \in L, x \in X$, $\{a/x\}$ is defined as an $\mathbf{L}$-set in X: $\{a/x\}(x) = a$, $\{a/x\}(y) = 0$, if $y \neq x$.

A binary $\mathbf{L}$-relation $\approx$ on X is an $\mathbf{L}$-equality if it satisfies: $\forall x, y, z \in X$, $(x \approx x) = 1$(reflexivity), $(x \approx y) = (y \approx x)$ (symmetry), $(x \approx y) \otimes (y \approx z) \leq (x \approx z)$(transitivity), and $(x \approx y) = 1$ implies $x = y$.

$I \in L^{X\times Y}$ is a binary $\mathbf{L}$-relation, and it is compatible with respect to $\approx_X$ and $\approx_Y$ if $I(x_1, y_1) \otimes (x_1 \approx_X x_2) \otimes (y_1 \approx_Y y_2) \leq I(x_2, y_2)$ for any $x_i \in X, y_i \in Y (i = 1, 2)$. Analogously, $A \in L^X$ is compatible with respect to $\approx_X$ if $A(x_1) \otimes (x_1 \approx_X x_2) \leq A(x_2)$. An $\mathbf{L}$-set $A \in L^{\langle X,\approx\rangle}$ is called an $\approx$-singleton if there exists $x_0 \in X$, such that $A(x) = (x \approx x_0)$ for any $x \in X$.

An **L**-order on X with an **L**-equality relation $\approx$ is a binary **L**-relation $\preceq$ which is compatible with respect to $\approx$ and satisfies: $\forall x, y, z \in X$ $(x \preceq x) = 1$(reflexivity), $(x \preceq y) \wedge (y \preceq x) \leq (x \approx y)$ (antisymmetry), $(x \preceq y) \otimes (y \preceq z) \leq (x \preceq z)$ (transitivity). A set X equipped with an **L**-order $\preceq$ and an **L**-equality $\approx$ is called an **L**-ordered set $\langle\langle X, \approx\rangle, \preceq\rangle$.

These notions are generalizations of the classical notions. Indeed, if **L**=**2**, **L**-order $\preceq$, **L**-equality $\approx$ coincide with the classical order $\leq$ and equality $=$.

For $A, B \in L^X$, we define $S(A, B) = \bigwedge_{x \in X} A(x) \to B(x)$, $(A \approx B) = \bigwedge_{x \in X} A(x) \leftrightarrow B(x)$, and $(A \preceq B) = S(A, B)$, thus $\langle\langle L^X, \approx\rangle, \preceq\rangle$ is an **L**-ordered set, see Example 1. We write $A \subseteq B$, if $S(A, B) = 1$.

Example 1. This is [1] Example 6(1). For $\emptyset \neq M \subseteq L^X$, we obtain that $\langle\langle M, \approx\rangle, S\rangle$ is an **L**-ordered set. In fact, reflexivity and antisymmetry are trivial, we have to prove transitivity and compatibility. Transitivity: $S(A, B) \otimes S(B, C) \leq S(A, C)$ holds if and only if $S(A, B) \otimes S(B, C) \leq A(x) \to C(x)$, i.e., $\forall x \in X$, $A(x) \otimes S(A, B) \otimes S(B, C) \leq C(x)$, and it is true since $A(x) \otimes S(A, B) \otimes S(B, C) \leq A(x) \otimes (A(x) \to B(x)) \otimes (B(x) \to C(x)) \leq C(x)$. In the similarly way, we also prove Compatibility: $S(A, B) \otimes (A \approx A') \otimes (B \approx B') \leq S(A', B')$.

For $S(A, B)$, Lemma 1 will be used in the paper, see [3].

Lemma 1. *(1)* $S(A, \bigcap_{i \in I} B_i) = \bigwedge_{i \in I} S(A, B_i)$, *(2)* $A(x) \otimes S(A, B) \leq B(x)$.

Suppose X and Y are two sets with **L**-equalities $\approx_X$ and $\approx_Y$, respectively. An **L**-Galois connection ([1]) between $\langle X, \approx_X\rangle$ and $\langle Y, \approx_Y\rangle$ is a pair $\langle {}^\uparrow, {}^\downarrow \rangle$ of mappings ${}^\uparrow : L^{\langle X, \approx_X\rangle} \to L^{\langle Y, \approx_Y\rangle}$, ${}^\downarrow : L^{\langle Y, \approx_Y\rangle} \to L^{\langle X, \approx_X\rangle}$, and satisfying the following conditions:

$S(A_1, A_2) \leq S(A_2^\uparrow, A_1^\uparrow)$, $S(B_1, B_2) \leq S(B_2^\downarrow, B_1^\downarrow)$,
$A \subseteq A^{\uparrow\downarrow}$, and $B \subseteq B^{\downarrow\uparrow}$ for any $A, A_1, A_2 \in L^X, B, B_1, B_2 \in L^Y$.

A mapping $C : L^X \to L^Y$ is an **L**-closure operator, if for $A, B \in L^X$, we have (1) $A \subseteq C(A)$, (2) $S(A, B) \leq S(C(A), C(B))$, and (3) $C(C(A)) = C(A)$.

3 Approximable Concepts Introduced by Zhang

As showed in Introduction, Zhang, P. Hitzler, Shen considered a special form of Chu space in [17, 23, 24] as follows.

Definition 2. *A Chu space P is a triple $P = (P_o, \models_P, P_a)$, where P_o is a set of objects and P_a is a set of attributes. The satisfaction relation $\models_P$ is a subset of $P_o \times P_a$. A mapping from a Chu space $P = (P_o, \models_P, P_a)$ to a Chu space $Q = (Q_o, \models_Q, Q_a)$ is a pair of functions (f_a, f_o) with $f_a : P_a \to Q_a$ and $f_o : Q_o \to P_o$ such that for any $x \in P_a$ and $y \in Q_o$, $f_o(y) \models_P x$ iff $y \models_Q f_a(x)$.*

With respect to a Chu space $P = (P_o, \models_P, P_a)$, two functions can be defined:

$\alpha : \mathcal{P}(P_o) \rightarrow \mathcal{P}(P_a)$ with $X \rightarrow \{a \mid \forall x \in X\ x \models_P a\}$,
$\omega : \mathcal{P}(P_a) \rightarrow \mathcal{P}(P_o)$ with $Y \rightarrow \{o \mid \forall y \in Y\ o \models_P y\}$.
α, ω form a pair of Galois connection between $\mathcal{P}(P_o)$ and $\mathcal{P}(P_a)$ ([15]).

A subset $A \subseteq P_a$ is called an intent of a formal concept if it is a fixed point of $\alpha \circ \omega$, i.e., $\alpha(\omega(A)) = A$. In [23], A is also called a (formal) concept (of attributes).

If A is a concept, for every subset $B \subseteq A$, we have $B \subseteq \alpha(\omega(B)) \subseteq \alpha(\omega(A)) = A$ ([24]).

Dually, an extent of a formal concept, or a (formal) concept (of objects) also defined in [24].

Zhang pointed out in [24] that in FCA, a Chu space is called a formal context, but "Chu" carries with it the notion of morphism, to form a category. On the other hand, FCA provides the notion of concept, intrinsic to a Chu space.

As a generalization of the notion of concept, Zhang and Shen introduced the notion of approximable concept in [23]. A subset $A \subseteq P_a$ is an approximable concept (of attributes) if for every finite subset $X \subseteq A$, we have $\alpha(\omega(X)) \subseteq A$. Clearly, every concept is an approximable concept, but the converse is false (see Example 2).

Example 2. In [23], Zhang and Shen gave the following example, to show an approximable concept is not a concept in general.

P	$\uparrow t$	$\uparrow b$	$\uparrow *$	$\uparrow 0$	$\uparrow 1$	$\uparrow 2$	$\cdots$	$\uparrow -1$	$\uparrow -2$	$\cdots$
t	×	×	×	×	×	×	$\cdots$	×	×	$\cdots$
b		×					$\cdots$			$\cdots$
*		×	×				$\cdots$			$\cdots$
0		×		×			$\cdots$	×	×	$\cdots$
1		×		×	×		$\cdots$	×	×	$\cdots$
$\vdots$		$\vdots$		$\vdots$	$\vdots$		$\vdots$	$\vdots$	$\vdots$	$\vdots$
-1		×					$\cdots$	×	×	$\cdots$
-2		×					$\cdots$		×	$\cdots$
$\vdots$		$\vdots$		$\vdots$	$\vdots$		$\vdots$	$\vdots$	$\vdots$	$\vdots$

Given $S = \{b, \cdots, -2, -1, 0, 1, 2, \cdots, t, *\}$, with the order $b < \cdots < -2 < -1 < 0 < 1 < 2 < \cdots < t$, and $b < * < t$. Then S is a complete lattice which is not algebraic.

$\forall x \in S$, let $\uparrow x = \{y \mid x \leq y\}$, $\times$ indicates that $y \in \uparrow x$, we obtain a Chu space as follows.

$P_a = \{\uparrow x \mid x \in S\}$, $P_o = \{x \mid x \in S\}$, $x \models \uparrow y$, if $x \in \uparrow y$, then $P = (P_o, \models, P_a)$ is a Chu space.

(1) $\{\uparrow i \mid i \leq 0$, or $i \geq 0\} \cup \{\uparrow b\}$ is an approximable concept, not a concept.

(2) Clearly $* \in \alpha(\omega(\{\uparrow i \mid i \geq 0\}))$, and but for any finite subset X of $\{\uparrow i \mid i \geq 0\}$, we have $* \notin \alpha(\omega(X))$. $\{\uparrow i \mid i \geq 0\}$ is a family of concept.

4 Approximable Concepts in Fuzzy Setting

4.1 Definitions

In [1, 5], suppose X and Y are two sets with **L**-equalities $\approx_X$ and $\approx_Y$, respectively; I an **L**-relation between X and Y which is compatible with respect to $\approx_X$ and $\approx_Y$. A pair $\langle {}^{\uparrow}, {}^{\downarrow} \rangle$ of mappings was defined as:

$${}^{\uparrow} : L^X \to L^Y, \text{ for } A \in L^X, \ A^{\uparrow}(y) = \bigwedge_{x \in X} A(x) \to I(x, y).$$

and $${}^{\downarrow} : L^Y \to L^X, \text{ for } B \in L^Y, \ B^{\downarrow}(x) = \bigwedge_{y \in Y} B(y) \to I(x, y).$$

Then $\langle X, Y, I \rangle$ is a formal **L**-context; $\langle A, B \rangle$ is a concept in fuzzy setting, if $A = A^{\uparrow\downarrow}$, $B = B^{\downarrow\uparrow}$. That is, A is an extent of a concept, B is an intent of a concept; or A is a concept of objects, B is a concept of attributes. $\beta(X, Y, I) = \{\langle A, B \rangle \mid \langle A, B \rangle \text{ is a concept } \}$ is a formal concept lattice.

As a generalization, we introduced the notion of an approximable concept in fuzzy setting ([9, 12, 13]). According to the reviewer's suggestion of [9], there exist two choices for the definition of an approximable concept. We adopted one kind (Definition 4) in [9]. In the section, we discuses the other kind (Definition 5), and prove the equivalence between the two definitions.

Given two **L**-ordered sets $(X, \approx_X)$, $(Y, \approx_Y)$, and I is an **L**-relation. Let $P_o = (X, \approx_X)$, $P_a = (Y, \approx_Y)$, and $\models$ induced by the **L**-relation I, that is to say, $(x \models y) = (xIy)$. We obtain a Chu space $P = ((X, \approx_X), \models, (Y, \approx_Y))$ in fuzzy setting, and $\alpha = {}^{\uparrow}$, $\omega = {}^{\downarrow}$, i.e.,

$$\alpha : L^X \to L^Y, \text{ for } A \in L^X, \ \alpha(A)(y) = A^{\uparrow}(y) = \bigwedge_{x \in X} A(x) \to I(x, y).$$

$$\omega : L^Y \to L^X, \text{ for } B \in L^Y, \ \omega(B)(x) = B^{\downarrow}(x) = \bigwedge_{y \in Y} B(y) \to I(x, y).$$

Definition 3. *Suppose $H \in L^X$, if $\{x \in X \mid H(x) > 0\}$ is finite, then H is called finite.*

Clearly if **L**=**2**, $\{x \in X \mid H(x) > 0\} = \{x \in X \mid H(x) = 1\}$, is the same with the finite set in classical set theory.

In [9], we defined the notion of an approximable concept,

Definition 4. *Given $A \in L^X$, if for each finite $H \in L^X$, we have $(H \preceq A) \leq (\omega(\alpha(H)) \preceq A)$, i.e., $S(H, A) \leq S(\omega(\alpha(H)), A)$, then A is called to be an extent of a formal fuzzy approximable concept. A is also called an (formal fuzzy) approximable concept (of objects).*

Dually, a set $A \in L^Y$ is an intent of a formal fuzzy approximable concept, if for each finite $H \in L^Y$, we have $(H \preceq A) \leq (\alpha(\omega(H)) \preceq A)$, i.e., $S(H, A) \leq S(\alpha(\omega(H)), A)$. A is also called an (formal fuzzy) approximable concept (of attributes). We will use the symbol $\mathcal{A}\ (Y, I)$ to denote the set of all approximable concepts A (of attributes).

Since for each finite $H \in L^X$, we have $H \subseteq \omega(\alpha(H))$, that is to say, $H(x) \leq \omega(\alpha(H))(x)$ for every $x \in X$. So we obtain $H(x) \to A(x) \geq \omega(\alpha(H))(x) \to A(x)$. Thus $S(H, A) \geq S(\omega(\alpha(H)), A)$ for every $A \in L^X$.

In the similar way, for each finite $H \in L^Y$, and $A \in L^Y$, we also have $S(H, A) \geq S(\alpha(\omega(H)), A)$.

By the above discussion, we obtain an equivalent definition,

Defintion 4$'$. *$A \in L^X$ is called an extent of a formal fuzzy approximable concept, if for each finite $H \in L^X$, we have $(H \preceq A) = (\omega(\alpha(H)) \preceq A)$, i.e., $S(H, A) = S(\omega(\alpha(H)), A)$.*

Dually, an **L**-*set $A \in L^Y$ is called an intent of a formal fuzzy approximable concept, if for each finite $H \in L^Y$, we have $(H \preceq A) = (\alpha(\omega(H)) \preceq A)$, i.e., $S(H, A) = S(\alpha(\omega(H)), A)$.*

The second choice for the definition of an approximable concept is Definition 5.

Definition 5. *Given $A \in L^X$, if for each finite $H \in L^X$, and $H \subseteq A$, we have $\omega(\alpha(H)) \subseteq A$, then A is called to be an extent of a formal fuzzy approximable concept. A is also called an (formal fuzzy) approximable concept (of objects).*

Dually, a set $A \in L^Y$ is an intent of a formal fuzzy approximable concept, if for each finite $H \in L^Y$, and $H \subseteq A$, we have $\alpha(\omega(H)) \subseteq A$. A is also called an (formal fuzzy) approximable concept (of attributes).

From the one direction, we have

Lemma 2. *Suppose A is an approximable concept in the sense of Definition 4, then A is also an approximable concept in the sense of Definition 5.*

Proof. It is clearly. □

From the other direction, suppose A is an approximable concept in the sense of Definition 5, for $F = \{A(y_0)/y_0\}$, clearly we have $F \subseteq A$. Thus $\alpha(\omega(F)) \subseteq A$. So we obtain $\alpha(\omega(F))(y) \leq A(y)$ for every $y \in Y$.

$$\omega(F)(x) = \bigwedge_{y \in Y} F(y) \to I(x, y) = A(y_0) \to I(x, y_0).$$

and

$$\alpha(\omega(F))(y) = \bigwedge_{x \in X} \omega(F)(x) \to I(x, y) = \bigwedge_{x \in X} (A(y_0) \to I(x, y_0)) \to I(x, y).$$

By Definition 5, we have, $\bigwedge_{x \in X} (A(y_0) \to I(x, y_0)) \to I(x, y) \leq A(y)$.

Lemma 3. *Suppose A is an approximable concept in the sense of Definition 5, then A is also an approximable concept in the sense of Definition 4.*

Proof. Suppose A is an approximable concept in the sense of Definition 5.

(1) For simplicity, we may assume $H = \{a/y_0\}$, then

$$S(H, A) = \bigwedge_{y \in Y} H(y) \to A(y) = a \to A(y_0).$$

$$\omega(H)(x) = \bigwedge_{y \in Y} H(y) \to I(x, y) = a \to I(x, y_0).$$

and

$$\alpha(\omega(H))(y) = \bigwedge_{x \in X} \omega(H)(x) \to I(x, y) = \bigwedge_{x \in X} [a \to I(x, y_0)] \to I(x, y).$$

(2) We have to prove A is an approximable concept in the sense of Definition 4, it suffices to prove $S(H, A) \leq S(\alpha(\omega(H)), A)$. i.e.,

$$a \to A(y_0) \leq \bigwedge_{y \in Y} [\bigwedge_{x \in X} (a \to I(x, y_0)) \to I(x, y)] \to A(y) \quad (*).$$

(3) To prove (*), it suffices to prove

$$a \to A(y_0) \leq [\bigwedge_{x \in X} (a \to I(x, y_0)) \to I(x, y)] \to A(y) \text{ holds for every } y \in Y.$$

It is valid, since

$$a \to A(y_0) \leq [A(y_0) \to I(x, y_0)] \to [a \to I(x, y_0)]$$
$$\leq [(a \to I(x, y_0)) \to I(x, y)] \to [(A(y_0) \to I(x, y_0)) \to I(x, y)].$$

Thus, we obtain,

$$[(a \to I(x, y_0)) \to I(x, y)] \otimes [a \to A(y_0)] \leq [(A(y_0) \to I(x, y_0)) \to I(x, y)].$$

So we have

$$[(a \to I(x, y_0)) \to I(x, y)] \otimes [a \to A(y_0)] \leq \bigwedge_{x \in X} [(A(y_0) \to I(x, y_0)) \to I(x, y)],$$

that is to say,

$$[a \to A(y_0)]$$
$$\leq [(a \to I(x, y_0)) \to I(x, y)] \to \bigwedge_{x \in X} [(A(y_0) \to I(x, y_0)) \to I(x, y)]$$
$$\leq \bigwedge_{x \in X} [(a \to I(x, y_0)) \to I(x, y)] \to \bigwedge_{x \in X} [(A((y_0) \to I(x, y_0)) \to I(x, y)]$$
$$\leq \bigwedge_{x \in X} [(a \to I(x, y_0)) \to I(x, y)] \to A(y).$$

By this, (*) holds. Hence we obtain $S(H, A) \leq S(\alpha(\omega(H)), A)$. That is, A is an approximable concept in the sense of Definition 4. □

So we obtain,

Proposition 1. *Definition 4 and Definition 5 are equivalent.*

The following proposition gives a representation of an approximable concept.

Proposition 2. *Suppose A is an approximable concept in $\mathcal{A}(Y.I)$, then*

$$A(y) = \bigvee_{\textit{finite } H \in L^Y} S(H, A) \otimes \alpha(\omega(H))(y).$$

Proof. Since A is an approximable concept, so for each finite $H \in L^Y$, we have $S(H, A) \leq S(\alpha(\omega(H)), A)$, thus we have,

$$S(H, A) \otimes \alpha(\omega(H))(y) \leq S(\alpha(\omega(H)), A) \otimes \alpha(\omega(H))(y) \leq A(y).$$

So we obtain, $\bigvee_{\text{finite } H \in L^Y} S(H, A) \otimes \alpha(\omega(H))(y) \leq A(y)$.

On the other hand, for $\{A(y)/y\} \in L^Y$, since $H \subseteq \alpha(\omega(H))$, we obtain

$$S(\{A(y)/y\}, A) \otimes \alpha(\omega(\{A(y)/y\}))(y) = 1 \otimes \alpha(\omega(\{A(y)/y\}))(y)$$
$$= \alpha(\omega(\{A(y)/y\}))(y) \geq \{A(y)/y\}(y) = A(y).$$

Furthermore we have,

$$\bigvee_{\text{finite } H \in L^Y} S(H, A) \otimes \alpha(\omega(H))(y) \geq A(y).$$ Hence the equation holds. □

4.2 Main Results

In [10, 11, 13], we introduced the notions of a directed set, a way-below relation, a continuous lattice, an algebraic lattice in fuzzy setting.

Suppose $M \subseteq L^X$, for an **L**-set $\mathcal{U}$ in M, two operators were defined in [3], for every $x \in X$, $\bigcup\mathcal{U}(x) = \bigvee_{A\in M} \mathcal{U}(A) \otimes A(x)$, $\bigcap\mathcal{U}(x) = \bigwedge_{A\in M} \mathcal{U}(A) \to A(x)$.

Clearly, $\bigcup\mathcal{U}$ and $\bigcap\mathcal{U}$ are generalizations of the union and the intersection of a system of sets in the classical case.

We gave the notion of a directed set in fuzzy setting, that is, Definition 6.

Definition 6. *Suppose $\mathcal{U}$ is an **L**-set in M, then $\mathcal{U}$ is called a directed **L**-set, if for any $A, B \in M$, there exists $C \in M$, we have $\mathcal{U}(A) \leq \mathcal{U}(C) \otimes S(A, C)$, $\mathcal{U}(B) \leq \mathcal{U}(C) \otimes S(B, C)$.*

Clearly, when **L**=**2**, if $A \in \mathcal{U}$, $B \in \mathcal{U}$, by Definition 6, there exists $C \in \mathcal{U}$, such that $A \subseteq C, B \subseteq C$.

For any directed **L**-set $\mathcal{U}$ in M, we replace $\bigcup\mathcal{U}$ by $\bigsqcup\mathcal{U}$, i.e., for every $x \in X$, $\bigsqcup\mathcal{U}(x) = \bigvee_{A\in M} \mathcal{U}(A) \otimes A(x)$.

Clearly, $\bigsqcup\mathcal{U}$ is a generalization of a directed union of a system of sets in the classical case.

By means of the notion of directed **L**-sets, we defined the notion of way-below relation. For $A, B \in M$, and $\mathcal{U}$ is any directed **L**-set in M, let

$$W(A, B) = \bigwedge_{\{\mathcal{U} \text{ is directed}\}} S(B, \bigsqcup\mathcal{U}) \to \bigvee_{E\in M} \mathcal{U}(E) \otimes S(A, E)$$

In fact, when **L**=**2**, the above definition coincides with the definition of way below relation according to the definition in [15].

Furthermore, we gave the notion of compact elements in fuzzy setting.

Definition 7. *Suppose $A \in M$, A is compact with respect to $\bigsqcup$ in M, if for any directed **L**-set $\mathcal{U}$ in M, we have $S(A, \bigsqcup\mathcal{U}) \to \bigvee_{E\in M} \mathcal{U}(E) \otimes S(A, E) = 1$.*

In fact, when **L**=**2**, the above definition coincides with the definition of a compact element according to the definition in [15].

An **L**-ordered set $\langle\langle M, \approx\rangle, \preceq\rangle$ is a completely algebraic lattice, if

(1) M is closed for $\bigsqcup$ and $\bigcap$. Moreover, for any $A \in M$, there exists a compact directed **L**-set $\mathcal{U}$ in M, such that $A = \bigsqcup\mathcal{U}$.

(2) for any directed **L**-set $\mathcal{U}$ in M, $\sup\mathcal{U}$ is a $\approx$-singleton,

(3) for any **L**-set $\mathcal{U}^*$ in M, $\inf\mathcal{U}^*$ is a $\approx$-singleton.

In what follows, we will adopt Definition 4, use the symbol $\mathcal{A}(Y, I)$ to denote the set of all approximable concepts A (of attributes), and prove that $\mathcal{A}(Y, I)$ is a completely algebraic lattice **L**-ordered sets in the sense of [3].

Suppose $A, B \in \mathcal{A}(Y, I)$, we define $(A \preceq B) = \bigwedge_{y\in Y} A(y) \to B(y)$, and $(A \approx B) = (A \preceq B) \wedge (B \preceq A)$. From Example 1, we know that $\langle\langle \mathcal{A}(Y, I), \approx\rangle, \preceq\rangle$ is an **L**-ordered set. That is,

Lemma 4. *$\langle\langle \mathcal{A}(Y,I), \approx\rangle, \preceq\rangle$ is an* **L***-ordered set.*

On the one hand, suppose $\mathcal{U}$ is a directed **L**-set in $\mathcal{A}(Y,I)$, let $D = \bigsqcup\mathcal{U}$, we have

Lemma 5. *D is an approximable concept.*

Proof. Suppose $\mathcal{U}$ is a directed **L**-set in $\mathcal{A}(Y,I)$, let $D = \bigsqcup\mathcal{U}$, we have to prove D is an approximable concept.

For each finite $H \in L^Y$, for every $A \in \mathcal{A}(Y,I)$, since A is an approximable concept, we have $S(H,A) = S(\alpha(\omega(H)),A)$. Thus

$$S(H,D) \geq \bigvee_{A\in\mathcal{A}(Y,I)} \mathcal{U}(A) \otimes S(H,A),$$

and

$$S(\alpha(\omega(H)),D) \geq \bigvee_{A\in\mathcal{A}(Y,I)} \mathcal{U}(A) \otimes S(\alpha(\omega(H)),A).$$

By the above analysis, we obtain

$$S(H,D) \geq \bigvee_{A\in\mathcal{A}(Y,I)} \mathcal{U}(A) \otimes S(H,A)$$
$$= \bigvee_{A\in\mathcal{A}(Y,I)} \mathcal{U}(A) \otimes S(\alpha(\omega(H)),A) \text{ (since } A \text{ is an approximable concept)}$$
$$\leq S(\alpha(\omega(H)),D) \leq S(H,D). \text{ (since } H \subseteq \alpha(\omega(H)))$$

So, we have

$$S(H,D) = S(\alpha(\omega(H)),D) = \bigvee_{A\in\mathcal{A}(Y,I)} \mathcal{U}(A) \otimes S(\alpha(\omega(H)),A)$$
$$= \bigvee_{A\in\mathcal{A}(Y,I)} \mathcal{U}(A) \otimes S(H,A).$$

i.e., D is an approximable concept. □

Lemma 5 shows that $\mathcal{A}(Y,I)$ is closed for the operator $\bigsqcup$.
Lemma 6 shows that for any directed **L**-set $\mathcal{U}$ in $\mathcal{A}(Y,I)$, $\sup\mathcal{U}$ is a $\approx$-singleton.

Lemma 6. *$\sup\mathcal{U}$ is a $\approx$-singleton.*

Proof. Since $\mathcal{U}$ is a directed **L**-set in $\mathcal{A}(Y,I)$, by Lemma 5, $D = \bigsqcup\mathcal{U}$ is an approximable concept. i.e., $D \in \mathcal{A}(Y,I)$.

We have to show $\sup\mathcal{U}$ is a $\approx$-singleton, it suffices to prove

$$(U(\mathcal{U})(D)) \bigwedge (LU(\mathcal{U})(D)) = 1.$$

(1) By the definition of D, we have $\mathcal{U}(A) \otimes A(y) \leq D(y)$, which is equivalent to $\mathcal{U}(A) \leq A(y) \to D(y)$. Thus we have $1 \leq \mathcal{U}(A) \to (A \preceq D)$, so we obtain $U(\mathcal{U})(D) \geq 1$.

(2) By the definition of L, we have to show $1 \leq (U(\mathcal{U}))(A) \to (D \preceq A)$ for any $A \in \mathcal{A}(Y,I)$. It is sufficient to prove $(U(\mathcal{U}))(A) \leq (D \preceq A)$ for any $A \in \mathcal{A}(Y,I)$.

Since $(D \preceq A) = S(D,A) = \bigwedge_{y\in Y} D(y) \to A(y)$, we have to prove $(U(\mathcal{U}))(A) \leq D(y) \to A(y)$ for each $y \in Y$. i.e., $D(y) \otimes (U(\mathcal{U}))(A) \leq A(y)$ holds for each $y \in Y$. That is, for each $y \in Y$, we have

$$\bigvee_{C\in\mathcal{A}(Y,I)} \mathcal{U}(C)\otimes C(y)\otimes (U(\mathcal{U}))(A) \le A(y).$$

It is true, since

$$\begin{aligned}
&\mathcal{U}(C)\otimes C(y)\otimes (U(\mathcal{U}))(A)\\
&= \mathcal{U}(C)\otimes C(y)\otimes \bigwedge_{B\in\mathcal{A}(Y,I)} \mathcal{U}(B)\to (B\preceq A)\\
&\le \mathcal{U}(C)\otimes C(y)\otimes [\mathcal{U}(C)\to (C\preceq A)]\\
&= \mathcal{U}(C)\otimes C(y)\otimes [\mathcal{U}(C)\to S(C,A)]\\
&\le C(y)\otimes S(C,A)\le A(y).
\end{aligned}$$

By the above proof, we obtain that $(LU(\mathcal{U}))(D) = 1$. Hence Lemma 6 holds. □

On the other hand, for any **L**-set $\mathcal{U}^*$ in $\mathcal{A}(Y,I)$, where $\mathcal{U}^*$ may be not directed. Let $E = \bigcap \mathcal{U}^*$, then we have

Lemma 7. *E is an approximable concept.*

Proof. For each finite $H\in L^Y$, and each $A\in\mathcal{A}(Y,I)$, we have $S(H,A)\le S(\alpha(\omega(H)),A)$. i.e.,

$$\bigwedge_{y\in Y} H(y)\to A(y) \le \bigwedge_{y\in Y} \alpha(\omega(H))(y)\to A(y).$$

By this, we have

$$\begin{aligned}
&\bigwedge_{A\in\mathcal{A}(Y,I)} \mathcal{U}^*(A)\to [\bigwedge_{y\in Y} H(y)\to A(y)]\\
&\le \bigwedge_{A\in\mathcal{A}(Y,I)} \mathcal{U}^*(A)\to [\bigwedge_{y\in Y} \alpha(\omega(H))(y)\to A(y)].
\end{aligned}$$

We thus have

$$\begin{aligned}
&\bigwedge_{y\in Y} H(y)\to [\bigwedge_{A\in\mathcal{A}(Y,I)} \mathcal{U}^*(A)\to A(y)]\\
&\le \bigwedge_{y\in Y} \alpha(\omega(H))(y)\to [\bigwedge_{A\in\mathcal{A}(Y,I)} \mathcal{U}^*(A)\to A(y)].
\end{aligned}$$

So, we obtain

$$\bigwedge_{y\in Y} H(y)\to E(y) \le \bigwedge_{y\in Y} \alpha(\omega(H))(y)\to E(y).$$

This implies that $S(H,E)\le S(\alpha(\omega(H)),E)$. i.e., E is an approximable concept. □

Lemma 7 shows that $\mathcal{A}(Y,I)$ is closed for the operator $\bigcap$.

Lemma 8. *inf* $\mathcal{U}^*$ *is a* $\approx$*-singleton.*

The proof see [9].

Lemma 9. *Suppose* $A\in\mathcal{A}(Y,I)$*, then there exists a directed* **L***-set* $\mathcal{U}_A$ *in* $\mathcal{A}(Y,I)$*, such that* $A = \bigsqcup \mathcal{U}_A$.

The proof see [9].

Lemma 9 will be used to prove that $\mathcal{A}(Y, I)$ is algebraic.
For $A \in \mathcal{A}(A, I)$, by Lemma 9, $A = \bigsqcup \mathcal{U}_A$, where $\mathcal{U}_A$ is a directed **L**-set in $\mathcal{A}(Y, I)$. On the other hand, by Lemma 6, we show that $\sup \mathcal{U}$ is a $\approx$-singleton for any directed **L**-set $\mathcal{U}$ in $\mathcal{A}(Y, I)$; by Lemma 7, $\mathcal{A}(Y, I)$ is closed for $\bigcap$, and by Lemma 8, $\inf \mathcal{U}^*$ is a $\approx$-singleton for any **L**-set $\mathcal{U}^*$ in $\mathcal{A}(Y, I)$,. Furthermore, Lemmas 4, 5, 6, 7, 8 show that $\langle\langle \mathcal{A}(Y, I), \approx\rangle, \preceq\rangle$ is a completely lattice **L**-ordered set. So we obtain,

Proposition 3. *$\langle\langle \mathcal{A}(Y, I), \approx\rangle, \preceq\rangle$ is a completely lattice* **L***-ordered set with two operators $\bigsqcup$ and $\bigcap$.*

In the end of the section, we will show the approximable concept lattice is algebraic.

Lemma 10. *For each finite $H \in L^Y$, $\alpha(\omega(H))$ is compact with respect to $\bigsqcup$ in $\mathcal{A}(Y, I)$.*

The proof see [9].

Proposition 4. *$\mathcal{A}(Y, I)$ is algebraic.*

Proof. By Lemmas 8, 10. □

Note 1. In [9], suppose $(V, \leq)$ is an algebraic completely lattice, $K(\ll)$ is the set of all compact elements, and $\ll$ is the way below relation ([15]), see Section 2.1. We also constructed a Chu space in the fuzzy sense (i.e., the objects, the attributes, the satisfaction relation are in fuzzy setting), such that it's approximable concepts of attributes is isomorphic to $(V, \leq)$ in the sense of [1] Definition 3.

5 Generalized Approximable Concepts

As showed in Introduction, R. Bělohlávek and Stanislav Krajči gave the generalization of concept lattice, respectively, see [1, 6].

In [18, 19], Stanislav Krajči obtained a common platform for both of them, and proved all complete lattices are isomorphic to the generalized concept lattices.

We introduce some main notions from [18, 19].

Suppose L is a poset, C and D are two supremum-complete upper-semilattices. i.e., there exists $\sup X = \bigvee X$ for each subset of C or D (in fact, C, D are complete lattices). Let $\bullet : C \times D \to L$ be monotone and left-continuous in both their arguments, that is to say,

1a) $c_1 \leq c_2$ implies that $c_1 \bullet d \leq c_2 \bullet d$ for all $c_1, c_2 \in C$ and $d \in D$.
1b) $d_1 \leq d_2$ implies that $c \bullet d_1 \leq c \bullet d_2$ for all $c \in C$ and $d_1, d_2 \in D$.
2a) If $c \bullet d \leq \iota$ holds for $d \in D, \iota \in L$ and for all $c \in X \subseteq C$, then $\sup X \bullet d \leq l$.
2b) If $c \bullet d \leq \iota$ holds for $c \in C, \iota \in L$ and for all $d \in Y \subseteq D$, then $c \bullet \sup Y \leq l$.

Let A and B be non-empty sets and R be L-fuzzy relation on their Cartesian product, $R : A \times B \to L$. Stanislav Krajči defined two mappings as follows,

(1) $\nearrow$: ${}^{B}D \to^{A} C$, if $g : B \to D$, then $\nearrow (g) : A \to C$, where $\nearrow (g)(a) = \sup\{c \in C \mid \forall b \in B, c \bullet g(b) \leq R(a,b)\}$.

(2) $\swarrow$: ${}^{A}C \to^{B} D$, if $f : A \to C$, then $\swarrow (f) : B \to D$, where $\swarrow (f)(b) = \sup\{d \in D \mid \forall a \in A, f(a) \bullet d \leq R(a,b)\}$.

In [18, 19], Stanislav Krajči introduced a generalized concept lattice.

Based on the common platform, we give a generalization of an approximable concept, i.e., a generalized approximable concept.

The notions of a directed set, an algebraic lattice were introduced in Section 2.1. In the section, because the definition is not symmetric, similarly, we also give the notions of a up-directed set, a left-algebraic lattice.

Definition 8. *Suppose $h : B \to D$, if there exists $\{b_i \mid i \in I\} \subseteq B$, where I is a finite index, such that $h(b_i) \neq 0$, and $h(b) = 0$ for all $b \in B, b \neq b_i$, then h is called finite.*

Definition 9. *Suppose $g : B \to D$, g is a generalized approximable concept, if for each finite $h \leq g$, we have $\swarrow\nearrow (h) \leq g$.*

The collection of all generalized approximable concepts denoted by $\mathcal{A}$. In the first part, we will show that $(\mathcal{A}, \leq)$ is a left-algebraic lattice.

When L, C, D are finite, the notions of a generalized approximable concept and a generalized concept are identical.

Lemma 11. *Suppose $g \in \mathcal{A}$, $\{\swarrow\nearrow (h) \mid$ finite $h \leq g\}$ is up-directed.*

Proof. For $g \in \mathcal{A}$, suppose h_1, h_2 are finite, and $h_1, h_2 \leq g$, we have $\swarrow\nearrow (h_1) \leq \swarrow\nearrow (h_1 \vee h_2)$, $\swarrow\nearrow (h_2) \leq \swarrow\nearrow (h_1 \vee h_2)$. where $(h_1 \vee h_2)(a) = h_1(a) \vee h_2(a)$. Thus $h_1 \vee h_2$ is also finite, and $h_1 \vee h_2 \leq g$. □

Lemma 12. *Suppose $g \in \mathcal{A}$, we have $g = \sup\{\swarrow\nearrow (h) \mid$ finite $h \leq g\}$.*

Proof. It is trivial. □

By Lemmas 11, 12, we have g is the supremum of a up-directed set.

Lemma 13. *Suppose h is finite, then $\swarrow\nearrow (h)$ is compact.*

Proof. It is trivial. □

Lemma 14. *Suppose $\{g_i \mid i \in I\}$ is a up-directed set of generalized approximable concepts, then $\bigvee_{i \in I} g_i$ is also a generalized approximable concept.*

Proof. For each finite $h \leq \bigvee_{i \in I} g_i$, there exist $g_1, g_2, \cdots, g_m$, such that $h(b_i) \leq g_i(b_i)$; and for every $b \in B$, $b \neq b_i$, $h(b) = 0$.

Since $\{g_i \mid i \in I\}$ is up-directed, there exists g_{i_0}, such that $h \leq g_{i_0}$. Furthermore, g_{i_0} is a generalized approximable concept, we have $\swarrow\nearrow (h) \leq g_{i_0} \leq \bigvee_{i \in I} g_i$, which implies that $\bigvee_{i \in I} g_i$ is a generalized approximable concept. □

Lemma 15. *Suppose* $\{g_i \mid i \in I\}$ *is a set of generalized approximable concepts, then* $\bigwedge_{i\in I} g_i$ *is also a generalized approximable concept.*

Proof. It is trivial. □

Proposition 5. $(\mathcal{A}, \leq)$ *is left-algebraic.*

Proof. By Lemmas 11, 12, 13, 14, 15. □

Proposition 5 shows that all generalized approximable concepts form a left-algebraic lattice. Conversely, in the second part, suppose $(P, \leq)$ is a left-algebraic lattice, we will construct a generalized approximable concept lattice which is isomorphic to $(P, \leq)$.

The elements of P denoted by x, y, and the elements of $K(\ll)$ denoted by p, q, where $K(\ll)$ is the set of all compact elements.

Let $A = P$, $B = K(\ll)$, and $R(x,p) : A \times B \rightarrow L$ indicates the degree of p belonging to x. By Proposition 3, we obtain a generalized approximable concept lattice $(\mathcal{A}, \leq)$.

In what follows, We will prove that $(P, \leq)$ is isomorphic to $(\mathcal{A}, \leq)$.

Suppose $e \in D, p \in K(\ll)$, we define a mapping $\{e/p\} : K(\ll) \rightarrow D$, where $(\{e/p\})(p) = e$; $(\{e/p\})(q) = 0$, if $q \neq p$.

Similarly, for $m \in C$, $x \in P$, we also define a mapping $\{m/x\} : P \rightarrow C$, where $\{m/x\}(x) = m$; and $\{m/x\}(y) = 0$, if $y \neq x$.

Lemma 16. *(1)* $\nearrow (\{e/p\})(x) = \sup\{c \in C \mid c \bullet e \leq R(x,p)\}$,
(2) $\swarrow (\{m/x\})(p) = \sup\{d \in D \mid m \bullet d \leq R(x,p)\}$.

Proof. (1) $\nearrow (\{e/p\})(x) = \sup\{c \in C \mid \forall q \in K(\ll), c \bullet (\{e/p\})(q) \leq R(x,q)\}$ $= \sup\{c \in C \mid c \bullet e \leq R(x,p)\}$.

(2) It is analogous. □

Proposition 6. *Suppose* $g : K(\ll) \rightarrow D$ *is a generalized approximable concept,* $p \in K(\ll)$, $g(p) \neq 0$, *then we have* $g(p) = 1$.

Proof. For $g : K(\ll) \rightarrow D$, and $p \in K(\ll)$, $g(p) \neq 0$. Let $e = g(p) \in D$, we obtain a mapping $\{e/p\} : K(\ll) \rightarrow D$ as defined above.

By Lemma 16, let $x = p \in K(\ll) \subseteq P$, we have

$\nearrow (\{e/p\})(p) = \sup\{c \in C \mid c \bullet e \leq R(p,p)\} = 1$.

$\swarrow\nearrow (\{e/p\})(p) = \sup\{d \mid \nearrow (\{e/p\})(p) \bullet d \leq R(p,p)\} = 1$.

Since $\{d/p\} \leq g$, and g is a generalized approximable concept, we have $\swarrow\nearrow (\{e/p\}) \leq g$. Thus $\swarrow\nearrow (\{e/p\})(p) \leq g(p)$. So we obtain $g(p) = 1$. □

By this, for a generalized approximable concept g, we define:
$x_g = \vee\{p \mid g(p) = 1\}$. On the other hand, for every $x \in P$, since P is left-algebraic, $x = \vee\{\downarrow x \cap K(\ll)\}$. We may define $g_x : K(\ll) \rightarrow D$, such that $g_x(p) = 1$ for every $p \ll x$. Then g_x is a generalized approximable concept. Thus we obtain an isomorphism between P and generalized approximable concept lattice $\mathcal{A}$. Thus P and $\mathcal{A}$ is isomorphic.

6 Conclusions

The paper consist of two parts. In the first part, as a generalization of approximable concept, we introduced the notion of an approximable concept in fuzzy setting, discussed the equivalence of two definitions of an approximable concept, and proved that all approximable concepts form an algebraic completely lattice of **L**-ordered sets. Furthermore, we obtained the isomorphism between them in [9]. In the second part, we gave another generalization of approximable concept, that is, generalized approximanle concept, and proved that all generalized approximable concept lattices are isomorphic to the algebraic complete lattices.

Acknowledgements

This work was supported by the National Natural Science Foundation of China (Grant No. 10471035/A010104) and Natural Science Foundation of Shandong Province (Grant No. 2003ZX13).

The authors would like to thank the Referees and Prof. R. Bělohlávek for their valuable suggestions and comments, which have helped improve this paper significantly.

References

1. Bělohlávek, R.: Concept lattices and order in fuzzy logic. Annals of Pure and Applied Logic 128, 277–298 (2004)
2. Bělohlávek, R.: Fuzzy Galois connections. Math. Logic Quart. 45(4), 497–504 (1999)
3. Bělohlávek, R.: Fuzzy relational systems. In: Foundations and Principles. Foundations and Principles, Kluwer, Dordrecht (2002)
4. Bělohlávek, R., Sklenař, V., Zacpal, J.: Crisply generated fuzzy concepts: reducing the number of concepts in formal concept analysis. In: Proc. 5th Int. Conf. on Recent Advances in Soft Computing, RASC 2004, Nottingham, United Kingdom, 16–18 December, 2004, pp. 524–529 (2004) pp. 63 (extended abstract) (full paper on the included CD)
5. Bělohlávek, R., Vychodil, V.: What is a fuzzy concept lattice? In: Proc. CLA 2005, 3rd Int. Conference on Concept Lattices and Their Applications, Olomouc, Czech Republic, pp. 34–45 (2005) `http://sunsite.informatik.rwth-aachen.de/Publications/CEUR-WS/Vol-162/`
6. Bělohlávek, R., Vychodil, V.: Reducing the size of fuzzy concept lattices by hedges. In: FUZZ-IEEE 2005, The IEEE International Conference on Fuzzy Systems, Reno (Nevada, USA), May 22-25, 2005, pp. 663–668 (2005)
7. Ben Yahia, S., Jaoua, A.: Discovering knowledge from fuzzy concept lattice. In: Kandel, A., Last, M., Bunke, H. (eds.) Data Mining and Computational Intelligence, pp. 169–190. Physica-Verlag (2001)
8. Chen, X.Y., Li, Q.G.: Formal topology, Chu space and approximable concept, In: Proc. CLA, 3rd Int. Conference on Concept Lattices and Their Applications, Olomouc, Czech Republic, pp. 158-165. (2005), `http://sunsite.informatik.rwth-aachen.de/Publications/CEUR-WS/Vol-162/`

9. Chen, X.Y., Li, Q.G., Deng, Z.K.: Chu space and approximable concept lattice in fuzzy setting. In: FUZZ-IEEE 2007, The IEEE International Conference on Fuzzy Systems, London, July 24-26 (2007)
10. Chen, X.Y.: Continuous lattice of L-sets, submitted
11. Chen, X.Y., Li, Q.G., Deng, Z.K.: Way-below relation in fuzzy setting (Abstract). In: Proc. International Symposium on Domain Theory 2006, Hunan University, Changsha, P. R. China, June 2-6, 2006, pp. 22–25 (2006)
12. Chen, X.Y., Li, G.Q., Long, F., Deng, Z.K.: Generalizations of approximation concept lattice. In: Yahia, S.B., Nguifo, E.M. (eds.) Proceedings of CLA 2006: The 4th international conference on Concept Lattice and Their Applications, Yasmine Hammamet, Tunisia, October-November 2006, pp. 231–242 (2006)
13. Chen, X.Y.: Domain, Approximable Concept Lattice, Rough Sets and Topology, PhD thesis, Hunan University, P. R. China (2007)
14. Ganter, B., Wille, R.: Formal concept analysis. Springer, Heidelberg (1999)
15. Gierz, G., Hofmann, K.H., Keimel, K., Lawson, J.D., Mislove, M., Scott, D.S.: A compendium of continuous lattices. Springer, Berlin Heidelberg, New York (1980)
16. Goguen, J.: L-fuzzy sets. J. Math. Anal. Appl. 18, 145–174 (1967)
17. Hitzler, P., Zhang, G.Q.: A Cartesian Closed Category of Approximable Concept Structures. In: Wolff, K.E., Pfeiffer, H.D., Delugach, H.S. (eds.) ICCS 2004. LNCS (LNAI), vol. 3127, pp. 170–185. Springer, Heidelberg (2004)
18. Krajči, S.: The basic theorem on generalized concept lattice. In Bělohlávek, V.R. (ed.) CLA 2004, Ostrava, Proceedings of the 2nd International Workshop, pp. 25-33(2004) ISBN 80-248-0597-9
19. Krajči, S.: A generalized concept lattice. Logic Journal of IGPL 2005 13(5), 543–550 (2005)
20. Krajči, S.: Cluster based efficient generation of fuzzy concepts. Neural Network World 13(5), 521–530 (2003)
21. Pratt, V.: Chu spaces as a semantic bridge between linear logic and mathematics. Theoretical Computer Science 294, 439–471 (2003)
22. Vickers, S.: Topology via Logic. Cambridge Univ. Press, Cambridge (1989)
23. Zhang, G.Q., Shen, G.Q.: Approximable concepts, Chu spaces, and information systems. In: De Paiva, V., Pratt, V. (eds.) Theory and Applications of Categories Special Volume on Chu Spaces: Theory and Applications, vol. 17(5), pp. 80–102 (2006), see: `http://newton.cwru.edu/publications.html`
24. Zhang, G.Q.: Chu space, concept lattices, and domains. Electronic Notes in Theoretical Computer Science 83, 17 (2004)

Rule Validation of a Meta-classifier Through a Galois (Concept) Lattice and Complementary Means

Mohamed Aoun-Allah[1] and Guy Mineau[2]

[1] Imam University, Computer Science Dept.,
11681 Riyadh, P.O. Box 84880, Kingdom of Saudi Arabia
[2] Laval University, Computer Science and Software Engineering Dept.,
Laboratory of Computational Intelligence (LCI/LIC),
Quebec City, Quebec, G1K 7P4, Canada
{Mohamed.Aoun-Allah,Guy.Mineau}@ift.ulaval.ca,
http://w3.ift.ulaval.ca/~moaoa,
http://www.ift.ulaval.ca/~LIC/guy/guy.htm

Abstract. In this work we are interested in the problem of mining very large distributed databases. We propose a distributed data mining technique which produces a *meta-classifier* that is both predictive and descriptive. This meta-classifier is made of a set of classification rules, which can be refined then validated. The refinement step, proposes to remove from the meta-classifier rules that according to their confidence coefficient, computed by statistical means, would not have a good prediction capability when used with new objects. The validation step uses some samples to fine-tune rules in the rule set resulted from the refinement step. This paper deals especially with the validation process. Indeed, we propose two validation techniques: the first one is very simple and the second one uses a Galois lattice. A detailed description of these processes is presented in the paper, as well as the experimentation proving the viability of our approach.

1 Introduction

We witness nowadays an explosion of electronic data. Indeed, almost everything (grocery, medical file, car repair history, etc.) is recorded on databases for future analysis. To perform this analysis, many centralized data mining tools exist. But with increasingly bigger databases, these tools become very time-consuming.

Distributed data mining tools are created as an alternative to centralized ones for inherently distributed data or in order to speed-up processing time. So, from a set of individual databases $\{DB_i\}$, we propose to produce a meta-classifier $R = \cup_i C_i$, where each C_i is a classifier made of a rule set. This set of rules R could be further refined then validated. The refinement step, proposes to remove from R rules that according to their confidence coefficient, computed by statistical means, would not have a good prediction capability when used with new objects. The validation step, which is the core of this paper, uses some

S. Ben Yahia et al. (Eds.): CLA 2006, LNAI 4923, pp. 123–138, 2008.

samples to fine-tune rules in the rule set resulted from the refinement step. In this paper, we propose two validation techniques. The first one uses some samples extracted from the distributed databases in order to compute the error rate for each rule. This error rate is then used to decide whether the rule should be kept in the final meta-classifier. The second validation process uses a Galois lattice where more sophisticated fine-tuning is conducted.

The paper proceeds as follows. Section 2 briefly enumerates some existing distributed data mining techniques. In Section 3, we introduce the proposed technique where we detail both the simple and the Galois lattice based validation processes. Then, we present in Section 4 experiments conducted to assess both techniques and which prove the viability of our method. We finally present a conclusion and our future work.

2 State of the Art

A rule set can be used both as a predictive and descriptive tool. Therefore, the technique we developed produces rule sets. In the literature purely predictive techniques such as *bagging* [1], *boosting* [2], *stacking* [3], and the *arbiter* and *combiner* methods [4] are found, but since they are not descriptive, they are deemed to be irrelevant to this paper.

Also, as we present in this paper a technique developed in a *distributed data mining* perspective, we will ignore some other non relevant techniques as the *Ruler System* [5] that was developed for the aggregation of several decision trees built on the same data set in a centralized system, the *Distributed Learning System* [6] developed in a context of information management system that builds a distributed learning system, and the *Fragmentation Approach* [7] which uses probalistic rules.

The closest existing techniques to ours are:

- the MIL algorithm [8] [9];
- the *Distributed Rule Learner* technique [10]; and
- a mixture of the last two techniques [11].

These techniques produce a set of disjoint cover rules, i.e., any object triggers one and only one rule, contrarily to our technique. Lifting up that constraint eliminates the processing time that is required to choose between these "competing" rules. However, we then need to referee between "conflicting" rules. Section 3 below shows how one can at a very low computing cost, implement an effective refereeing mechanism through a majority vote (see below). Therefore, as none of these techniques need to validate their rule set, we will not detail them in this paper. The interested reader may get more details on these techniques in [12], [13], and [14].

3 The Proposed Meta-classifier

The proposed algorithm goes roughly through two tasks: a distributed one achieved by "miner agents" which have to mine distributed databases on remote sites and to extract useful information, producing base classifiers C_i $\forall i \in [1, n]$, and a centralized task achieved by a "collector agent" which is responsible of aggregating information gathered by miner agents in order to produce the meta-classifier. Hereafter, we detail the tasks of these two types of agents.

3.1 The Tasks of a Miner Agent

We have already detailed the task of a miner agent in previous papers [13] [15] [12]. In what follows we summarize them through the algorithm of Fig 1.

```
Do, by a miner agent Am_i, working on database DB_i on a remote site:

 1. Apply on DB_i a classification algorithm producing a set of rules
    with disjoint cover. The produced set is: R_i = {r_ik | k ∈ [1..m_i]} where
    m_i is the number of rules;
 2. Compute for each r_ik a confidence coefficient c_{r_ik};
 3. Extract a random sample S_i from DB_i.
```

Fig. 1. Algorithm showing the tasks of a miner agent

The algorithm of Fig. 1 shows that a miner agent produces a set of classification rules, called *base classifier*, then computes for each rule a confidence coefficient. This coefficient, as its name suggests, reflects the confidence that we have in each rule based on some statistic means [13]. The sample S_i is used later by collector agent, in the process of rule validation. The size of this sample should be very small (about 50 objects) in order to reduce the amount of data traveling from the database site to the collector agent site.

3.2 The Tasks of a Collector Agent

The tasks of a collector agent are detailed in Fig. 2. This algorithm shows that a collector agent starts by aggregating all rules in order to produce the meta-classifier, R. Indeed, the simple aggregation of base classifiers has a good predictive capacity [13] [14].

The main problem that turns up from using the meta-classifier R as a descriptive tool could be the number of rules proposed to the user in order to explain the predicted class of a new object. In fact, the aggregation of x rules from n databases produces in the worst case nx rules. The subsequent steps (the refinement and the validation steps) are proposed to fix this drawback. In fact, the refinement step, proposes to remove from R rules that according to their confidence coefficient would not have a good prediction capability when used with new objects. The resulting meta-classifier is the set R_t. The validation step, which is the core of this paper, uses some samples to fine-tune rules in R_t by identifying those that actually have poor prediction performance on the

```
Do, by a collector agent CA, on the central site:
 1. Main step: Create the primary meta-classifier R as follows:
    R = ∪_{i=1...n} R_i where n is the number of sites
 2. Optional refinement step: From R, eliminate rules which have a
    confidence coefficient lower than a certain threshold t (determined
    empirically) and produce R_t:
    R_t = {r_ik ∈ R | c_{r_ik} ≥ t};
 3. Optional validation step:
    (a) Create S as follows: S = ∪_{i=1...n} S_i;
    (b) Use S to validate rules in R_t.
```

Fig. 2. Algorithm showing the tasks of a collector agent

samples. The validation process is detailed hereafter where we propose two techniques. The first one uses a Galois lattice, while the second one simply computes a new error rate considering S as a test set.

3.3 The Use of Set *R* as a Meta-classifier

The set R represents the aggregation of all base classifiers ($R = \cup_i R_i$). This rule set is used as a predictive model as well as a descriptive one. From a predictive point of view, the predicted class of a new object is the class predicted by a majority vote of all the rules that cover it, where the rules are weighted by their confidence coefficients[1].

It is to be noted that any object can be covered by at most n rules –knowing that n is the number of sites. The number of rules is not exactly equal to n because the confidence coefficient determination process could fail in certain circumstances, due to a lack of cover, and consequently the rule in question would be ignored. Besides, by gathering the sets R_i, a rule can appear in more than one base classifier. In this case, only one occurrence of the rule is kept by assigning it with a confidence coefficient equal to the mean of the confidence coefficients of its various occurrences.

From a descriptive point of view, the rules that cover an object propose our explanation as to why the object belongs to the class, even in the case of a tie of the simple and/or the weighted majority vote. As the whole system is developed as support to decision-making, the rules covering an object may be proposed to the user who could then judge, from his expertise, of their relevance. Presenting to a decision maker more than one rule in order to explain the class of an object may have its advantages since this provides a larger and more complete view of the "limits" of each class. We bring to mind, that in machine learning, the limit

[1] However, in a tie situation, we propose to carry out a simple majority vote. In rare cases, when the simple majority vote leads to a tie, we choose the majority class in the different training sets.

which defines separation between various classes is generally not unique nor clear cut, and consequently, several rules producing the same class can represent the "hyper-planes" separating the various classes, providing various views on the data.

3.4 The Use of a Galois Lattice to Validate Rules in a Meta-classifier

This validation process starts by creating a binary relation $\mathcal{I}$ defined over $R_t \times S$ where, at each intersection $(r_i,\ s_j)$, we find 0 if r_i does not cover s_j, 1 otherwise (See (1)).

$$\mathcal{I} = \{< r, s, f(r,s) > | \quad r \in R_t, s \in S, f(r,s) \in \{0,1\}\} \tag{1}$$

This binary relation is used as a context in order to build a Galois lattice $\mathcal{G}$. Consequently, each formal concept $(Rules, Objects)$ of $\mathcal{G}$ contains maximal pairs (closed sets) of objects and rules covering them. Thus, the obtained lattice represents a preset hierarchy of generalization/specialization of maximal pairs of rules and objects that they cover. The produced meta-classified is the set of rules $R_t^{\mathcal{G}}$.

Terminology. In order to simplify the presentation of the algorithm that uses $\mathcal{G}$ to validate the rules, we present some notation and terminology:

1. Our algorithm manipulates only binary databases denoting by "+" and "−" (*positive* and *negative* class) the two classes of the data set. Nevertheless, it could be extended to handle multiple class systems.
2. We note by *cpt* a concept of the lattice, R_{cpt} its extension and O_{cpt} its intention.
3. We call a *positive rule* (resp. a *negative rule*), a rule having the positive (resp. negative) class as conclusion.
4. We borrow some notation and terminology from [16], as we call the *least concept* the bottom most concept of the lattice. It contains no rules and all the objects. Dually, we call the *largest concept* the upper most concept of the lattice. It contains all the rules and no objects.
5. We note by $NbRules(ARuleSet)$ (resp. $NbObjects(AnObjectSet)$) the function that returns the number of rules in the rule set $ARuleSet$ (resp. objects in the set of objects $AnObjectSet$).
6. We note by $RulesOfTheClass(ARuleSet, clas)$ the function that returns rules of the set $ARuleSet$ belonging to the class $clas$ ($clas \in \{+,-\}$). The result is a set of rules.
7. We note by $ObjectsOfTheClass(AnObjectSet, clas)$ the function that returns objects of the set $AnObjectSet$ belonging to the class $clas$. The result is a set of objects.
8. $NbObjects(ObjectsOfTheClass(O_{cpt},+))$ (resp. $NbObjects(ObjectsOfTheClass(O_{cpt},-)))$ is abbreviated by $NbObj_{cpt}^{+}$ (resp. $NbObj_{cpt}^{-}$).

9. $NbRules(RulesOfTheClass(R_{cpt}, +))$ (resp. $NbRules(RulesOfTheClass(R_{cpt}, -)))$ is abbreviated by $NbRules^{+}_{cpt}$ (resp. $NbRules^{-}_{cpt}$).

How to Use a Concept Lattice. We bring to mind that we use a concept lattice in order to validate the rules of R_t that statistically (according to the confidence coefficient) should behave well when faced with new objects. This validation consists in choosing among rules that do not correctly predict the class of objects of S those to keep in the final meta-classifier $R_t^{\mathcal{G}}$. In other words, it consists in identifying rules to delete from R_t so that each conflicting rule is assessed. The successful rules are assigned to $R_t^{\mathcal{G}}$.

To achieve its task, the validation algorithm starts by identifying concepts having conflicting rules. To do this, we compute for each concept the number of positive rules ($NbRules^{+}_{cpt}$), the number of negative rules ($NbRules^{-}_{cpt}$), the number of objects belonging to the positive class ($NbObj^{+}_{cpt}$), and the number of objects belonging to the negative class ($NbObj^{-}_{cpt}$). Then we associate to each concept a label ("+", "−" or "?") according to the majority of rules. The label "?" is associated to a concept if the number of positive rules equals the number of negative rules (See (2))

$$Label(cpt) = \begin{cases} + \text{ If } NbRules^{+}_{cpt} > NbRules^{-}_{cpt} \\ - \text{ If } NbRules^{+}_{cpt} < NbRules^{-}_{cpt} \\ ? \text{ Otherwise} \end{cases} \quad (2)$$

We have to note that the labeling of concepts can be done during the construction of the lattice, at a negligible cost. Once the labeling of concepts is done, we could resume the algorithm as follows:

1. Go through the lattice from the least to the largest concept.
2. At the first concept containing rules in conflict, we identify rules belonging to the minority class that we will call them *problematic rules*. These rules are positive rules if the concept is labeled negative and vice-versa. If the concept is labeled "?", all its rules are considered problematic (See (3)). In other words, problematic rules are rules that should not be in the concept and hence, eventually, should not appear in the final meta-classifier.

$$PrbRules(cpt) = \begin{cases} RulesOfTheClass(R_{cpt}, +) \text{ If } Label(cpt) = - \\ RulesOfTheClass(R_{cpt}, -) \text{ If } Label(cpt) = + \\ R_{cpt} \quad \text{If } Label(cpt) = ? \end{cases} \quad (3)$$

3. In order to assess the impact of suppressing a rule from the lattice, we associate to each concept a cost function. Lets explain this function through an example. Figure 3(b) details the class distribution of the concept of Fig 3(a). We can easily notice that rule F is a problematic one since it is in conflict with rules C and E.

   It is clear that suppressing rule F from this concept has a positive effect or at least no effect, since the majority of objects and the majority of rules

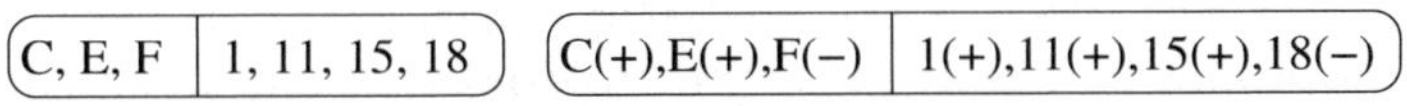

(a) Example of a concept (b) Class distribution in the concept

Fig. 3. An example of a concept from the lattice

belong to the positive class whereas F is a negative rule. And this is the case of all problematic rules where by definition they are rules belonging to the minority class. Consequently, the cost function for suppressing a problematic rule is computed through out all concepts that contain it. Fortunately, the lattice restrains our exploration of concepts where F appears, since only the superconcepts of the bottom most concept containing the first occurrence of the rule under consideration must be processed (See Fig. 4).

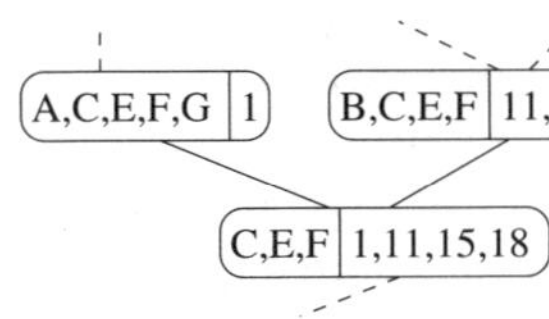

Fig. 4. Example of superconcepts of the one of Fig. 3

4. The cost function that we propose represents the gain of objects correctly classified minus the loss of objects incorrectly classified when the label of the concept changes (designated by function MP). In other words, suppose that we suppress from the concept of Fig. 3 the rule C or E. In that case the label of the concept passes from "+" to "?". The cost of this action according to our function is $-3+1=-2$ since objects 1, 11 and 15 are no longer correctly classified and the object 18 gains in classification since we consider that the class "?" is closer to "−" than "+" is close to "−".
5. For presentation purposes, we denote by *cpt′* the concept *cpt* abated by problematic rules. Hence, our cost function is defined as follows:

$$Cost_{MP}(cpt) = \begin{cases} 0 & \text{If } Label(cpt) = Label(cpt') \\ NbObj^{-}_{cpt} - NbObj^{+}_{cpt} & \text{If } (Label(cpt), Label(cpt')) \in \\ & \{(+,-),(+,?),(?,-)\} \\ NbObj^{+}_{cpt} - NbObj^{-}_{cpt} & \text{If } (Label(cpt), Label(cpt')) \in \\ & \{(-,+),(-,?),(?,+)\} \end{cases}$$

6. This cost is iteratively repeated on the remaining of the lattice by suppressing the same rule already identified as problematic. If the final cost over the lattice is positive, thus it is advantageous to eliminate the rule from the

concept and vice versa. If the final cost equals zero it is neither advantageous nor disadvantageous to keep or to ignore the rule. Actually, keeping or ignoring these rules (that we call "*limit rules*") produces two variants of cost function according to the decision taken about these rules.

7. The process of identifying problematic rules and computing the result of the cost function is also done over all possible combinations of problematic rules in one concept. So $2^p - 1$ rule subsets could be assessed if p is the number of problematic rules (obviously, the empty set is ignored). In other words, if in one concept there is more than one problematic rule, we compute the powerset of problematic rules and then for each set from the powerset the process described above is conducted.

Cost Functions. The first variant of the cost function is deduced from the use of the label "?" where it can be considered as an intermediate class between the "+" and "−". This function is designated by $MPQM$. Thus when the label of a concept goes from "−" to "+" by removing rules identified as problematic, the cost function returns twice the difference between positive and negative objects and vice versa. Whereas, when "?" appears as the label of the concept before or after removing problematic rules, the cost function returns a simple difference.

Another variant of cost function could be proposed by considering only the sign of the difference. Thus this binary function (designated by function BIN) return only $+1$, -1 or 0 according to the sign of the difference and the change of label.

The last variant proposed is deduced from the BIN function when considering "?" as an intermediate class. Thus, when the label changes going from "+" to "−" or inversely, the $BINQM$ function returns ± 2, otherwise it returns ± 1 or 0 if there is no label change.

Advantages of Using the Galois Lattice. The Hass diagram of the Galois lattice constitutes a hierarchy of generalization/specialization of the rules and objects that they cover. This hierarchy presents various advantages. We enumerate a few of them hereafter:

- If there exist two rules covering the same object but predicting different classes, called rules in conflict, these rules will necessarily appear in at least one concept of the lattice. In these concepts we find all rules in conflict, and the objects that they cover.
- In order to delete a rule r from the lattice, we do not need to visit each concept. We must just find the first concept that contains the first occurrence of r when we go upwards in the lattice. Then, all the occurrences of this rule will be in concepts that are superconcepts of the latter one. This is due to the structure of the lattice, where when we go up from the least concept to the largest one, the extension of a concept (i.e., rules) is enriched while intention (i.e., objects) is impoverished. In other words, when we go up in the lattice, the number of rules increases thus together they will cover fewer objects.

- Rules in a concept are coded by their numbers. Thus their treatment is very fast since the coverage of each rule is already computed and stored in the lattice. The only information that we may need, the conclusion of the rule, could be obtained instantly.

3.5 The Use of Samples as a Test Set to Validate Rules in a Meta-classifier

The main idea of this validation process is the following: rules that come out of the filtering process are those having a high confidence coefficient and thus they should demonstrate a good predictive accuracy when tackling new objects. Consequently, when these rules are assessed on a new test set, we can ignore those that do not satisfy our expectations.

For that, this validation process uses S as a test set and computes for each rule of R_t a new error rate called $E_r^{\mathcal{S}}(S)$ [17] [12] [15]. Since we assume that we should keep only very good rules, this validation process ignores rules having an error rate $E_r^{\mathcal{S}}(S)$ greater than a threshold $t_{\mathcal{S}}$. The produced meta-classifier is the rule set $R_t^{\mathcal{S}}$ (See (4)).

$$R_t^{\mathcal{S}} = \{r_{ik} \in R_t \mid E_{r_{ik}}^{\mathcal{S}}(S) \leq t_{\mathcal{S}}\} \tag{4}$$

4 Experiments

To assess the performance of our meta-learning method, we conducted a battery of tests in order to assess its prediction rate (accuracy) and its size (i.e., the number of rules it produces). We compared it to a C4.5 algorithm built on the whole data set, i.e., the aggregation of the distributed databases. This C4.5, produces the rule set R', which is used as a reference for its accuracy rate since we assumed in the introduction that it is impossible to gather all these bases onto the same site, and this, either because of downloading time, or because of the difficulty to learn from the aggregated base because of its size. The set R' is supposed to represent the ideal case because, theoretically, when the size of data increases, the resulting classifier will be more representative of the data set.

Exp. 1: Find the best threshold t for producing the meta-classifier R_t.
Exp. 2: Find the best pair $(t, Costfunction)$ for producing $R_t^{\mathcal{G}}$.
Exp. 3: Find the best pair $(t, t_{\mathcal{S}})$ for producing $R_t^{\mathcal{S}}$.
Exp. 4: Compare R, R_t, $R_t^{\mathcal{G}}$ and $R_t^{\mathcal{S}}$ to R' on the basis of accuracy.
Exp. 5: Compare R, R_t, $R_t^{\mathcal{G}}$ and $R_t^{\mathcal{S}}$ to R' on the basis of the size of their rule set.

All these experiments were run on ten data sets: adult, chess end-game (King+Rook versus King+Pawn), Crx, house-votes-84, ionosphere, mushroom, pima-indians-diabetes, tic-tac-toe, Wisconsin Breast Cancer (BCW)[18] and Wisconsin Diagnostic Breast Cancer (WDBC), taken from the UCI repository [19].

The size of these data sets varies from 351 to 45222 objects. Furthermore, in order to get more realistic data sets, we introduced noise in the ten aforementioned databases, and this by reversing the class attribute[2] of successively 10%, 20%, 25% and 30% of each data set objects. Hence, since for each data set we have, in addition to the original set, 4 other noisy sets, the total number of databases used for our tests is 50.

In order to simulate distributed data sets, we did the following. We divided each database into a test set with proportion of 1/4. This data subset was used as a test set. The remaining data subset (of proportion 3/4), was divided in its turn randomly into 2, 3, 4 or 5 data subsets in order to simulate distributed databases. The size of these bases was chosen to be disparate and in such a way that there was a significant difference between the smallest and the biggest data subset. Figure 5 shows an example of such subdivision.

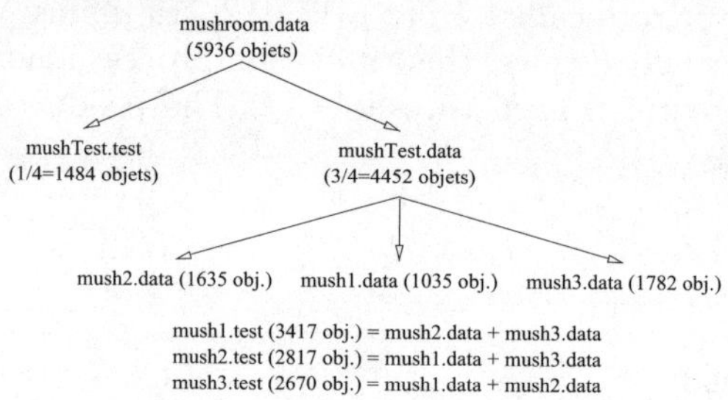

Fig. 5. Example of subdivision for a database from the UCI

For the construction of the base classifiers we used C4.5 release 8 [20] which produces a decision tree that is then directly transformed into a set of rules. For the concept lattice construction we used the algorithm proposed in [21].

4.1 Experiment 1: Best Parameter t for R_t

In order to find the best t for R_t, we tried all values ranging from 0.95 to 0.20, with decrements of 0.05 and 0.01. The analysis of results that we got for R_t over the 50 data sets show that the minimum error rate of R_t is obtained in almost all the cases with the threshold $t = 0.95$ or $t = 0.01$. When this is not the case, the error rate of R_t, with $t = 0.95$ or $t = 0.01$, is worse than the minimum error rate by no more than 0.1% except for the Pima-Indians original database where the difference is 1%.

The choice of a high threshold (such as 0.95) suggests that keeping only rules with a high value of the confidence coefficient produces good results. Moreover, a threshold as low as 0.01 signifies that the aggregation of almost all the rules produces also good results thanks to the weighted majority vote. So weighting

[2] Please note that i) all data sets have a binary class attribute ii) we deleted from these data sets objects with missing values .

the rules by this confidence coefficient seems to be quite sufficient to provide our method with a satisfactory accuracy rate.

In order to choose between these two thresholds, we draw a table of the number of tests for which they produced the minimum error rate. In the case where the best accuracy is obtained with t' which was neither 0.01 nor 0.95, we associated t' with the closer of the two thresholds 0.01 or 0.95.

Table 1. Number of occurrences of the minimum of R_t error rate with $t = 0.01$ and $t = 0.95$

Databases	Min with 0.01	Min with 0.95	Constant error rate
Original databases	6	3	2
10% noisy DB	3	3	4
20% noisy DB	4	1	5
25% noisy DB	3	5	3
30% noisy DB	4	3	3
TOTAL	**20**	15	17

From table 1 we can choose the threshold 0.01 as the best one since it produces the minimum error rate of R_t in the majority of the cases.

4.2 Experiment 2: Best Pair (t, $CostFunction$) for $R_t^{\mathcal{G}}$

We start by identifying the threshold t that produces the minimum error rate of $R_t^{\mathcal{G}}$ when considering the four cost functions (MP, $MPQM$, BIN and $BINQM$) presented above with for each one a little variant by taking or ignoring limit rules for a total of 8 functions. We conducted the same tests as with R_t. We found that $t = 0.01$ produces a score of 124 occurrences of the minimum, whereas, $t = 0.95$ produces a score of 116 occurrences. Consequently, we can assume that the threshold $t = 0.01$ is the best for $R_t^{\mathcal{G}}$ too. This is an interesting result, since it proves that the proposed technique (thanks to the weighted majority vote between conflicting rules) is robust faced to poor rules.

Once threshold t fixed, we have to find the best cost function. To do so, we analyze Table 2 below. The best function is clearly "$MPQM$" or "MP" with limit rules (LR). Functions "$BINQM$" and "BIN" with limit rules produce also good results but they are slightly less efficient from a prediction point of view than their competitors.

The most interesting conclusion that we can draw from these results is that limit rules are very important and should be kept in the final meta-classifier, even if the cost function did not succeeded to prove it. We recall to mind that a cost function returns the value 0 for a limit rule.

4.3 Experiment 3: Best Pair (t, t_S) for R_t^S

For each value of threshold t (listed in §4.1), we assessed the refinement process using S as a test set with threshold t_S equal respectively to 2%, 5% and 10%.

Table 2. Number of bases for which cost functions produce the lowest R_t^G error rate

	Original DB	10% noisy DB	20% noisy DB	25% noisy DB	30% noisy DB	Σ
$MPQM$ with LR	4	5	4	3	2	**18**
$MPQM$ without LR	3	4	1	3	2	13
MP with LR	4	5	4	4	2	**19**
MP without LR	3	4	1	3	2	13
$BINQM$ with LR	6	2	4	1	3	16
$BINQM$ without LR	3	3	1	4	2	13
BIN with LR	6	2	3	1	4	16
BIN without LR	3	3	1	4	2	13

In order to find the best pair (t, t_S) for R_t^S, we started by finding the threshold t producing most frequently the minimum error when using the three values of threshold t_S (2%, 5% and 10%). Then, we chose the best t_S –i.e., producing the lowest error rate– among the three values above.

To find the best t we conducted the same analysis as previously. The analysis showed that the thresholds that most frequently produce the minimum error rate for R_t^S are $t = 0.01$ and $t = 0.95$. As we did for R_t, we drew tables showing the number of times that these two values of t produce the lowest error rate when t_S equals respectively 2%, 5% and 10% (See Tab. 3).

Table 3. Total of the number of occurrences of the lowest error rate of R_t^S with $t_S \in \{2\%, 5\%, 10\%\}$ and $t \in \{0.01, 0.95\}$ computed over the 50 data sets

	Min with 0.01	Min with 0.95	Constant error rate
$t_S = 2\%$	30	9	13
$t_S = 5\%$	18	13	22
$t_S = 10\%$	17	15	21
Total	**65**	37	56

Based on Table 3, it is obvious to conclude that $t = 0.01$ is the best threshold since it frequently produces the lowest error with the three values of t_S. This result confirms our previous choices and results and it proves once again the power of the weighted majority vote against poor rules.

Once threshold t fixed, we have to find threshold t_S. To do so, we analyze Table 4, which presents the number of bases for which t_S produced the minimum error rate of R_t^S. A simple look to Table 4 shows that the best value of t_S is 5%; nevertheless, $t_S = 10\%$ gives almost the same number of bases.

The very low number of bases obtained with $t_S = 2\%$ indicates that a very tight threshold t_S will exclude some interesting rules which decreases the prediction capacity of R_t^S. In contrast, a slightly loose threshold as $t_S = 10\%$ could accept in R_t^S some bad rules which also decreases the prediction capacity of the meta-classifier.

Table 4. Number of bases for which thresholds t_S produce the lowest R_t^S error rate

	Original DB	10% noisy DB	20% noisy DB	25% noisy DB	30% noisy DB	Σ
$t_S = 2\%$	5	4	3	2	3	17
$t_S = 5\%$	7	8	10	10	9	**44**
$t_S = 10\%$	6	7	10	10	9	42

4.4 Experiment 4: Compare R, R_t, $R_t^{\mathcal{G}}$ and R_t^S to R' on the Basis of Accuracy

Once parameters for R_t, $R_t^{\mathcal{G}}$ and R_t^S has been found, we can compare the prediction performance of our meta-classifiers R, R_t, $R_t^{\mathcal{G}}$ and R_t^S to the classifier R' used as reference since it is the ideal case. To do so, we compare R' with its error rate confidence interval (lower and upper bounds) computed at 95% confidence to our meta-classifiers using their respective optimal parameters, over the original databases and over the noisy databases.

As a detailed citation of these results needs more than a few pages, we will restrain our presentation to the most important results. Indeed, we can resume our observations to the following:

1. The prediction performance of R is very comparable to that of R' since in 34 cases over 50, the error rate of R is statistically comparable (with 95% confidence) to that of R'. It worsens only in 5 cases but it improves in 11 cases.
2. The sets R and R_t have sensibly the same error rate except for 6 cases where in two of them the error rate of R_t is worse than that of R by no more than 0.1% (which is not a significant difference) and in the 4 other cases R_t do better than R with an error rate difference ranging from 1.1% to 3%. Hence, we can conclude that R_t, in general, does predict as well as R or better.
3. Globally, R_t and $R_t^{\mathcal{G}}$ present comparable prediction performance. Indeed, over the 50 data sets, $R_t^{\mathcal{G}}$ presents exactly the same error rate over 30 data sets, better error rate for 11 data sets with a difference of at most 4.7% and a worse error rate for only 9 cases with a difference of at most 2.1%.
4. The prediction performance of $R_t^{\mathcal{G}}$ is slightly better than the one of R_t^S since in 38 cases over 50, the difference of error rates of these rule sets is null or less than 1%. The error rate of $R_t^{\mathcal{G}}$ is better in 8 cases by a difference ranging from 1.1% to 4.3%. It is worst in 4 cases by a difference ranging from 1.2% to 4.2%.
5. When databases are very noisy, our meta-classifiers R, R_t, $R_t^{\mathcal{G}}$ and R_t^S produce better error rates (statistically, with confidence of 95%) than R'.

On the light of these results, we can conclude that the proposed meta-classifiers (R, R_t, $R_t^{\mathcal{G}}$ and R_t^S) present globally comparable prediction performance, nonetheless, $R_t^{\mathcal{G}}$ is slightly better than R_t^S which proves that a tight fine-tuning of rules using a Galois lattice is better than a simple filter using a threshold. Moreover, the error rates of our meta-classifiers are comparable to R', but that of our meta-classifiers outperform R' as noise increases in the data set.

The reader should notice that even if R_t or $R_t^{\mathcal{G}}$ or $R_t^{\mathcal{S}}$ does not significantly improve R in terms of its error rate, applying the threshold t then validating rules offers some advantages, like decreasing the meta-classifier size (as we will see below).

4.5 Experiment 5: Compare R, R_t, $R_t^{\mathcal{G}}$ and $R_t^{\mathcal{S}}$ to R' on the Basis of the Size of Their Rule Set

Table 5 presents the number of rules in the classifiers: R', R, R_t, $R_t^{\mathcal{G}}$ and $R_t^{\mathcal{S}}$. It is clear from this table that R, R_t, $R_t^{\mathcal{G}}$ and $R_t^{\mathcal{S}}$ have a relatively low number of rules which is in certain circumstances inferior that the number of rules in our reference classifier R'. This result is very encouraging since our meta-classifier can be seen as neither more difficult nor easier to interpret than R'.

Table 5. The number of rules in each rule set

	Adult	BCW	Chess	Crx	Iono.	Mush.	Pima.	Tic.	Vote	Wdbc
R'	523	10	31	25	7	24	21	69	5	11
R	592	50	54	23	11	11	30	77	10	18
R_t	482	33	54	20	9	11	26	64	6	17
$R_t^{\mathcal{G}}$	469	32	46	19	9	11	26	61	6	17
$R_t^{\mathcal{S}}$	408	31	46	16	8	11	13	51	5	15

Moreover, we can easily observe that the refinement step as well as the validation steps are very useful since they can reduce the number of rules in R significantly. For instance, the rule set size reduction of $R_t^{\mathcal{G}}$ is up to 40% with Vote data set (36%, 21% and 21% for respectively BCW, Adult and Tic-Tac-Toe data sets). The reduction of the size of $R_t^{\mathcal{S}}$ reaches the 57% with Pima-Indians data set. Besides, it is clear from Table 5 that the validation process using $\mathcal{S}$ as a test set eliminates more rules than the one using a Galois lattice. Unfortunately, the high reduction in rule set size of $R_t^{\mathcal{S}}$ comes with a slightly decrease in prediction performance compared to $R_t^{\mathcal{G}}$, as we saw previously.

5 Conclusion

We presented in this paper a distributed data mining technique that goes globally through two steps. The first one is to mine in a distributed manner, each data set, then in a centralized site information gathered from the first process is aggregated. The resulting meta-classifier is as a rule set that can be refined and validated. This paper dealt especially with the validation step where we proposed two validation processes: A simple one and a Galois lattice based one.

We have demonstrated in this paper that concept lattices could be very useful in DDM. Indeed, from a prediction point of view a validated meta-classifier (i.e., rules of the meta-classifier are validated by a concept lattice) performs as well as or even better, than a classifier built on the whole data set, which is used as a reference, depending on the level of noise in it. Moreover, from a description point of view (i.e., number of rules in the classifier), the size of validated meta-classifier is usually comparable to that of the reference centralized classifier. When we compare the simple rule sets aggregation with this set refined then validated by a Galois lattice, a significant decrease of the set of rules (up to 40%) could be observed.

The simple validation process proposed in this paper do almost as good as the Galois lattice based one: the prediction performance of $R_t^{\mathcal{G}}$ is slightly better than R_t^S but the size of the latter rule set is lower than the size of the first one. These results are very interesting since we offer to a decision maker two validation techniques where the first one optimizes the meta-classifier size and the second one optimizes its prediction capabilities.

Currently, we are working on the design of a smart validation process that dynamically chooses the more appropriate validation process to use, according to the prevailing circumstances pertaining to the nature of the data set. This work will be presented in a forthcoming paper.

References

1. Breiman, L.: Bagging predictors. Machine Learning 24(2), 123–140 (1996)
2. Schapire, R.E.: The strength of weak learnability. Machine Learning 5(2), 197–227 (1990)
3. Tsoumakas, G., Vlahavas, I.: Distributed Data Mining of Large Classifier Eensembles. In: Vlahavas, I., Spyropoulos, C. (eds.) Proceedings Companion Volume of the Second Hellenic Conference on Artificial Intelligence, Thessaloniki, Greece, pp. 249–256 (April 2002)
4. Prodromidis, A.L., Chan, P.K., Stolfo, S.J.: Meta-learning in distributed data mining systems: Issues and approaches. In: Kargupta, H., Chan, P. (eds.) Advances in Distributed and Parallel Knowledge Discovery, ch. 3, part II, pp. 81–113. AAAI Press MIT Press, Menlo Park, Cambridge (2000)
5. Fayyad, U.M., Djorgovski, S.G., Weir, N.: Automating the analysis and cataloging of sky surveys. In: Advances in Knowledge Discovery and Data Mining, pp. 471–493. AAAI Press/The MIT Press, Menlo Park, California (1996)
6. Sikora, R., Shaw, M.: A Computational Study of Distributed Rule Learning. Information Systems Research 7(2), 189–197 (1996)
7. Wüthrich, B.: Probabilistic knowledge bases. IEEE Transactions on Knowledge and Data Engineering 7(5), 691–698 (1995)
8. Williams, G.J.: Inducing and Combining Decision Structures for Expert Systems. PhD thesis, The Australian National University (January 1990)
9. Hall, O.L., Chawla, N., Bowyer, W.K.: Decision tree learning on very large data sets. In: IEEE International Conference on Systems, Man, and Cybernetics (october 1998), vol. 3, pp. 2579–2584 (1998)
10. Provost, F.J., Hennessy, D.N.: Scaling up: Distributed machine learning with cooperation. In: Thirteenth National Conference on Artificial Intelligence (AAAI-1996), pp. 74–79 (1996)

11. Hall, O.L., Chawla, N., Bowyer, W.K.: Learning rules from distributed data. In: Workshop on Large-Scale Parallel KDD Systems (KDD99). Also in RPI, CS Dep. Tech. Report 99-8, 77–83 (1999)
12. Aounallah, M., Mineau, G.: Rule confidence produced from disjoint databases: a statistically sound way to regroup rules sets. In: IADIS international conference, Applied Computing 2004, Lisbon, Portugal, vol. 31, pp. II–27–II31 (2004)
13. Aounallah, M., Mineau, G.: Le forage distribué des données: une méthode simple, rapide et efficace. Revue des Nouvelles Technologies de l'Information, extraction et gestion des connaissances RNTI-E-6(1), 95–106 (2006)
14. Aounallah, M., Mineau, G.: Distributed Data Mining: Why Do More Than Aggregating Models. In: Twentieth International Join Conference on Artificial Intelligence (IJCAI 2007), Hyderabad, India (January 2007)
15. Aounallah, M., Quirion, S., Mineau, G.: Forage distribué des données : une comparaison entre l'agrégation d'échantillons et l'agrégation de règles. Revue des Nouvelles Technologies de l'Information, extraction et gestion des connaissances RNTI-E-3(1), 43–54 (2005)
16. Ganter, B., Wille, R.: Formal Concept Analysis: Mathematical Foundations. Springer, Heidelberg (1999)
17. Mineau, G.W., Aounallah, M., Quirion, S.: Distributed Data Mining vs. In: Tawfik, A.Y., Goodwin, S.D. (eds.) Canadian AI 2004. LNCS (LNAI), vol. 3060, pp. 454–460. Springer, Heidelberg (2004)
18. Mangasarian, O.L., Wolberg, W.H.: Cancer diagnosis via linear programming. SIAM News 23(5), 1–18 (1990)
19. Blake, C., Merz, C.: UCI repository of machine learning databases (1998), `http://www.ics.uci.edu/$\sim$mlearn/MLRepository.html`
20. Quinlan, J.R.: Improved use of continuous attributes in C4.5. Journal of Artificial Intelligence Research 4, 77–90 (1996)
21. Valtchev, P., Missaoui, R., Lebrun, P.: A partition-based approach towards building Galois (concept) lattices. Technical Report 2000-08, Département d'Informatique, UQAM, Montréal (CA) (August 2000)

Graded LinClosure and Its Role in Relational Data Analysis*

Radim Belohlavek[1,2] and Vilem Vychodil[1,2]

[1] Dept. Systems Science and Industrial Engineering
T. J. Watson School of Engineering and Applied Science
Binghamton University–SUNY, PO Box 6000, Binghamton, NY 13902–6000, USA
{rbelohla,vychodil}@binghamton.edu
[2] Dept. Computer Science, Palacky University, Olomouc
Tomkova 40, CZ-779 00 Olomouc, Czech Republic

Abstract. We present graded extension of the algorithm LinClosure. Graded LinClosure can be used to compute degrees of semantic entailment from sets of fuzzy attribute implications. It can also be used together with graded extension of Ganter's NextClosure algorithm to compute non-redundant bases of data tables with fuzzy attributes. We present foundations, the algorithm, and illustrative examples.

1 Introduction

Fuzzy logic is a formal framework for dealing with a particular type of imprecision. A key idea of fuzzy logic is a graded approach to truth in which we allow for truth degrees other than 0 (falsity) and 1 (full truth). This enables us to consider truth of propositions to various degrees, e.g., proposition "Peter is old" can be assigned a degree 0.8, indicating that "Peter is old" is almost (fully) true. One way of looking at the proposition "Peter is old" being true to degree 0.8 is that it expresses a *graded attribute* "being old to degree 0.8" of the *object* "Peter". When dealing with multiple graded attributes, we often need to determine their dependencies. In [2,5] we have introduced fuzzy attribute implications as particular dependencies between graded attributes in data sets representing objects and their graded attributes (so-called data tables with fuzzy attributes). A fuzzy attribute implication can be seen as a rule of the form "$A \Rightarrow B$", saying "for each object from the data set: if the object has all graded attributes from A, then it has all graded attributes from B". We have proposed several ways to compute, given an input data set represented by a data tables with fuzzy attributes, a minimal set of dependencies describing all dependencies which are valid (true) in the table, see [7] for a survey.

In this paper we focus on computational aspects of one of the algorithms proposed so far. Namely, we show how to compute fixed points of certain fuzzy

* Supported by grant No. 1ET101370417 of GA AV ČR, by grant No. 201/05/0079 of the Czech Science Foundation, and by institutional support, research plan MSM 6198959214.

S. Ben Yahia et al. (Eds.): CLA 2006, LNAI 4923, pp. 139–154, 2008.

closure operators that appear in algorithms from [2,11]. We introduce an extended version of the LINCLOSURE algorithm which is well known from database systems [22]. Compared to the original LINCLOSURE, our extended algorithm, called a GRADED LINCLOSURE (shortly, a GLINCLOSURE) is more versatile (this is discussed in Section 3) while having the same asymptotic time complexity as LINCLOSURE. This is an important feature since it is often the case that fuzzy logic extensions of classical algorithms proposed in the literature are of significantly higher time complexity than their classical counterparts. Since there is a close relationship between dependencies in data tables with fuzzy attributes and data tables over domains with similarity relations, one can also use GLINCLOSURE for computing functional dependencies in data tables over domains with similarity relations. The latter naturally appear in an extension of Codd's relational model which takes into account similarities on domains, see [8,9].

2 Preliminaries and Motivation

In this section we present preliminaries of fuzzy logic and basic notions of fuzzy attribute implications which will be used in further sections. More details can be found in [1,17,19,21,23] and [2,5,7]. In Section 2.2 we also present motivations for developing LINCLOSURE in fuzzy setting.

2.1 Fuzzy Logic and Fuzzy Set Theory

Since fuzzy logic and fuzzy sets are developed using general structures of truth degrees, we first introduce structures of truth degrees which are used in our approach. Our basic structures of truth degrees will be so-called complete residuated lattices with hedges, see [1,17,19,20,23]. A complete residuated lattice with hedge is an algebra $\mathbf{L} = \langle L, \wedge, \vee, \otimes, \rightarrow, {}^*, 0, 1\rangle$ such that

(i) $\langle L, \wedge, \vee, 0, 1\rangle$ is a complete lattice with 0 and 1 being the least and greatest element of L, respectively;
(ii) $\langle L, \otimes, 1\rangle$ is a commutative monoid (i.e., $\otimes$ is commutative, associative, and for each $a \in L$ we have $a \otimes 1 = 1 \otimes a = a$);
(iii) $\otimes$ and $\rightarrow$ satisfy so-called adjointness property: $a \otimes b \leq c$ iff $a \leq b \rightarrow c$ is true for each $a, b, c \in L$;
(iv) hedge * is a unary operation $^*: L \rightarrow L$ satisfying, for each $a, b \in L$: (1) $1^* = 1$; (2) $a^* \leq a$; (3) $(a \rightarrow b)^* \leq a^* \rightarrow b^*$; (4) $a^{**} = a^*$.

Operations $\otimes$ and $\rightarrow$ are (truth functions of) "fuzzy conjunction" and "fuzzy implication". Hedge * is a (truth function of) logical connective "very true" and properties of hedges have natural interpretations [19,20]. A common choice of $\mathbf{L}$ is a structure with $L = [0, 1]$ (real unit interval), $\wedge$ and $\vee$ being minimum and maximum, $\otimes$ being a left-continuous t-norm with the corresponding $\rightarrow$. Three most important pairs of adjoint operations on $[0, 1]$ are Łukasiewicz, Gödel, and Goguen (product), see [1] for details. Complete residuated lattices include also finite structures of truth degrees. For instance, one can put $L = \{a_0 =$

$0, a_1, \ldots, a_n = 1\} \subseteq [0,1]$ ($a_0 < \cdots < a_n$) with $\otimes$ given by $a_k \otimes a_l = a_{\max(k+l-n,0)}$ and the corresponding $\rightarrow$ given by $a_k \rightarrow a_l = a_{\min(n-k+l,n)}$. Such an $\mathbf{L}$ is called a finite Łukasiewicz chain. Another possibility is a finite Gödel chain which consists of L and restrictions of Gödel operations on $[0,1]$ to L. A special case of a complete residuated lattice with hedge is the two-element Boolean algebra $\mathbf{2}$ (structure of truth degrees of classical logic). Two boundary cases of hedges are (i) identity, i.e. $a^* = a$ ($a \in L$); (ii) so-called globalization [25]:

$$a^* = \begin{cases} 1 & \text{if } a = 1, \\ 0 & \text{otherwise.} \end{cases} \tag{1}$$

Moreover, for each $\mathbf{L}$ we consider a derived truth function $\ominus$ defined by

$$a \ominus b = a \otimes ((a \rightarrow b)^* \rightarrow 0). \tag{2}$$

For * being globalization, $a \ominus b$ simplifies as follows:

$$a \ominus b = \begin{cases} 0 & \text{if } a \leq b, \\ a & \text{else.} \end{cases} \tag{3}$$

Remark 1. Note that the derived truth function $\ominus$ can be seen as a particular subtraction of truth degrees because $a \otimes ((a \rightarrow b)^* \rightarrow 0)$ can be described as a degree to which "a is true and it is not (very) true that b is greater than a". The meaning of $\ominus$ as a type of subtraction is apparent especially in case of globalization, see (3). Due to (3), the result of $a \ominus b$ is 0 if b is greater than a. We will comment on the purpose of $\ominus$ later on.

Until otherwise mentioned, we assume that $\mathbf{L}$ denotes a complete residuated lattice (with hedge *) which serves as a structure of truth degrees. Using $\mathbf{L}$, we define the following notions. An $\mathbf{L}$-set (a fuzzy set) A in universe U is a mapping $A\colon U \rightarrow L$, $A(u)$ being interpreted as "the degree to which u belongs to A". If U is a finite universe $U = \{u_1, \ldots, u_n\}$ then an $\mathbf{L}$-set A in U can be denoted by $A = \{{}^{a_1}\!/u_1, \ldots, {}^{a_n}\!/u_n\}$, meaning that $A(u_i)$ equals a_i ($i = 1, \ldots, n$). For brevity, we introduce the following convention: we write $\{\ldots, u, \ldots\}$ instead of $\{\ldots, {}^{1}\!/u, \ldots\}$, and we also omit elements of U whose membership degree is zero. For example, we write $\{u, {}^{0.5}\!/v\}$ instead of $\{{}^{1}\!/u, {}^{0.5}\!/v, {}^{0}\!/w\}$, etc. Let $\mathbf{L}^U$ denote the collection of all $\mathbf{L}$-sets in U. Denote by $|A|$ the cardinality of the support set of A, i.e. $|A| = |\{u \in U \mid A(u) > 0\}|$. The operations with $\mathbf{L}$-sets are defined componentwise. For instance, the union of $\mathbf{L}$-sets $A, B \in \mathbf{L}^U$ is an $\mathbf{L}$-set $A \cup B$ in U such that $(A \cup B)(u) = A(u) \vee B(u)$ ($u \in U$). Due to (3), for * being globalization, we get

$$(A \ominus B)(u) = \begin{cases} 0 & \text{if } A(u) \leq B(u), \\ A(u) & \text{else.} \end{cases} \tag{4}$$

Remark 2. Fuzzy set $A \ominus B$ can be interpreted as follows. A degree $(A \ominus B)(u)$ to which $u \in U$ belongs to $A \ominus B$ is a truth degree to which "u belongs to A and u does not belong to B at least to which it belongs to A". Think of A as of a fuzzy set assigning to each $u \in U$ a threshold degree $A(u)$. Then if $B(u)$ exceeds the threshold given by $A(u)$, element u will not be present in the resulting fuzzy set $A \ominus B$ (it means that $(A \ominus B)(u) = 0$), i.e. "u will be removed". If $B(u)$ does not exceed the threshold, we have $(A \ominus B)(u) = A(u)$, i.e. "u will be preserved".

For $a \in L$ and $A \in \mathbf{L}^U$, we define an **L**-set $a \otimes A$ (a-multiple of A) by $(a \otimes A)(u) = a \otimes A(u)$, for each $u \in U$. Binary **L**-relations (binary fuzzy relations) between U and V can be thought of as **L**-sets in $U \times V$. For $A, B \in \mathbf{L}^U$, we define $S(A, B) \in L$ by

$$S(A, B) = \bigwedge_{u \in U} (A(u) \to B(u)). \tag{5}$$

$S(A, B)$ is called a subsethood degree of A in B and it generalizes the classical subsethood relation $\subseteq$ in a fuzzy setting. In particular, if **L** (structure of truth degrees) is **2** (two-element Boolean algebra), then **2**-sets coincide in an obvious manner with (characteristic functions of) ordinary sets. Also, in case of $\mathbf{L} = \mathbf{2}$ we have that $S(A, B) = 1$ iff $A \subseteq B$. For general **L**, we write $A \subseteq B$ iff $S(A, B) = 1$; and $A \subset B$ iff $S(A, B) = 1$ and $A \neq B$. As a consequence of properties of $\to$ and $\bigwedge$, we get that $A \subseteq B$ iff $A(u) \leq B(u)$ for each $u \in U$, see [1,17,21].

A fuzzy closure operator with hedge * (shortly, a fuzzy closure operator) [3] on a set U is a mapping $C : \mathbf{L}^U \to \mathbf{L}^U$ satisfying, for each $A, B_1, B_2 \in \mathbf{L}^U$: $A \subseteq C(A)$, $S(B_1, B_2)^* \leq S(C(B_1), C(B_2))$, and $C(A) = C(C(A))$.

2.2 Fuzzy Attribute Implications

Let Y denote a *finite set of attributes.* Each **L**-set $M \in \mathbf{L}^Y$ of attributes can be seen as a set of graded attributes because M prescribes, for each attribute $y \in Y$, a degree $M(y) \in L$. A *fuzzy attribute implication* (*over attributes* Y) is an expression $A \Rightarrow B$, where $A, B \in \mathbf{L}^Y$ are fuzzy sets of attributes. Fuzzy attribute implications (FAIs) represent particular data dependencies. The intuitive meaning we wish to give to $A \Rightarrow B$ is: "if it is (very) true that an object has all (graded) attributes from A, then it has also all (graded) attributes from B". Formally, for an **L**-set $M \in \mathbf{L}^Y$ of attributes, we define a *truth degree* $||A \Rightarrow B||_M \in L$ *to which* $A \Rightarrow B$ *is true in* M by

$$||A \Rightarrow B||_M = S(A, M)^* \to S(B, M), \tag{6}$$

with $S(\cdots)$ defined by (5). The degree $||A \Rightarrow B||_M$ can be understood as follows: if M (semantic component) represents presence of attributes of some object, i.e. $M(y)$ is truth degree to which "the object has the attribute $y \in Y$", then $||A \Rightarrow B||_M$ is the truth degree to which "if the object has all attributes from A, then it has all attributes from B", which corresponds to the desired interpretation of $A \Rightarrow B$. Note also that the hedge * present in (6) serves as a modifier of interpretation of $A \Rightarrow B$ and plays an important technical role. If * is globalization, i.e. if * is defined by (1), then $||A \Rightarrow B||_M = 1$ (i.e., $A \Rightarrow B$ is fully true in M) iff we have:

$$\text{if } A \subseteq M, \text{ then } B \subseteq M. \tag{7}$$

More information about the role of hedges can be found in [2,5,7]. See also [24] for a related approach.

Let T be a set of fuzzy attribute implications. An **L**-set $M \in \mathbf{L}^Y$ is called a *model of* T if, for each $A \Rightarrow B \in T$, $||A \Rightarrow B||_M = 1$. The set of all models of T will be denoted by $\mathrm{Mod}(T)$, i.e.

$$\mathrm{Mod}(T) = \{M \in \mathbf{L}^Y \mid \text{for each } A \Rightarrow B \in T\text{: } ||A \Rightarrow B||_M = 1\}. \tag{8}$$

A *degree* $||A \Rightarrow B||_T$ *to which* $A \Rightarrow B$ *semantically follows from* T is defined by

$$||A \Rightarrow B||_T = \textstyle\bigwedge_{M \in \mathrm{Mod}(T)} ||A \Rightarrow B||_M. \tag{9}$$

Described verbally, $||A \Rightarrow B||_T$ is defined to be the degree to which "$A \Rightarrow B$ is true in each model of T". Hence, degrees $||\cdots||_T$ defined by (9) represent degrees of semantic entailment from T. Let us note that degrees $||\cdots||_T$ can also be fully described via the (syntactic) concept of a *provability degree*, see [7,12].

The set $\mathrm{Mod}(T)$ of all models of T form a particular fuzzy closure system in Y, see [11] for details. Thus, for each **L**-set $M \in \mathbf{L}^Y$ we can consider its closure in $\mathrm{Mod}(T)$ which is then the least model of T containing M. The closure operator associated with $\mathrm{Mod}(T)$ can be described as follows. First, for any set T of FAIs and any **L**-set $M \in \mathbf{L}^Y$ of attributes define an **L**-set $M^T \in \mathbf{L}^Y$ of attributes by

$$M^T = M \cup \textstyle\bigcup\{S(A,M)^* \otimes B \mid A \Rightarrow B \in T\}. \tag{10}$$

Note that if * is globalization, (10) simplifies as follows:

$$M^T = M \cup \textstyle\bigcup\{B \mid A \Rightarrow B \in T \text{ and } A \subseteq M\}. \tag{11}$$

Using (10), for each $n \in \mathbb{N}_0$ we define a fuzzy set $M^{T_n} \in \mathbf{L}^Y$ of attributes by

$$M^{T_n} = \begin{cases} M & \text{for } n = 0 \\ (M^{T_{n-1}})^T & \text{for } n \geq 1. \end{cases} \tag{12}$$

Finally, we define an operator $cl_T \colon \mathbf{L}^Y \to \mathbf{L}^Y$ by

$$cl_T(M) = \textstyle\bigcup_{n=0}^{\infty} M^{T_n}. \tag{13}$$

The following assertion shows the importance of cl_T.

Theorem 1 (see [11]). *Let* **L** *be a finite residuated lattice with hedge,* T *be a set of fuzzy attribute implications. Then*

(i) cl_T *defined by* (13) *is a fuzzy closure operator;*
(ii) $cl_T(M)$ *is the least model of* T *containing* M*, i.e.* $cl_T(M) \in \mathrm{Mod}(T)$ *and, for each* $N \in \mathrm{Mod}(T)$*, if* $M \subseteq N$ *then* $cl_T(M) \subseteq N$*;*
(iii) $||A \Rightarrow B||_T = S(B, cl_T(A))$. □

Remark 3. Note that Theorem 1 (iii) says that degrees of semantic entailment from sets of fuzzy attribute implications can be expressed as subsethood degrees of consequents of FAIs into least models generated by antecedents of FAIs. Hence, a single model of T suffices to express the degree $||A \Rightarrow B||_T$, cf. definition (9). In other words, an efficient procedure for computing of closures $cl_T(\cdots)$ would give us an efficient procedure to compute degrees of semantic entailment.

Another area in which a closure operator similar to (13) appears is the computation of non-redundant bases of data tables with fuzzy attributes. A *data table with fuzzy attributes* is a triplet $\langle X, Y, I\rangle$ where X is a set of objects, Y is a finite set of attributes (the same as above), and $I \in \mathbf{L}^{X\times Y}$ is a binary $\mathbf{L}$-relation between X and Y assigning to each object $x \in X$ and each attribute $y \in Y$ a degree $I(x, y)$ to which "object x has attribute y". $\langle X, Y, I\rangle$ can be thought of as a table with rows and columns corresponding to objects $x \in X$ and attributes $y \in Y$, respectively, and table entries containing degrees $I(x, y)$. A row of a table $\langle X, Y, I\rangle$ corresponding to an object $x \in X$ can be seen as a set I_x of graded attributes (a fuzzy set of attributes) to which an attribute $y \in Y$ belongs to a degree $I_x(y) = I(x, y)$. Furthermore, a *degree* $||A \Rightarrow B||_{\langle X,Y,I\rangle}$ *to which* $A \Rightarrow B$ *is true in data table* $\langle X, Y, I\rangle$ is defined by

$$||A \Rightarrow B||_{\langle X,Y,I\rangle} = \textstyle\bigwedge_{x\in X} ||A \Rightarrow B||_{I_x}. \tag{14}$$

By definition, $||A \Rightarrow B||_{\langle X,Y,I\rangle}$ is a degree to which "$A \Rightarrow B$ is true in each row of table $\langle X, Y, I\rangle$", i.e. a truth degree of "for each object $x \in X$: if it is (very) true that x has all attributes from A, then x has all attributes from B". A set T of FAIs is called *complete in* $\langle X, Y, I\rangle$ if $||A \Rightarrow B||_T = ||A \Rightarrow B||_{\langle X,Y,I\rangle}$, i.e. if, for each $A \Rightarrow B$, a degree to which T entails $A \Rightarrow B$ coincides with a degree to which $A \Rightarrow B$ is true in $\langle X, Y, I\rangle$. If T is complete and no proper subset of T is complete, then T is called a *non-redundant basis of* $\langle X, Y, I\rangle$. Note that both the notions of a complete set and a non-redundant basis refer to a given data table with fuzzy attributes.

In order to describe particular non-redundant bases of data tables with fuzzy attributes we need to recall basic notions of formal concept analysis of data tables with fuzzy attributes [4,7]. Given a data table $\langle X, Y, I\rangle$, for $A \in \mathbf{L}^X$, $B \in \mathbf{L}^Y$ we define $A^{\uparrow} \in \mathbf{L}^Y$ and $B^{\downarrow} \in \mathbf{L}^X$ by

$$A^{\uparrow}(y) = \textstyle\bigwedge_{x\in X}(A(x)^* \to I(x, y)), \tag{15}$$

$$B^{\downarrow}(x) = \textstyle\bigwedge_{y\in Y}(B(y) \to I(x, y)). \tag{16}$$

Operators ${}^{\downarrow}, {}^{\uparrow}$ form so-called Galois connection with hedge, see [4]. The set of all fixed points of ${}^{\downarrow}, {}^{\uparrow}$ (so-called fuzzy concepts) hierarchically ordered by a subconcept-superconcept relation is called a *fuzzy concept lattice with hedge*, see [4,7]. A crucial role in determining a non-redundant basis of a given $\langle X, Y, I\rangle$ is played by an operator which is a modification of cl_T, see (13). The modified operator can be described as follows. For $M \in \mathbf{L}^Y$ put

$$M^{T^*} = M \cup \textstyle\bigcup\{S(A, M)^* \otimes B \mid A \Rightarrow B \in T \text{ and } A \neq M\}. \tag{17}$$

If * is globalization, (17) is equivalent to

$$M^{T^*} = M \cup \textstyle\bigcup\{B \mid A \Rightarrow B \in T \text{ and } A \subset M\}. \tag{18}$$

We can now define an operator cl_{T^*} in much the same way as cl_T:

$$M^{T^*_n} = \begin{cases} M & \text{for } n = 0 \\ (M^{T^*_{n-1}})^{T^*} & \text{for } n \geq 1, \end{cases} \tag{19}$$

$$cl_{T^*}(M) = \bigcup_{n=0}^{\infty} M^{T^*_n} . \tag{20}$$

For cl_{T^*} defined by (19), we have the following

Theorem 2 (see [2,5,7]). *Let* $\mathbf{L}$ *be a finite residuated lattice with globalization,* $\langle X, Y, I \rangle$ *be a data table with fuzzy attributes. Then there is* T *such that*

(i) cl_{T^*} *is a fuzzy closure operator;*
(ii) *a set of FAIs defined by* $\{P \Rightarrow P^{\downarrow\uparrow} \mid P = cl_{T^*}(P) \text{ and } P \neq P^{\downarrow\uparrow}\}$ *is a non-redundant basis of* $\langle X, Y, I \rangle$. □

Remark 4. From Theorem 2 it follows that for * being globalization a non-redundant basis of $\langle X, Y, I \rangle$ is determined by particular fixed points of cl_{T^*}, namely, by fuzzy sets $P \in \mathbf{L}^Y$ of attributes such that $P = cl_{T^*}(P)$ and $P \neq P^{\downarrow\uparrow}$. The basis given by Theorem 2 is also a minimal one, i.e. each set of fuzzy attribute implications which is complete in $\langle X, Y, I \rangle$ has at least the same number of FAIs as the basis given by 2, see [5]. Notice that we have not specified the set T of fuzzy attribute implications which is used by cl_{T^*}. A detailed description of that set is outside the scope of our paper, see [2,5,7]. An approach for general hedges has been presented in [6,10]. Let us just mention that T is computationally tractable. Fuzzy sets of attributes satisfying $P = cl_{T^*}(P)$ and $P \neq P^{\downarrow\uparrow}$ will occasionally be referred to as *pseudo-intents* or *pseudo-closed fuzzy set of attributes.* This is for the sake of consistency with [2,5,7], cf. also [14,15,16,18].

3 Graded LinClosure

Throughout this section, we assume that $\mathbf{L}$ is a finite linearly ordered residuated lattice with globalization, see (1). Structure $\mathbf{L}$ represents a finite linear scale of truth degrees.

Problem Setting. Given a fuzzy set $M \in \mathbf{L}^Y$ of attributes we wish to compute its closures $cl_T(M)$ and $cl_{T^*}(M)$ defined by (13) and (20), respectively. First, note that cl_T and cl_{T^*} differ only in non-strict/strict fuzzy set inclusions "$\subseteq$" and "$\subset$" used in (11) and (18). A direct method to compute $cl_T(M)$ and $cl_{T^*}(M)$, which is given by the definitions of cl_T and cl_{T^*}, leads to an algorithm similar to Closure which is known from database systems [22]. In more detail: for a given M, we iterate through all FAIs in T and for each $A \Rightarrow B \in T$ we test if $A \subseteq M$ ($A \subset M$); if so, we add B to M (i.e., we set $M := M \cup B$) and repeat the process until M cannot be enlarged; the resulting M is the closure under cl_T (cl_{T^*}) of the original fuzzy set M. Clearly, this procedure is sound. Let n be the number of attributes in Y and p be the number of FAIs in T. In the worst case, we have to make p^2 iterations in order to compute the closure because in each loop through all FAIs in T there can be only one $A \Rightarrow B$ such that $A \subseteq M$ ($A \subset M$) and $B \nsubseteq M$ (i.e., only one FAI from T causes M to be enlarged). Moreover, for each $A \Rightarrow B$ we need n steps to check the non-strict/strict subsethood $A \subseteq M$ ($A \subset M$). To sum up, the complexity of this algorithm is $O(np^2)$, where n is the number of attributes and p is the number of FAIs from T (cf. [22]).

In this section we present an improved version of the algorithm, so-called GLINCLOSURE (GRADED LINCLOSURE), which computes $cl_T(M)$ and $cl_{T^*}(M)$ with complexity $O(n)$, where n is the size of the input. GLINCLOSURE uses each FAI from T only once and allows us to check the non-strict/strict inclusions $A \subseteq M$ ($A \subset M$) which appear in (11) and (18) in a constant time. Our algorithm results by extending LINCLOSURE [22] so that

(i) we can use *fuzzy sets of attributes* instead of classical sets (this brings new technical problems with efficient comparing of truth degrees and checking of strict inclusion, see below);
(ii) we can use the algorithm also to compute *systems of pseudo-intents* (i.e., fixed points of cl_{T^*}), and thus non-redundant bases (the original LINCLOSURE [22] cannot be used to compute pseudo-intents [16], it can only compute fixed points of the classical counterpart of cl_T), this also brings technical complications since we have to maintain a "waitlist of attributes" which can possibly be updated (or not) in future iterations.

In what follows we present a detailed description of the algorithm and analysis of its complexity.

Input and Output of the Algorithm. The input for GLINCLOSURE consists of a set T of fuzzy attribute implications over Y, a fuzzy set $M \in \mathbf{L}^Y$ of attributes, and a flag *PCLOSED* $\in$ {*false*, *true*}. The meaning of *PCLOSED* is the following. If *PCLOSED* is set to *true*, the output of GLINCLOSURE is $cl_{T^*}(M)$ (the least fixed point of cl_{T^*} which contains M); if *PCLOSED* is set to *false*, the output of GLINCLOSURE is $cl_T(M)$ (the least fixed point of cl_T, i.e. the least model of T, which contains M).

Representation of Graded Attributes. During the computation, we represent fuzzy sets ($\mathbf{L}$-sets) of attributes in Y by ordinary sets of tuples $\langle y, a\rangle$, where $y \in Y$ and $a \in L$. Namely, a fuzzy set $\{{}^{a_1}/y_1, {}^{a_2}/y_2, \ldots, {}^{a_n}/y_n\}$ ($a_1 \neq 0, \ldots, a_n \neq 0$) will be represented by an ordinary set $\{\langle y_1, a_1\rangle, \langle y_2, a_2\rangle, \ldots, \langle y_n, a_n\rangle\}$. We will use both notations $A(y) = a$ and $\langle y, a\rangle \in A$. Whenever we consider $\langle y, a\rangle$, we assume $a \neq 0$. From the implementational point of view, we may represent fuzzy sets of attributes by lists of tuples $\langle y, a\rangle$ instead of sets of such tuples. In such a case, we write $(\langle y_1, a_1\rangle, \langle y_2, a_2\rangle, \ldots, \langle y_n, a_n\rangle)$ instead of $\{\langle y_1, a_1\rangle, \langle y_2, a_2\rangle, \ldots, \langle y_n, a_n\rangle\}$.

Quick Test of Subsethood. We avoid repeated testing of inclusions in (11) and (18) analogously as in the original LINCLOSURE. For each fuzzy attribute implication $A \Rightarrow B$ from T we keep a record of the number of attributes due to which A is not contained in the constructed closure. If this number reaches zero, we get that $A \subseteq M$ and we can process $A \Rightarrow B$, see (11). This suffices to check non-strict subsethood which is needed to compute fixed points of cl_T. In order to check strict subsethood which is needed to compute $cl_{T^*}(M)$, we need to have a quick test to decide if $A \subset M$ provided we already know that $A \subseteq M$. The test can be done with the following notion of *cardinality* of fuzzy sets. Take a fixed monotone injective mapping $f_{\mathbf{L}}\colon L \to [0,1]$. That is, $f_{\mathbf{L}}$ is injective, and for each $a, b \in L$, if $a \leq b$ then $f_{\mathbf{L}}(a) \leq f_{\mathbf{L}}(b)$. For each fuzzy set $M \in \mathbf{L}^Y$ of attributes we define a number $\mathrm{card}(M) \in [0, \infty)$ by

$$\mathrm{card}(M) = \textstyle\sum_{\langle y,a\rangle \in M} f_{\mathbf{L}}(a). \tag{21}$$

For instance, if L is a subset of $[0,1]$, we can put $f_{\mathbf{L}}(a) = a$ ($a \in L$), and thus $\mathrm{card}(M) = \sum_{\langle y,a\rangle \in M} a$.

Lemma 1. *Let $A, B \in \mathbf{L}^Y$ s. t. $A \subseteq B$. Then $A \subset B$ iff* $\mathrm{card}(A) < \mathrm{card}(B)$. □

Remark 5. Note that checking strict inclusion in fuzzy setting is more difficult that in the ordinary case where one can decide it simply by comparing numbers of elements in both sets. In fuzzy setting, we can have fuzzy sets which have the same number of elements belonging to the sets (to a non-zero degree) but the sets may not be equal. For instance, if $L = [0,1]$, then $A = \{^{0.7}/y, {}^{0.5}/z\}$ and $B = \{^{0.9}/y, {}^{0.5}/z\}$ both contain two elements (to a non-zero degree), i.e. $|A| = |B| = 2$ (see preliminaries). Hence, the values of $|\cdots|$ alone are not sufficient to decide $A \subset B$ provided that $A \subseteq B$. This is why we have introduced "cardinalities" by (21).

Data Structures Used During the Computation:

NEWDEP is a fuzzy set of attributes which is the closure being constructed;
CARDND is the cardinality of *NEWDEP* given by (21);
COUNT$[A \Rightarrow B]$ is a nonnegative integer indicating the number of attributes from A such that $A(y) >$ *NEWDEP*(y), *COUNT*$[A \Rightarrow B] = 0$ means that A is a subset of *NEWDEP*;
CARD$[A \Rightarrow B]$ is a number indicating cardinality of A, it is used to decide if A is strictly contained in *NEWDEP* when *COUNT*$[A \Rightarrow B]$ reaches zero;
UPDATE is a fuzzy set of attributes which are waiting for update;
WAITLIST is a list of (pointers to) fuzzy sets of attributes which can be added to *NEWDEP* as soon as *NEWDEP* will increase its cardinality; this is necessary if *PCLOSED* = *true*, *WAITLIST* is not used if *PCLOSED* = *false*;
LIST$[y]$ is an attribute-indexed collection of (pointers to) FAIs from T such that $A \Rightarrow B$ is referenced in *LIST*$[y]$ iff $A(y) > 0$;
SKIP$[y][A \Rightarrow B] \in \{$*false*, *true*$\}$ indicates whether attribute y has already been updated in $A \Rightarrow B$ (*SKIP*$[y][A \Rightarrow B] =$ *true*) or not (*SKIP*$[y][A \Rightarrow B] =$ *false*); *SKIP* is necessary in graded setting to avoid updating an attribute twice which may result in incorrect values of *COUNT*$[A \Rightarrow B]$;
DEGREE$[y][A \Rightarrow B]$ represents a degree in which attribute y is contained in A;
() denotes the empty list.

Note that *NEWDEP*, *UPDATE*, *COUNT*, and *LIST* play a similar role as in LinClosure, cf. [22]. The algorithm is depicted in Fig. 1. Let us mention that the algorithm has two basic parts:

Input: a set T of FAIs over Y, a fuzzy set $M \in \mathbf{L}^Y$ of attributes, and a flag $PCLOSED \in \{false, true\}$

Output: $cl_{T^*}(M)$ if $PCLOSED = true$, or $cl_T(M)$ if $PCLOSED = false$

Initialization:

```
if M = ∅ and PCLOSED = true:
   return ∅
NEWDEP := M
for each A ⇒ B ∈ T:
   if A = ∅:
      NEWDEP := NEWDEP ∪ B
   else:
      COUNT[A ⇒ B] := |A|
      CARD[A ⇒ B] := card(A)
      for each ⟨y, a⟩ ∈ A:
         add A ⇒ B to LIST[y]
         DEGREE[y][A ⇒ B] := a
         SKIP[y][A ⇒ B] := false
UPDATE := NEWDEP
CARDND := card(NEWDEP)
WAITLIST := ()
```

Computation:

```
while UPDATE ≠ ∅:
   choose ⟨y, a⟩ ∈ UPDATE
   UPDATE := UPDATE − {⟨y, a⟩}
   for each A ⇒ B ∈ LIST[y] such that
            SKIP[y][A ⇒ B] = false and DEGREE[y][A ⇒ B] ≤ a:
      SKIP[y][A ⇒ B] = true
      COUNT[A ⇒ B] := COUNT[A ⇒ B] − 1
      if COUNT[A ⇒ B] = 0 and
            (PCLOSED = false or CARD[A ⇒ B] < CARDND):
         ADD := B ⊖ NEWDEP
         CARDND := CARDND + Σ_{⟨y,a⟩∈ADD} (f_L(a) − f_L(NEWDEP(y)))
         NEWDEP := NEWDEP ∪ ADD
         UPDATE := UPDATE ∪ ADD
         if PCLOSED = true and ADD ≠ ∅:
            while WAITLIST ≠ ():
               choose B ∈ WAITLIST
               remove B from WAITLIST
               ADD := B ⊖ NEWDEP
               CARDND := CARDND + Σ_{⟨y,a⟩∈ADD} (f_L(a) − f_L(NEWDEP(y)))
               NEWDEP := NEWDEP ∪ ADD
               UPDATE := UPDATE ∪ ADD
      if COUNT[A ⇒ B] = 0 and PCLOSED = true and
            CARD[A ⇒ B] = CARDND:
         add B to WAITLIST
return NEWDEP
```

Fig. 1. Graded LinClosure

(i) *initialization of data structures* (lines 1–16), and

(ii) *main computation* (lines 17–38).

Denote by k the number of truth degrees, i.e. $k = |L|$. Denote by n the length of the input which is the sum of attributes which belong to left-hand sides of FAIs to non-zero degrees, i.e.: $n = \sum_{A \Rightarrow B \in T} |A|$. The initialization can be done with time complexity $O(n)$. Of course, the initialization depends on data structures we choose for representing *LIST*, *SKIP*, *DEGREE*, *COUNT*, and *CARD*. In next section we propose an efficient structure encompassing the information from *LIST*, *SKIP*, *DEGREE*, *COUNT*, and *CARD* whose initialization takes $O(kn)$ steps. Since k (number of truth degrees) is a multiplicative constant (size of $\mathbf{L}$ is fixed and does not depend on the length of the input), the initialization is linearly dependent on the length of the input, i.e. it is indeed in $O(n)$.

In the second part (computation), each graded attribute $\langle y, a \rangle$ is considered at most once for update. This is ensured in a similar way as in the ordinary case, see [22]. For each fuzzy attribute implication, the value of *COUNT* reaches zero at most once. Then, the computation which follows (lines 24–27 or lines 32–35) is linearly dependent on the size of the left-hand side of the processed fuzzy attribute implication. This again depends on the representation of fuzzy sets and operations with fuzzy sets. If we represent fuzzy sets by a list of pairs of the form $\langle y, a \rangle$ where $y \in Y$ and $a \in L$ which is moreover sorted by the attributes (in some fixed order), we can perform all necessary operations in linear time proportional to the size of the fuzzy set. Thus, using analogous arguments as in case of the original LinClosure [22], we get that GLinClosure works with asymptotic time complexity $O(n)$. Note that lines 28–37 are not present in the ordinary LinClosure. This is because if we intend to compute fixed points of cl_{T^*}, a graded attribute can be scheduled for updating (added to *UPDATE*) only if we know that the left-hand side of FAI is strictly contained in *NEWDEP*. This is checked at line 36.

Remark 6. If $\mathbf{L}$ (our structure of truth degrees) is a two-element Boolean algebra, i.e. if $L = \{0, 1\}$, GLinClosure with *PCLOSED* set to *false* produces the same results as LinClosure [22] (the only difference is that our algorithm allows also for FAIs of the form $\{\} \Rightarrow B$ whereas the original LinClosure does not). From this point of view, GLinClosure is a generalization of LinClosure. GLinClosure is more versatile (even in crisp case): GLinClosure can be used to compute pseudo-intents (and thus a non-redundant basis of data tables with fuzzy attributes) which cannot be done with the original LinClosure (without additional modifications).

4 Implementation Details, Examples, and Remarks

As mentioned before, the efficiency of an implementation of GLinClosure is closely connected with data structures. The information contained in *LIST*, *SKIP*, *DEGREE*, *COUNT*, and *CARD* can be stored in a single efficient data

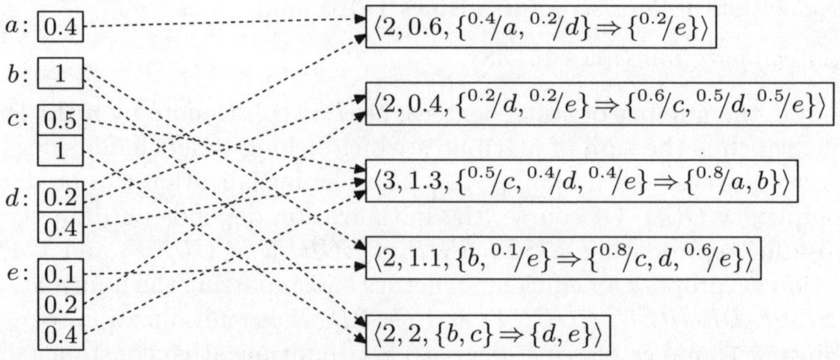

Fig. 2. T-structure encompassing *LIST*, *SKIP*, *DEGREE*, *COUNT*, and *CARD*

structure. This structure, called a *T-structure*, is a particular attribute-indexed vector of lists of pointers to structures carrying values from *COUNT* and *CARD*. We illustrate the construction of a T-structure by an example. Consider a set T of FAIs which consists of the following fuzzy attribute implications:

$$\varphi_1\colon \{\} \Rightarrow \{{}^{0.4}\!/a, {}^{0.1}\!/d\}, \qquad \varphi_4\colon \{{}^{0.5}\!/c, {}^{0.4}\!/d, {}^{0.4}\!/e\} \Rightarrow \{{}^{0.8}\!/a, b\},$$
$$\varphi_2\colon \{{}^{0.4}\!/a, {}^{0.2}\!/d\} \Rightarrow \{{}^{0.2}\!/e\}, \qquad \varphi_5\colon \{b, {}^{0.1}\!/e\} \Rightarrow \{{}^{0.8}\!/c, d, {}^{0.6}\!/e\},$$
$$\varphi_3\colon \{{}^{0.2}\!/d, {}^{0.2}\!/e\} \Rightarrow \{{}^{0.6}\!/c, {}^{0.5}\!/d, {}^{0.5}\!/e\}, \qquad \varphi_6\colon \{b, c\} \Rightarrow \{d, e\}.$$

Since φ_1 is of the form $\{\} \Rightarrow B$, its right-hand side is added to *NEWDEP* and the implication itself is not contained in *LIST* and other structures. The other formulas, i.e. $\varphi_2, \ldots, \varphi_6$, are used to build a new T-structure which is depicted in Fig. 2. The T-structure can be seen as consisting of two main parts. First, a set of records encompassing information about the FAIs, *COUNT*, and *CARD*. For each FAI φ_i, we have a single record, called a T-record, of the form $\langle COUNT[\varphi_i], CARD[\varphi_i], \varphi_i\rangle$, see Fig. 2 (right). Second, an attribute-indexed vector of lists containing truth degrees and pointers to T-records, see Fig. 2 (left). A list which is indexed by attribute $y \in Y$ will be called a y-list. The aim of this part of the structure is to keep information about the occurrence of graded attributes that appear in left-hand sides of FAIs from T. In more detail, a y-list contains truth degree $a \in L$ iff there is at lest one $A \Rightarrow B \in T$ such that $0 \neq A(y) = a$. Moreover, if a y-list contains a as its element, then it is connected via pointer to all T-records $\langle m, n, C \Rightarrow D\rangle$ such that $C(y) = a$. Because of the computational efficiency, each y-list is sorted by truth degrees in the ascendant manner. Note that pointers between elements of lists Fig. 2 (left) and T-records Fig. 2 (right) represent information in *SKIP* ($SKIP[y][A \Rightarrow B] = \mathit{false}$ means that pointer from element $A(y)$ of y-list to T-record of $A \Rightarrow B$ is present). As one can see, a T-structure can be constructed by a sequential updating of the structure with time complexity $O(kn)$, where n is the size of the input (each graded attribute is considered once) and k is the number of truth degrees (this is an overhead needed to keep y-lists sorted). In the following examples, we will use a convenient notation for writing T-structures which correspond in an

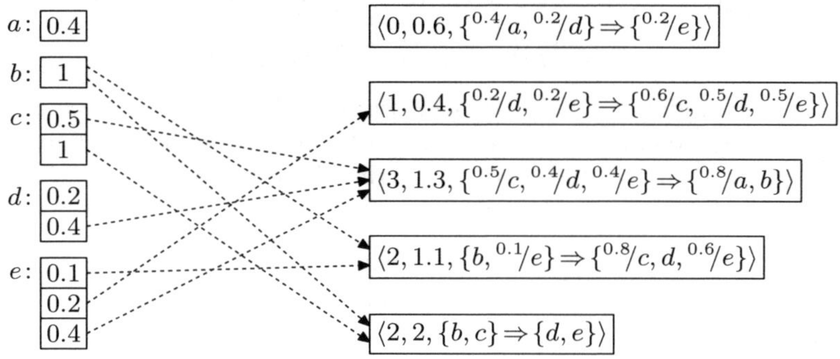

Fig. 3. T-structure before processing the first FAI

obvious way with graphs of the from of Fig. 2. For example, instead of Fig. 2, we write:

a: $[(0.4, \langle 2, 0.6, \varphi_2\rangle)]$
b: $[(1, \langle 2, 2, \varphi_6\rangle, \langle 2, 1.1, \varphi_5\rangle)]$
c: $[(0.5, \langle 3, 1.3, \varphi_4\rangle), (1, \langle 2, 2, \varphi_6\rangle)]$
d: $[(0.2, \langle 2, 0.4, \varphi_3\rangle, \langle 2, 0.6, \varphi_2\rangle), (0.4, \langle 3, 1.3, \varphi_4\rangle)]$
e: $[(0.1, \langle 2, 1.1, \varphi_5\rangle), (0.2, \langle 2, 0.4, \varphi_3\rangle), (0.4, \langle 3, 1.3, \varphi_4\rangle)]$

Example 1. Consider T which consists of $\varphi_1, \ldots, \varphi_6$ as above in this section. Let $M = \{{}^{0.2}\!/d\}$, and $\mathit{PCLOSED} = \mathit{false}$. After the initialization (line 16 of the algorithm), we have $\mathit{NEWDEP} = \{{}^{0.4}\!/a, {}^{0.2}\!/d\}$ and $\mathit{UPDATE} = (\langle a, 0.4\rangle, \langle d, 0.2\rangle)$. Recall that during the update, values of *COUNT* and *SKIP* are changed. Namely, values of *COUNT* may be decremented and values of *SKIP* are changed to *true*. The latter update is represented by removing pointers from the T-structure. After the update of $\langle a, 0.4\rangle$ and $\langle d, 0.2\rangle$, the T-record $\langle 0, 0.6, \varphi_2 = \{{}^{0.4}\!/a, {}^{0.2}\!/d\} \Rightarrow \{{}^{0.2}\!/e\}\rangle$ of φ_2 is processed because we have $\mathit{COUNT}[\varphi_2] = 0$ (see the first item of the T-record). At this point, the algorithm is in the following state:

b: $[(1, \langle 2, 2, \varphi_6\rangle, \langle 2, 1.1, \varphi_5\rangle)]$ $\qquad$ $\mathit{ADD} = (\langle e, 0.2\rangle)$
c: $[(0.5, \langle 3, 1.3, \varphi_4\rangle), (1, \langle 2, 2, \varphi_6\rangle)]$ $\qquad$ $\mathit{NEWDEP} = \{{}^{0.4}\!/a, {}^{0.2}\!/d, {}^{0.2}\!/e\}$
d: $[(0.4, \langle 3, 1.3, \varphi_4\rangle)]$ $\qquad$ $\mathit{UPDATE} = (\langle e, 0.2\rangle)$
e: $[(0.1, \langle 2, 1.1, \varphi_5\rangle), (0.2, \langle 1, 0.4, \varphi_3\rangle), (0.4, \langle 3, 1.3, \varphi_4\rangle)]$

The corresponding T-structure is depicted in Fig. 3.
As a further step of the computation, an update of $\langle e, 0.2\rangle$ is performed and then the T-record $\langle 0, 0.4, \varphi_3 = \{{}^{0.2}\!/d, {}^{0.2}\!/e\} \Rightarrow \{{}^{0.6}\!/c, {}^{0.5}\!/d, {}^{0.5}\!/e\}\rangle$ of φ_3 is processed:

b: $[(1, \langle 2, 2, \varphi_6\rangle, \langle 1, 1.1, \varphi_5\rangle)]$ $\qquad$ $\mathit{ADD} = (\langle c, 0.6\rangle, \langle d, 0.5\rangle, \langle e, 0.5\rangle)$
c: $[(0.5, \langle 3, 1.3, \varphi_4\rangle), (1, \langle 2, 2, \varphi_6\rangle)]$ $\qquad$ $\mathit{NEWDEP} = \{{}^{0.4}\!/a, {}^{0.6}\!/c, {}^{0.5}\!/d, {}^{0.5}\!/e\}$
d: $[(0.4, \langle 3, 1.3, \varphi_4\rangle)]$ $\qquad$ $\mathit{UPDATE} = (\langle c, 0.6\rangle, \langle d, 0.5\rangle, \langle e, 0.5\rangle)$
e: $[(0.4, \langle 3, 1.3, \varphi_4\rangle)]$

Right after the update of $\langle c, 0.6\rangle$, $\langle d, 0.5\rangle$, and $\langle e, 0.5\rangle$, the algorithm will process the T-record of φ_4. After that, we have the following situation:

b: $[(1, \langle 2, 2, \varphi_6\rangle, \langle 1, 1.1, \varphi_5\rangle)]$ $ADD = (\langle a, 0.8\rangle, \langle b, 1\rangle)$
c: $[(1, \langle 2, 2, \varphi_6\rangle)]$ $NEWDEP = \{{}^{0.8}\!/a, b, {}^{0.6}\!/c, {}^{0.5}\!/d, {}^{0.5}\!/e\}$
$UPDATE = (\langle a, 0.8\rangle, \langle b, 1\rangle)$

Then, $\langle a, 0.8\rangle$ is updated. Notice that this update has no effect because the T-structure no longer contains attributes of the form $\langle a, x\rangle$ waiting for update (the a-list is empty). After the update of $\langle b, 1\rangle$, the T-record $\langle 0, 1.1, \varphi_5 = \{b, {}^{0.1}\!/e\} \Rightarrow \{{}^{0.8}\!/c, d, {}^{0.6}\!/e\}\rangle$ of φ_5 is processed. We arrive to:

c: $[(1, \langle 1, 2, \varphi_6\rangle)]$ $ADD = (\langle c, 0.8\rangle, \langle d, 1\rangle, \langle e, 0.6\rangle)$
$NEWDEP = \{{}^{0.8}\!/a, b, {}^{0.8}\!/c, d, {}^{0.6}\!/e\}$
$UPDATE = (\langle c, 0.8\rangle, \langle d, 1\rangle, \langle e, 0.6\rangle)$

The algorithm updates $\langle c, 0.8\rangle$, $\langle d, 1\rangle$, $\langle e, 0.6\rangle$ however such updates are all without any effect because the d-list and e-list are already empty, and the c-list contains a single record with $1 \not\leq 0.8$ (see the condition at line 20 of the algorithm). Thus, the T-structure remains unchanged, $UPDATE$ is empty, and the procedure stops returning the value of $NEWDEP$ which is $\{{}^{0.8}\!/a, b, {}^{0.8}\!/c, d, {}^{0.6}\!/e\}$.

Example 2. In this example we demonstrate the role of the $WAITLIST$. Let T be a set of FAIs which consists of

$$\psi_1\colon \{{}^{0.2}\!/a\} \Rightarrow \{{}^{0.6}\!/a, {}^{0.3}\!/c\}, \qquad \psi_3\colon \{{}^{0.6}\!/a, {}^{0.3}\!/c\} \Rightarrow \{b\},$$
$$\psi_2\colon \{{}^{0.3}\!/c\} \Rightarrow \{{}^{0.2}\!/b\}, \qquad \psi_4\colon \{{}^{0.6}\!/a, b, {}^{0.3}\!/c\} \Rightarrow \{d\}.$$

Moreover, we consider $M = \{{}^{0.3}\!/a\}$ and $PCLOSED = \mathit{true}$. After the initialization (line 16), we have $NEWDEP = \{{}^{0.3}\!/a\}$, $CARDND = 0.3$ ($f_{\mathbf{L}}$ is identity), $UPDATE = (\langle a, 0.3\rangle)$, $WAITLIST = ()$, and the T-structure is the following:

a: $[(0.2, \langle 1, 0.2, \psi_1\rangle), (0.6, \langle 3, 1.9, \psi_4\rangle, \langle 2, 0.9, \psi_3\rangle)]$
b: $[(1, \langle 3, 1.9, \psi_4\rangle)]$
c: $[(0.3, \langle 3, 1.9, \psi_4\rangle, \langle 2, 0.9, \psi_3\rangle, \langle 1, 0.3, \psi_2\rangle)]$

The computation continues with the update of $\langle a, 0.3\rangle$. During that, the T-record $\langle 1, 0.2, \psi_1\rangle$ will be updated to $\langle 0, 0.2, \psi_1\rangle$. Since $CARD[\psi_1] = 0.2 < 0.3 = CARDND$, the left-hand side of ψ_1 is strictly contained in $NEWDEP$, and the algorithm processes $\langle 0, 0.2, \psi_1 = \{{}^{0.2}\!/a\} \Rightarrow \{{}^{0.6}\!/a, {}^{0.3}\!/c\}\rangle$, i.e. we get to

a: $[(0.6, \langle 3, 1.9, \psi_4\rangle, \langle 2, 0.9, \psi_3\rangle)]$ $ADD = (\langle a, 0.6\rangle, \langle c, 0.3\rangle)$
b: $[(1, \langle 3, 1.9, \psi_4\rangle)]$ $NEWDEP = \{{}^{0.6}\!/a, {}^{0.3}\!/c\}$
c: $[(0.3, \langle 3, 1.9, \psi_4\rangle, \langle 2, 0.9, \psi_3\rangle, \langle 1, 0.3, \psi_2\rangle)]$ $CARDND = 0.9$
$UPDATE = (\langle a, 0.6\rangle, \langle c, 0.3\rangle)$

After the update of $\langle a, 0.6\rangle$, we have:

b: $[(1, \langle 2, 1.9, \psi_4\rangle)]$
c: $[(0.3, \langle 2, 1.9, \psi_4\rangle, \langle 1, 0.9, \psi_3\rangle, \langle 1, 0.3, \psi_2\rangle)]$

Then, the algorithm continues with updating $\langle c, 0.3\rangle$. The T-record $\langle 2, 1.9, \psi_4\rangle$ is updated to $\langle 1, 1.9, \psi_4\rangle$ and removed from the c-list. In the next step, the T-record $\langle 1, 0.9, \psi_3\rangle$ is updated to $\langle 0, 0.9, \psi_3\rangle$. At this point, we have $CARD[\psi_3] = 0.9 = CARDND$, i.e. we add fuzzy set $\{b\}$ of attributes (the right-hand side of

ψ_3) to the *WAITLIST*. Finally, $\langle 1, 0.3, \psi_2\rangle$ is updated to $\langle 0, 0.3, \psi_2\rangle$ which yields the following situation: the T-structure consists of b: $[(1, \langle 1, 1.9, \psi_4\rangle)]$, $ADD = (\langle b, 0.2\rangle)$, $NEWDEP = \{{}^{0.6}/a, {}^{0.2}/b, {}^{0.3}/c\}$, $CARDND = 1.1$, and $UPDATE = (\langle b, 0.2\rangle)$. Since *ADD* is nonempty, the algorithm continues with flushing the *WAITLIST* (lines 28–35). After that, the new values are set to $NEWDEP = \{{}^{0.6}/a, b, {}^{0.3}/c\}$, $CARDND = 1.9$, and $UPDATE = (\langle b, 0.2\rangle, \langle b, 1\rangle)$. The process continues with updating $\langle b, 0.2\rangle$ (no effect) and $\langle b, 1\rangle$. Here again, we are in a situation where $CARD[\psi_4] = 1.9 = CARDND$, i.e. $\{d\}$ is added to the *WAITLIST*, only this time, the computation ends because *UPDATE* is empty, i.e. $\{d\}$ will not be added to *NEWDEP*. Thus, the resulting value being returned is $\{{}^{0.6}/a, b, {}^{0.3}/c\}$.

5 Conclusions

We have shown an extended version of the LinClosure algorithm, so-called Graded LinClosure (GLinClosure). Our algorithm can be used in case of graded as well as binary attributes. Even for binary attributes, GLinClosure is more versatile than the original LinClosure (it can be used to compute systems of pseudo-intents) but it has the same asymptotic complexity $O(n)$. Future research will focus on further algorithms for formal concept analysis of data with fuzzy attributes.

References

1. Belohlavek, R.: Fuzzy Relational Systems: Foundations and Principles. Kluwer, Academic/Plenum Publishers, New York (2002)
2. Belohlavek, R., Chlupova, M., Vychodil, V.: Implications from data with fuzzy attributes. In: AISTA 2004 in Cooperation with the IEEE Computer Society Proceedings, p. 5 (2004)
3. Belohlavek, R., Funiokova, T., Vychodil, V.: Fuzzy closure operators with truth stressers. Logic Journal of IGPL 13(5), 503–513 (2005)
4. Belohlavek, R., Vychodil, V.: Reducing the size of fuzzy concept lattices by hedges. In: FUZZ-IEEE 2005, The IEEE International Conference on Fuzzy Systems, Reno (Nevada, USA), May 22–25, 2005, pp. 663–668 (2005) (proceedings on CD), abstract in printed proceedings, p. 44
5. Belohlavek, R., Vychodil, V.: Fuzzy attribute logic: attribute implications, their validity, entailment, and non-redundant basis. In: Liu, Y., Chen, G., Ying, M. (eds.) Fuzzy Logic, Soft Computing & Computational Intelligence: Eleventh International Fuzzy Systems Association World Congress, vol. I, pp. 622–627. Tsinghua University Press and Springer (2005)
6. Belohlavek, R., Vychodil, V.: Fuzzy attribute implications: Computing non-redundant bases using maximal independent sets". In: Zhang, S., Jarvis, R. (eds.) AI 2005. LNCS (LNAI), vol. 3809, pp. 1126–1129. Springer, Heidelberg (2005)
7. Belohlavek, R., Vychodil, V.: Attribute implications in a fuzzy setting. In: Missaoui, R., Schmidt, J. (eds.) Formal Concept Analysis. LNCS (LNAI), vol. 3874, pp. 45–60. Springer, Heidelberg (2006)

8. Belohlavek, R., Vychodil, V.: Functional dependencies of data tables over domains with similarity relations. In: Proc. IICAI 2005, pp. 2486–2504 (2005)
9. Belohlavek, R., Vychodil, V.: Data tables with similarity relations: Functional dependencies, complete rules and non-redundant bases. In: Li Lee, M., Tan, K.-L., Wuwongse, V. (eds.) DASFAA 2006. LNCS, vol. 3882, pp. 644–658. Springer, Heidelberg (2006)
10. Belohlavek, R., Vychodil, V.: Computing non-redundant bases of if-then rules from data tables with graded attributes. In: Zhang, Y.Q., Lin, T.Y. (eds.) Proc. IEEE-GrC 2006, pp. 205–210 (2006)
11. Belohlavek, R., Vychodil, V.: Properties of models of fuzzy attribute implications. In: Proc. SCIS & ISIS 2006: Joint 3rd International Conference on Soft Computing and Intelligent Systems and 7th International Symposium on advanced Intelligent Systems, Tokyo Institute of Technology, Japan Society for Fuzzy Theory and Intelligent Informatics, pp. 1880–3741 (2006)
12. Belohlavek, R., Vychodil, V.: Fuzzy attribute logic over complete residuated lattices. J. Exp. Theor. Artif. Intelligence 18(4), 471–480 (2006)
13. Carpineto, C., Romano, G.: Concept Data Analysis. Theory and Applications. J. Wiley, Chichester (2004)
14. Ganter, B.: Begriffe und Implikationen, manuscript (1998)
15. Ganter, B.: Algorithmen zur formalen Begriffsanalyse. In: Ganter, B., Wille, R., Wolff, K.E. (eds.) (Hrsg.): Beiträge zur Begriffsanalyse, pp. 241–254. B. I. Wissenschaftsverlag, Mannheim (1987)
16. Ganter, B., Wille, R.: Formal Concept Analysis. Mathematical Foundations. Springer, Berlin (1999)
17. Goguen, J.: The logic of inexact concepts. Synthese 18(9), 325–373 (1968)
18. Guigues, J.-L., Duquenne, V.: Familles minimales d'implications informatives resultant d'un tableau de données binaires. Math. Sci. Humaines 95, 5–18 (1986)
19. Hájek, P.: Metamathematics of Fuzzy Logic. Kluwer, Dordrecht (1998)
20. Hájek, P.: On very true. Fuzzy Sets and Systems 124, 329–333 (2001)
21. Klir, G.J., Yuan, B.: Fuzzy Sets and Fuzzy Logic. Theory and Applications. Prentice-Hall, Englewood Cliffs (1995)
22. Maier, D.: The Theory of Relational Databases. Computer Science Press, Rockville (1983)
23. Pavelka, J.: On fuzzy logic I, II, III. Z. Math. Logik Grundlagen Math. 25, 45–52, 119–134, 447–464 (1979)
24. Pollandt, S.: Fuzzy Begriffe. Springer, Berlin Heidelberg (1997)
25. Takeuti, G., Titani, S.: Globalization of intuitionistic set theory. Annals of Pure and Applied Logic 33, 195–211 (1987)

Yet Another Approach for Completing Missing Values

Leila Ben Othman and Sadok Ben Yahia

Faculty of Sciences of Tunis
Computer Science Department
Campus University, 1060 Tunis, Tunisia
sadok.benyahia@fst.rnu.tn

Abstract. When tackling real-life datasets, it is common to face the existence of scrambled missing values within data. Considered as "dirty data", it is usually removed during the pre-processing step of the KDD process. Starting from the fact that "making up this missing data is better than throwing it away", we present a new approach trying to complete the missing data. The main singularity of the introduced approach is that it sheds light on a fruitful synergy between generic basis of association rules and the topic of missing values handling. In fact, beyond interesting compactness rate, such generic association rules make it possible to get a considerable reduction of conflicts during the completion step. A new metric called *"Robustness"* is also introduced, and aims to select the robust association rule for the completion of a missing value whenever a conflict appears. Carried out experiments on benchmark datasets confirm the soundness of our approach. Thus, it reduces conflict during the completion step while offering a high percentage of correct completion accuracy.

Keywords: Data mining, Formal Concept Analysis, Generic Association Rule Bases, Missing Values Completion.

1 Introduction

The field of Knowledge Discovery in Databases (KDD) has recently emerged as a new research discipline, standing at the crossroads of statistics, machine learning, data management, and other areas. The central step within the overall KDD process is data mining — the application of computational techniques for the sake of finding patterns and models in data. Implicitly, such knowledge is supposed to be mined from "high" quality data. However, most real-life datasets encompass missing data, that is commonly considered as withdrawable during the KDD pre-processing step.

Thus, setting up robust mining algorithms handling "dirty" data is a compelling and thriving issue to be addressed towards knowledge quality improvement. In this respect, a review of the dedicated literature pointed out a determined effort from the Statistics community. This is reflected by the wealthy

S. Ben Yahia et al. (Eds.): CLA 2006, LNAI 4923, pp. 155–169, 2008.

harvest of works addressing the completing missing value issue, *e.g.*, Gibbs sampling [7,14], the Expectation Maximization [9] and Bound and Collapse [18] — to cite but a few. Based on the missing information principle [12], *i.e.*, *the value for replacement is one of the existing data*, the use of association rules seemed to be a promising issue *c.f,* the approaches presented in [4,10,17,22]. The driving idea is that association rules ideally describe conditional expectation of the missing values according to the observed data caught out by their premise parts. Within based association rule approaches, we shall mention those that present a robust itemset support counting procedure, *i.e.,* without throwing out missing data. They are based on pioneering works of [11,16] and those that proceed by acquiring knowledge under incompleteness [19,20]. The main difference between approaches presenting a completion process stands in the way of tackling the conflict problem, *i.e.,* when many values are candidates for the completion of a missing data. In addition, the inherent oversized lists of association rules that can be drawn is a key factor in hampering the efficiency of heuristics used to address the conflict problem.

In this paper, we propose a new approach, called $GBAR_{MVC}$, aiming to complete missing values based on generic basis of association rules. In fact, beyond interesting compactness rate, the use of such generic association rules proved to be fruitful towards efficiently tackling the conflict problem. In addition, a new metric called "*Robustness*" is introduced and aims to select the robust rule for the completion of a missing value whenever a conflict appears. Conducted experiments on benchmark datasets show a high percentage of correct completion accuracy.

The remainder of the paper is organized as follows. Section 2 sketches a thorough study of the related work to the completion of missing values using association rules. In Section 3, we introduce the $GBAR_{MVC}$ approach for completing missing values based on generic basis of association rules. Experimental results showing the soundness of our approach are presented in section 4. Finally, we conclude and outline avenues of future work.

2 Basic Definitions and Related Work

In this section, we present the general framework for the derivation of association rules and the related work dealing with the completion of missing values using the association rule technique.

2.1 Association Rules

Complete - Incomplete context: A table $\mathcal{D}$ is a non-empty finite set of tuples (or transactions), where each tuple T is characterized by a non-empty finite set of attributes, denoted by $\mathcal{I}$. Each attribute X_i is associated with a domain, denoted $dom(X_i)$, which defines the set of possible values for X_i. It may happen that some attribute values for a tuple are missing. A context with missing values is called *incomplete context*, otherwise, it is said to be *complete*. In the sequel, we denote a missing value by "?".

Extraction context: An extraction context is a triplet $\mathcal{K} = (\mathcal{O}, \mathcal{I}, \mathcal{R})$, where $\mathcal{O}$ represents a finite set of transactions, $\mathcal{I}$ is a finite set of items and $\mathcal{R}$ is a binary (incidence) relation (*i.e.*, $\mathcal{R} \subseteq \mathcal{O} \times \mathcal{I}$). Each couple $(o, i) \in \mathcal{R}$ expresses that the transaction $o \in \mathcal{O}$ contains the item $i \in \mathcal{I}$.

Example 1. Let us consider the complete context depicted by Figure 1 (Left). This context is defined by 4 attributes X_1, X_2, X_3 and X_4, such that $dom(X_1) = \{A, B\}$, $dom(X_2) = \{C, D\}$, $dom(X_3) = \{E, F, G\}$ and $dom(X_4) = \{H, I\}$. The associated extraction context is depicted in Figure 1 (Center), where each couple (attribute, value) is mapped to an item. Figure 1 (Right) represents the extraction context in which missing values were randomly introduced. It is important to mention that each missing value indicates the presence of one item among the missing ones.

The formalization of the association rule extraction problem was introduced by Agrawal *et al.* [1]. Association rule derivation is achieved from a set $\mathcal{FI}_{\mathcal{K}}$ of frequent itemsets [2].

Frequent itemset: The support of an itemset I is the percentage of transactions containing I. The support of I, denoted $supp(I)$, is defined as $supp(I) = |\{o \in \mathcal{O} | I \subseteq o\}|$. I is said to be frequent if $supp(I)$ is greater than or equal to a user-specified minimum support, denoted *minsup*.

Association rule: An association rule R is a relation between itemsets of the form $R: X \Rightarrow (Y\text{-}X)$, in which X and Y are frequent itemsets, and $X \subset Y$. Itemsets X and $(Y\text{-}X)$ are called, respectively, *premise* and *conclusion* of the rule R. Valid association rules are those whose confidence measure, Conf$(R)=\frac{supp(Y)}{supp(X)}$, is greater than or equal to a minimal threshold of confidence denoted *minconf*. If Conf$(R)=1$, then R is called *exact association rule*, otherwise it is called *approximative association rule* [13]. Even though support and confidence metrics are commonly used to assess association rule validity, the Lift metric [6] is becoming of common use. In fact, this statistical metric, presenting a finer assessment of the correlation between the *premise* and the *conclusion* parts, is defined as follows: Lift(R)$=\frac{supp(Y)}{supp(X) \times supp(Y-X)}$. Nevertheless, in practice, the number of valid association rules is very high. To palliate this problem, several solutions towards a lossless reduction were proposed. They mainly consist in extracting an informative reduced subset of association rules, commonly called *generic basis*.

2.2 Related Work

The intuition behind the association rules based approaches for completing missing values, is that association rules describe a dependency among data including missing ones. Hence, it should be possible to guess these values by exploiting discovered rules [22]. Interestingly enough, all these approaches can be split into two pools. With respect to Table 1, the first pool approaches begin by discarding missing data. Then, they try to complete missing ones where association rules discovered from only complete data, are of use. However, such approaches may lead to biased results, since such rules were discovered from a misleading data,

	X_1	X_2	X_3	X_4
1	A	C	E	H
2	B	C	E	I
3	A	C	E	H
4	A	D	F	I
5	B	C	F	I
6	B	C	F	H
7	A	D	G	I
8	B	D	G	I

	A	B	C	D	E	F	G	H	I
1	×		×		×			×	
2		×	×		×				×
3	×		×		×			×	
4	×			×		×			×
5		×	×			×			×
6		×	×			×		×	
7	×			×			×		×
8		×		×			×		×

	A	B	C	D	E	F	G	H	I
1	×		×		×			×	
2		×	×		×				×
3	×		×		?	?	?	×	
4	×			×		×		?	?
5		×	×			×			×
6	?	?	×			×		×	
7	×			×			×		×
8		×	?	?			×		×

Fig. 1. Left: Extraction complete context $\mathcal{K}$. **Center**: The associated complete transactional mapping. **Right**: Extraction incomplete context.

which considerably affects the efficiency of the completion process [20]. Starting from the fact that *"making up missing data is better than throwing out it away"*, approaches of a second pool were proposed. Such approaches focus on mining knowledge under incompleteness. Unfortunately, these approaches suffer from the handling prohibitive number of rules generated from frequent itemsets. As a result, conflict between rules will be difficult to manage and leads to an inefficient completion process. To palliate such a drawback, we propose a new approach based on the use of generic basis of association rules, that aims to complete missing values and reduce conflict during the completion step. In addition, our proposed approach falls within the second pool since it does not discard missing data.

Table 1. Characteristics of the surveyed approaches dealing with missing values completion

	Pool 1		Pool 2	
	Approach 1 [4]	Approach 2 [10]	Approach 3 [15]	Approach 4 [22]
Knowledge Discovery under incompleteness	No	No	Yes	Yes
Conflict resolution	-	reducing conclusion part's rule	*Score-VM* [15] *J-Measure* [21]	*Score* [22]
Generation of rules based on	relevant maximal rectangles	frequent itemsets	frequent itemsets	frequent itemsets

3 The $GBAR_{MVC}$ Approach

The limitations of the above surveyed approaches motivate us to propose a new approach mainly based on the use of generic basis of association rules. The main motivation is that such generic rules consist of a reduced subset of association rules, *i.e.*, fulfills the compactness property. Thus, our proposed approach presents theses main features:

1. It does not discard missing data
2. It takes into account the presence of missing values
3. It is based on the use of a generic basis of association rules

In the following, we begin by motivating the benefits of using such rules for the missing values completion.

3.1 Motivations

Let us consider an incomplete transaction depicted by Figure 2 (Left). We suppose that the missing value can be either the item "A" or the item "B". We try to compare the use of the set of association rules, denoted $\mathcal{AR}$, versus the use of the associated generic basis of association rules, denoted $\mathcal{GB}$. The set of $\mathcal{AR}$ is given by Figure 2 (Center), while the $\mathcal{GB}$ set is given by Figure 2 (Right). By putting the focus on rules concluding on items "A" and "B", we remark that all rules belonging to $\mathcal{AR}$ can be used to complete the missing value. However, it should be noticed that R_2, R_3 and R_4 are not interesting for the completion in comparison with R_1. In fact, R_1 can easily be interpreted as follows: "it is sufficient to have the item "D" in the transaction to use R_1 for completing the missing value". However, R_2, R_3 and R_4 present much more satisfiable constraints (materialized by more items in the *premise* parts) to be used for the completion. Moreover, when R_1 can not be used for the completion, it becomes unnecessary to check whether R_2, R_3 and R_4 could be used since they present the same constraint imposed by R_1. The same statement can be observed for the rule R_7, which is not interesting in comparison with both R_5 and R_6. However, if we consider the set of $\mathcal{GB}$, one can see that all rules qualified as not interesting do not appear in $\mathcal{GB}$. This fact is the fundamental characteristic of the generic basis of association rules. Such rules are composed in their *premise* part by a *minimal generator*, which represents less constraints to satisfy when completing a missing value. Hence, a generic basis of association rules presents the following advantages:

- It encompasses a minimal set of rules. These rules are the most interesting for the missing value completion, *i.e.*, non interesting (redundant) rules are discarded, since they do not materialize relevant knowledge.
- It presents less constraints to satisfy when completing a missing value.
- It shows a considerable reduction of conflicts between rules.

As pointed out in [3], defining generic association rules relies on the *Closure* operator and the key notion of *minimal generator* [13]. Thus, before introducing the completion approach, we shall show how these key notions are redefined in the case of an incomplete context.

?	C	D	E	F	G

R_1	D ⇒ A	R_5	E G ⇒ B
R_2	D F ⇒ A	R_6	C G ⇒ B
R_3	D G ⇒ A	R_7	C E G ⇒ B
R_4	D F G ⇒ A		

R_1	D ⇒ A
R_5	E G ⇒ B
R_6	C G ⇒ B

Fig. 2. **Left**: An incomplete transaction. **Center**: Association rules ($\mathcal{AR}$). **Right**: Generic basis of association rules ($\mathcal{GB}$).

3.2 Basic Definitions

Certain Transaction: A transaction T is said to be *Certain*, with respect to an itemset X, denoted $\mathcal{X}-Certain$, if T contains X. The set of *Certain* transactions is defined as follows :

$$\mathcal{X} - Certain = \{T \in \mathcal{D} \,|\, \forall i \in X \; i \text{ is present in } T\}.$$

Probable transaction: A transaction T is said to be *Probable*, with respect to an item i, denoted $i - Probable$, if i is missing in T.
Shadowed transaction: A transaction T is said to be *Shadowed* with respect to an itemset $(X \cup i)$ if T contains X, such that T is $i - Probable$. The set of *Shadowed* transactions relatively to an itemset $(X \cup i)$, denoted $(X, i) - shadowed$ is as follows: $(X, i) - Shadowed = \{T \in \mathcal{D} \,|\, T \in X - Certain \cap i - Probable\}$.

Example 2. Let us consider the incomplete context depicted by Figure 1 (Right). Transaction T_3 is considered as $AC - Certain$, since it contains AC and it is $E - Probable$, since E is missing. Transaction T_3 is then considered as $(AC, E) - Shadowed$.

In what follows, we recall the definition of the *Almost-Closure* operator [5].

Definition 1. ***(Almost-Closure)*** *The Almost-Closure operator of an itemset X, denoted $\mathcal{AC}(X)$, is defined as $\mathcal{AC}(X) = X \cup \{i \,|\, i \in \mathcal{I} \wedge supp(X) - supp(Xi) \leq \delta\}$ where δ is a positive integer representing the number of exceptions.*

This Definition points out that when an item $i \in \mathcal{AC}(X)$, then it is to say that this item is present in all transactions containing X with a bounded number of exceptions less than δ.

Example 3. Let us consider the complete context depicted by Figure 1 (Center). With respect to Definition 1, we have $\mathcal{AC}(AC) = ACEH$ with $\delta = 0$, *i.e.*, E and H exist in all transactions containing AC.

It is noteworthy that the *Almost-Closure* operator overlaps with that of *Closure* operator in a complete context for $\delta = 0$ [5]. The *Almost-Closure* was redefined to compute the δ-free sets[1] from an incomplete context [20]. We use this definition to introduce a *minimal generator* in an incomplete context. Then, we prove that with $\delta = 0$, the *Almost-Closure* does no longer correspond to the *Closure* operator like in a complete context. For this reason, in the remainder, we shall employ the *Pseudo-Closure* term to point out this distinction.

[1] A 0-free-set is also called minimal generator [5].

Definition 2. ***(Pseudo-Closure)*** *The Pseudo-Closure of an itemset X in an incomplete context, denoted $\mathcal{PC}(X)$, is defined as follows:*

$$\mathcal{PC}(X) = X \cup \{i \,|\, i \in \mathcal{I} \;\wedge\; supp(X) - supp(Xi) = |(X,i) - Shadowed|\,\}.$$

The idea of the *Pseudo-Closure* operator is to adopt an optimistic strategy. This involves a consideration of transactions containing X in which i is missing ($(X,i) - Shadowed$). These transactions are considered as transactions containing the item i.

Example 4. Let us consider the incomplete context depicted by Figure 1 (Right). We have $supp(AC) - supp(ACH) = 0$ which is equal to $|(AC,H) - Shadowed|$. Moreover, we have $supp(AC) - supp(ACE) = 1$ which represents the number of the transactions $(AC,E) - Shadowed$. Hence, $\mathcal{PC}(AC) = ACEH$.

Definition 3. ***(Minimal generator in an incomplete context)*** *An itemset g is said to be minimal generator in an incomplete context if it is not included in the Pseudo-Closure of any of its subsets of size $|g| - 1$.*

Proposition 1. *The Pseudo-closure is no more a Closure operator.*

Proof. By fulfilling the extensivity property, the *Closure* operator induces that each *minimal generator* and its associated *Pseudo-closed* itemset have the same support value. However, the *Pseudo-closure* adopts an optimistic strategy as presented in [20]. When computing the *Pseudo-closure* of an itemset X, if an item is missing, then it is considered as present. Thus, the *minimal generator* and its *Pseudo-closed* itemset do not necessarily have the same support value. Consequently, the *Pseudo-closure* in an incomplete context is not a *Closure* operator.■

However, it is important to mention that the Closure operator induces generic basis of exact association rules. For this reason, we use then a generic basis of pseudo-exact association rules since the Pseudo-closure operator is of use. In what follows, we adapt the definition of the generic basis of exact association rules introduced in [3] to an incomplete context. Such rules allow the selection of a generic subset of all association rules. Thus, the minimal set of rules is used for completing missing values, since it reduces conflict between rules during the completion step.

Definition 4. ***(Generic basis of pseudo-exact association rules)*** *Let $\mathcal{FPC}$ be the set of frequent Pseudo-closed itemsets extracted from an incomplete context. For each frequent Pseudo-closed itemset $c \in \mathcal{FPC}$, let $\mathcal{MG}_c$ be the set of its minimal generators. The generic basis of pseudo-exact association rules $\mathcal{G}B$ is defined as:*

$$\mathcal{G}B = \{R : g \Rightarrow (c \text{ - } g) \mid c \in \mathcal{FPC} \text{ and } g \in \mathcal{MG}_c \text{ and } g \neq c^{(2)}\}.$$

For the completion of the missing values, we use generic rules of the form *premise* $\Rightarrow (X_n, v_n)$, where *premise* is a conjunction of elements of the form (X_j, v_j), $n \neq j$ where (X_j, v_j) is considered as an item.

[2] The condition $g \neq c$ ensures discarding rules of the form $g \Rightarrow \emptyset$.

3.3 The Missing Values Completion $GBAR_{MVC}$

In the remainder, we present a missing value completion approach called $GBAR_{MVC}$[3]. This approach is based on the one hand, extracting the generic basis of pseudo-exact association rules from an incomplete context. On the other hand, we provide a new metric called *Robustness* that aims to select the robust rule for the completion of a missing value whenever a conflict appears. This new metric evaluates the degree of correlation between the premise and the conclusion of a rule materialized through the *Lift* measure [6] and it introduces the degree of assessment of the incomplete transaction. This assessment is materialized through the *Matching* measure. Below, we recall the notion of *consistently interpreting* a transaction by a rule [22] and we provide the definitions of the *Matching* and *Robustness* metrics.

Consistently interpreting [22]: A rule R : $premise \Rightarrow (X_n, v_n)$ is said to be *consistently interpreting* a transaction T presenting a missing value in the attribute X_n, if there is no element (X_j, v_j) in the premise of R that differs from the existing value of X_j in T.

Definition 5. *The* ***Matching*** *measure of a rule* R : $premise \Rightarrow (X_n, v_n)$ *with an incomplete transaction* T *is defined as follows :*

$$Matching(R,t) = \begin{cases} 0 & \text{if } R \text{ is not consistently interpreting } T \\ \frac{\sum matched(X_j, v_j)}{number\ of\ attributes} & \text{otherwise.} \end{cases}$$

where

$$matched(X_j, v_j) = \begin{cases} 0 \text{ if } X_j \text{ presents a missing value in } T \\ 1 \text{ otherwise.} \end{cases}$$

Example 5. Let us consider transaction T_6: $(X_1, ?)(X_2, C)(X_3, F)(X_4, H)$. Rule R_1 : $(X_2, D)(X_3, F) \Rightarrow (X_1, A)$ does not consistently interpret T_6, since the value of the attribute X_2 for T_6 is C, which is different from the value D related to attribute X_2 in the rule R_1. Thus, $Matching(R_1, T_6) = 0$. However, if we consider the example of R_2 : $(X_2, C)(X_3, F) \Rightarrow (X_1, B)$, we can affirm that $Matching(R_2, T_6) = \frac{1}{2}$ since (X_2, C) and (X_3, F) are present in T_6.

The main idea of our proposed approach is to select a rule that maximizes both the *Lift* and the *Matching* values. The *Lift* measure of a rule A ⇒ B is interesting for the completion issue since it describes the strength of the correlation between A and B, *i.e.*, the presence of the item A indicates an increase of the item B. The purpose of the *Matching* measure is to select the rule that corresponds best to the incomplete transaction. For example, if the hair color of a person is missing and we are faced by a conflict between these two rules: Bleu eyes ⇒ Blond hair and redheaded person ∧ clear skin ⇒ Red hair. Then, we tend to use the second rule since it presents a maximum matching. This is performed through the *Robustness* metric defined as follows:

[3] The acronym $GBAR_{MVC}$ stands for Generic Basis of Association Rules based approach for Missing Values Completion.

Definition 6. *The Robustness of an associative rule R for completing a missing transaction T is defined as follows:*

$$Robustness(R, T) = Matching(R, T) \times Lift(R).$$

In the remainder, we present the $GBAR_{MVC}$ algorithm, whose pseudo-code is given by Algorithm 1. The main steps of $GBAR_{MVC}$ algorithm are sketched by the following sequence :

- For each missing attribute X_n of an incomplete transaction T, select rules concluding on X_n and consistently interpreting T. We denote such rule set by $R_{probables}(T, X_n)$ (lines 3-7).
- If the set $R_{probables}(T, X_n)$ is empty, then there are no rules permitting the completion of X_n (lines 8-9).
- If all rules in $R_{probables}(T, X_n)$ conclude on the same value v, then v is used to complete the missing attribute value (lines 11-12).
- Otherwise, *i.e.*, $R_{probables}(T, X_n)$ lead to a conflict. Hence, we compute the *Robustness* value for all rules belonging to $R_{probables}(T, X_n)$ (lines 14-18).
- The rule presenting the highest *Robustness* value is used to complete the missing value on X_n (line 19).

4 Experimental Results

It was worth the effort to experience in practice the potential benefits of the proposed approach. Thus, we have implemented both $GBAR_{MVC}$ and AR_{MVC} [22] approaches in the C++ language using gcc version 3.3.1. Experiments were conducted on a Pentium IV PC with a 2.4 GHz and 512 MB of main memory, running Red Hat Linux. The set of *minimal generators* and their associated *Pseudo-closed* itemsets were extracted thanks to *MVminer* kindly provided by F. Rioult. For these experiments, we consider a complete database to act as a reference database, and we randomly introduce missing values per attribute with the following different rates : 5%, 10%, 15% and 20%. Benchmark datasets used for these experiments are from the UCI Machine Learning Repository[4]. Characteristics of these datasets are depicted by Table 2. During these experiments, we compared statistics yielded by $GBAR_{MVC}$ vs those of AR_{MVC}, by stressing on the following metrics :

- The percentage of missing values that an approach permits to complete.
- The *accuracy* : the percentage of correctly completed missing values.

Table 3 sketches the variation of the completion percentage and the *Accuracy* metric vs the percentage of the missing values variation of $GBAR_{MVC}$. From the reported statistics, we remark that the variation of incrusted missing values does not really affect the percentage of the completion. However, the higher the percentage of the missing values is, the lower the obtained *accuracy*. This decrease in

[4] http://www.ics.uci.edu/~mlearn/MLRepository.html

```
Algorithm : GBAR_MVC
Data: - K_MV : Incomplete context
      - GB : Generic basis of pseudo-exact association rules
Results: K_MV completed
Begin
    Foreach incomplete transaction T in K_MV do
        Foreach attribute X_n in T with a missing value do
            Foreach rule R in GB such that X_n appears in the conclusion
            do
                If R consistently interpreting T then
                    R_probables(T, X_n) = R_probables(T, X_n) ∪ R;
            If |R_probables(T, X_n)| = 0 then
                V_completion = ∅;
            Else
                If R_probables(T, X_n) concludes on the same value v then
                    V_completion = v;
                Else
                    max=0;
                    Foreach rule r in R_probables(T, X_n) do
                        r.Robustness=r.Matching× r.Lift;
                        If r.Robustness>max then
                            V_completion=r.conclusion;
            T.X_n = V_completion;
    return (K_MV completed);
End
```

Algorithm 1. $GBAR_{MVC}$ algorithm

of the percentage of the correctly completed missing values seems to be legitimate and quite expectable. This result can be explained by the following: the higher the incrusted number of missing values is, the worse the extracted rule quality. This fact considerably affects the *Accuracy* metric. Table 4 sketches the variation of the completion percentage and the *Accuracy* metric vs the variation of the *minsup* value. From Table 4, we can remak, as far as the *minsup* value increases, the percentage of the completion diminishes. On the contrary, in most cases by increasing the *minsup* value the accuracy value increases. In fact, rules with a higher *minsup* permit an accurate completion since they describe a more frequent expectation of the missing values according to the observed data. Table 5 sketches the statistics for the completion percentage and those of the *Accuracy* values obtained by $GBAR_{MVC}$ *vs* those pointed out by AR_{MVC} for a *minsup* value equal to 10%. For both approaches, as far as we lower the percentage of missing values, the number of rules considered during the completion step increases. However, those used by $GBAR_{MVC}$ is by far less than the rules used by AR_{MVC}. A careful scrutinize of these statistics permits to shed light on the following:

Table 2. Dataset characteristics

Dataset	Number of transactions	Number of items	Number of attributes
Mushroom	8124	128	23
Zoo	101	56	28
Tic-tac-toe	958	58	29
House-votes	435	36	18
Monks2	432	38	19

Table 3. Variation of the *percentage of completion* and the *Accuracy* metric of $GBAR_{MVC}$ *vs* the percentage of missing values variation for *minsup* value equal to 10%

Dataset	# missing values(%)	# of rules	Percentage of completion (%)	Accuracy (%)
Mushroom	5	28293	74	99
	10	27988	76	99
	15	27988	78	97
	20	24410	80	97
Zoo	5	824650	100	97
	10	756741	98	89
	15	626390	100	88
	20	547075	99	88
Tic-tac-toe	5	315094	100	91
	10	296222	100	89
	15	279915	100	87
	20	266022	100	60
House-votes	5	125909	91	95
	10	102310	93	90
	15	94246	92	87
	20	81162	92	82
Monks2	5	28325	100	83
	10	25402	100	65
	15	21790	100	71
	20	19741	100	63

- **Mushroom - House-Votes:** For these datasets, the *percentage of completion* of AR_{MVC} is better than $GBAR_{MVC}$. This result is not explained by the reduced number of rules presented by $GBAR_{MVC}$. This is can be justified by the *Score* metric used by AR_{MVC}. This metric allows the use of rules on which all items in the *premise* part are missing. Such rules are not used by $GBAR_{MVC}$. We considered them as non reliable for the completion.
- **Zoo - Tic-tac-toe - Monks2:** In the contrary of the previous datasets, we remark that $GBAR_{MVC}$ has permitted a high *percentage of completion*

Table 4. Variation of the *percentage of completion* and the *Accuracy* metric of $GBAR_{MVC}$ *vs* the variation of the *minsup* value for a number of missing values equal to 20%

Dataset	minsup (%)	Percentage of completion (%)	Accuracy (%)
Mushroom	10	80	97
	15	75	98
	20	71	98
	25	55	73
	30	53	57
Zoo	10	99	88
	15	96	78
	20	92	88
	25	89	92
	30	88	93
Tic-tac-toe	10	100	60
	15	100	70
	20	89	85
	25	86	96
	30	65	100
House-votes	10	92	82
	15	45	90
	20	76	79
	25	70	74
	30	33	50
Monks2	10	100	63
	15	100	63
	20	100	75
	25	100	83
	30	83	92

as well as AR_{MVC}. This statement is observed even with the reduced number of rules of $GBAR_{MVC}$ in comparison with rules of AR_{MVC}. This fact represents the advantage of $GBAR_{MVC}$, *i.e.*, rules of $GBAR_{MVC}$ are not redundant.

- For all datasets, $GBAR_{MVC}$ has permitted a better *Accuracy*. This better *Accuracy* result can be justified as follows:
  1. Rules produced by $GBAR_{MVC}$ are more reliable in presence of missing values. This is materialized thorough the *Pseudo-Closure* definition.
  2. It was shown in [22] that the *Accuracy* depends on the number of the extracted rules. However, AR_{MVC} generates a large number of rules, which affects considerably the completion *Accuracy*.

Finally, according to these experimental results, it should be mentioned that $GBAR_{MVC}$ presents a more accurate completion process. Moreover, this completion process is less affected by the rate of the introduced missing values than

Table 5. Evaluation of the *percentage of completion* and the *Accuracy* metric of AR_{MVC} *vs.* $GBAR_{MVC}$ for a *minsup* value equal to 10%

		Mushroom			
	Number of missing values (%)	5	10	15	20
AR_{MVC}	Percentage of completion (%)	50	92	99	80
	Accuracy (%)	42	58	64	66
	Number of rules	79461	77161	76830	68168
$GBAR_{MVC}$	Percentage of completion (%)	74	76	78	80
	Accuracy (%)	99	99	97	97
	Number of rules	28893	27988	27988	24410
		Zoo			
	Number of missing values(%)	5	10	15	20
AR_{MVC}	Percentage of completion(%)	100	100	100	100
	Accuracy (%)	55	57	55	66
	Number of rules	3898169	3842627	3761081	3293571
$GBAR_{MVC}$	Percentage of completion (%)	100	098	100	099
	Accuracy (%)	97	89	88	88
	Number of rules	824650	756741	626390	547075
		Tic-tac-toe			
	Number of missing values(%)	5	10	15	20
AR_{MVC}	Percentage of completion (%)	100	100	100	100
	Accuracy (%)	86	73	76	71
	Number of rules	632826	592115	554530	528343
$GBAR_{MVC}$	Percentage of completion (%)	100	100	100	100
	Accuracy (%)	91	89	87	60
	Number of rules	315094	296222	279915	266022
		House-votes			
	Number of missing values (%)	5	10	15	20
AR_{MVC}	Percentage of completion (%)	95	96	97	98
	Accuracy (%)	87	77	73	71
	Number of rules	387342	369180	335639	309617
$GBAR_{MVC}$	Percentage of completion (%)	91	93	92	92
	Accuracy (%)	95	90	87	82
	Number of rules	125909	102310	94246	81162
		Monks			
	Number of missing values (%)	5	10	15	20
AR_{MVC}	Percentage of completion (%)	100	100	100	100
	Accuracy (%)	76	67	65	60
	Number of rules	52660	45249	33815	34490
$GBAR_{MVC}$	Percentage of completion (%)	100	100	100	100
	Accuracy (%)	83	65	71	63
	Number of rules	28325	25402	21790	19741

AR_{MVC}. This efficiency can be explained by the strategy adopted during the completion step. In fact, based on generic bases of association rules, it permitted a considerable reduction of conflicts, leading to high rate of correct completion accuracy.

5 Conclusion and Future Work

In this paper, we proposed a new approach called $GBAR_{MVC}$, permitting the completion of the missing values. The main particularity of our proposed approach is that is based on the generic basis of association rules and a new metric called *Robustness*. Carried out experiments on benchmark datasets confirmed that $GBAR_{MVC}$ approach turns out to be very beneficial for resolving the challenge of completing missing values, specially at the pre-processing KDD step. In fact, $GBAR_{MVC}$ approach offers a high rate of correct completion accuracy and outperforms the approach proposed in [22]. The preliminary obtained results offer exciting additional alternatives avenues of future work. In fact, we are interested first, in tackling the "silence problem", *i.e.*, improving the percentage of completion. Second, it will be interesting to complete missing values by using the concept of disjunction-free-sets [8]. These sets allow the extraction of generalized rules with negative terms which could be interesting for the completion of missing values. Finally, our future work includes a further evaluation of the *Robustness* metric.

References

1. Agrawal, R., Imielinski, T., Swami, A.: Mining association rules between sets of items in large databases. In: Proceedings of the ACM-SIGMOD International Conference on Management of Data, Washington D. C., USA, pp. 207–216 (May 1993)
2. Agrawal, R., Srikant, R.: Fast algorithms for mining association rules. In: Bocca, J.B., Jarke, M., Zaniolo, C. (eds.) Proceedings of the 20th International Conference on Very Large Databases, Santiago, Chile, pp. 478–499 (1994)
3. Bastide, Y., Pasquier, N., Taouil, R., Lakhal, L., Stumme, G.: Mining Minimal Non-redundant Association Rules Using Frequent Closed Itemsets. In: Palamidessi, C., Moniz Pereira, L., Lloyd, J.W., Dahl, V., Furbach, U., Kerber, M., Lau, K.-K., Sagiv, Y., Stuckey, P.J. (eds.) CL 2000. LNCS (LNAI), vol. 1861, Springer, Heidelberg (2000)
4. Ben Yahia, S., Arour, K., Jaoua, A.: Completing missing values in databases using discovered association rules. In: Proceedings of the International Conference on Artificial and Computational Intelligence for Decision, Control and Automation in Engineering and Industrial Applications (ACIDCA 2000), Monastir, Tunisia, March 22-24, pp. 138–143 (2000)
5. Boulicaut, J.-F., Bykowski, A., Rigotti, C.: Approximation of frequency queries by means of free-sets. In: Zighed, A.D.A., Komorowski, J., Żytkow, J.M. (eds.) PKDD 2000. LNCS (LNAI), vol. 1910, pp. 75–85. Springer, Heidelberg (2000)

6. Brin, S., Motwani, R., Silverstein, C.: Beyond market baskets: Generalizing association rules to correlations. In: Proceedings of the International Conference on Management of Data (ACM SIGMOD), Tucson, Arizona, USA, May 13–15, pp. 265–276. ACM Press, New York (1997)
7. Buntine, W.L.: Operations for learning with graphical models. Journal of Artificial Intelligence 2, 159–225 (1994)
8. Bykowski, A., Rigotti, C.: A condensed representation to find frequent patterns. In: Proceedings of the ACM SIGMOD-SIGACT-SIGART symposium of Principles of Database Systems, Santa Barbara, USA, pp. 267–273.
9. Dempster, A.P., Laird, N.M., Rubin, D.B.: Maximum likelihood from incomplete data via the EM algorithm. Journal of the Royal Statistical Society 39(1), 1–38 (1977)
10. Jami, S., Jen, T., Laurent, D., Loizou, G., Sy, O.: Extraction de règles d'association pour la prédiction de valeurs manquantes. ARIMA journal (3), 103–124 (2005)
11. Kryszkiewicz, M.: Probabilistic approach to association rules in incomplete databases. In: Lu, H., Zhou, A. (eds.) WAIM 2000. LNCS, vol. 1846, pp. 133–138. Springer, Heidelberg (2000)
12. Little, R.J.A., Rubin, D.B.: Statistical analysis with missing data. Wiley, New York (2002)
13. Pasquier, N., Bastide, Y., Touil, R., Lakhal, L.: Discovering frequent closed itemsets. In: Beeri, C., Bruneman, P. (eds.) ICDT 1999. LNCS, vol. 1540, pp. 398–416. Springer, Heidelberg (1998)
14. Giudici, P., Castelo, R.: Improving Markov Chain Monte Carlo model search for data mining. Machine Learning 50(1–2), 127–158 (2003)
15. Ragel, A.: Exploration des bases incomplètes: Application à l'aide au prétraitement des valeurs manquantes. PhD Thesis, Université de Caen, Basse Normandie (December 1999)
16. Ragel, A., Crémilleux, B.: Treatment of missing values for association rules. In: Wu, X., Kotagiri, R., Korb, K.B. (eds.) PAKDD 1998. LNCS, vol. 1394, pp. 258–270. Springer, Heidelberg (1998)
17. Ragel, A., Crémilleux, B.: MVC - a preprocessing method to deal with missing values. Knowledge-Based System Journal 12(5–6), 285–291 (1999)
18. Ramoni, M., Sebastiani, P.: Bayesian inference with missing data using bound and collapse. Journal of Computational and Graphical Statistics 9(4), 779–800 (2000)
19. Rioult, F.: Knowledge discovery in databases containing missing values or a very large number of attributes. PhD Thesis, Université de Caen, Basse Normandie (November 2005)
20. Rioult, F., Crémilleux, B.: Condensed representations in presence of missing values. In: Proceedings of the International symposium on Intelligent Data Analysis, Berlin, Germany, pp. 578–588 (2003)
21. Smyth, P., Goodman, R.M.: An information theoretic approach to rule induction from databases. IEEE Trans. On Knowledge And Data Engineering 4, 301–316 (1992)
22. Wu, C., Wun, C., Chou, H.: Using association rules for completing missing data. In: Proceedings of 4th International Conference on Hybrid Intelligent Systems (HIS 2004), Kitakyushu, Japan, 5-8 December 2004, pp. 236–241. IEEE Computer Society Press, Los Alamitos (2004)

Efficient Generic Association Rules Based Classifier Approach

Ines Bouzouita and Samir Elloumi

Faculty of Sciences of Tunis,
Computer Science Department, 1060 Tunis, Tunisia
{ines.bouzouita,samir.elloumi}@fst.rnu.tn

Abstract. *Associative classification* is a promising new approach that mainly uses association rule mining in classification. However, most associative classification approaches suffer from the huge number of the generated classification rules which takes efforts to select the best ones in order to construct the classifier. In this paper, a new associative classification approach called GARC is proposed. GARC takes advantage of generic basis of association rules in order to reduce the number of association rules without jeopardizing the classification accuracy. Furthermore, since rule ranking plays an important role in the classification task, GARC proposes two different strategies. The latter are based on some interestingness measures that arise from data mining literature. They are used in order to select the best rules during classification of new instances. A detailed description of this method is presented, as well as the experimentation study on 12 benchmark data sets proving that GARC is highly competitive in terms of accuracy in comparison with popular classification approaches.

Keywords: Associative Classification, Generic Basis, Classification Rules, Generic association rules, Classifier, interestingness measures.

1 Introduction

In the last decade, a new approach called *associative classification* (AC) was proposed to integrate association rule mining and classification in order to handle large databases. Given a training data set, the task of an associative classification algorithm is to discover the classification rules which satisfy the user specified constraints denoted respectively by minimum support (*minsup*) and minimum confidence (*minconf*) thresholds. The classifier is built by choosing a subset of the generated classification rules that could be of use to classify new objects or instances. Many studies have shown that AC often achieves better accuracy than do traditional classification techniques [1,2]. In fact, it could discover interesting rules omitted by well known approaches such as C4.5 [3]. However, the main drawback of this approach is that the number of generated associative classification rules could be large and takes efforts to retrieve, prune, sort and select high quality rules among them. To overcome this problem, we propose a

S. Ben Yahia et al. (Eds.): CLA 2006, LNAI 4923, pp. 170–185, 2008.

new approach called GARC which uses generic bases of association rules [4,5]. The main originality of GARC is that it extracts the generic classification rules directly from a generic basis of association rules, in order to retain a small set of rules with higher quality and lower redundancy in comparison with current AC approaches. Moreover, a new score is defined by the GARC approach to find an effective rule selection during the class label prediction of a new instance, in the sake of reducing the error rate. This tackled issue is quite challenging, since the goal is to use generic rules while maintaining a high classifier accuracy.

The remainder of the paper is organized as follows. Section 2 briefly reports basic concepts of associative classification, scrutinizes related pioneering works by splitting them into two groups. Generic bases of association rules are surveyed in section 3. Section 4 presents our proposed approach, where details about classification rules discovery, building classifier and two different strategies to classify new instances are discussed. Experimental results and comparisons are given in section 5. Finally, section 6 concludes this paper and points out future perspectives.

2 Associative Classification

An association rule is a relation between itemsets having the following form: $R : X \Rightarrow Y - X$, where X and Y are frequent itemsets for a minimal support *minsup*, and $X \subset Y$. Itemsets X and $(Y - X)$ are called, respectively, *premise* and *conclusion* of the rule R. An association rule is valid whenever its strength metric, confidence$(R)=\frac{support(Y)}{support(X)}$, is greater than or equal to the minimal threshold of confidence *minconf*.

An associative classification rule (ACR) is a special case of an association rule. In fact, an ACR conclusion part is reduced to a single item referring to a class label. For example, in an ACR such as $X \Rightarrow c_i$, c_i must be a class label.

2.1 Basic Notions

Let us define the classification problem in an association rule task. Let D be a training set with n attributes (columns) $A_1, \ldots, A_n$ and $|D|$ rows. Let C be the list of class labels.

Definition 1. *An object or instance in D can be described as a combination of attribute names and values a_i and an attribute class denoted by c_i [6].*

Definition 2. *An item is described by an attribute name and a value a_i [6].*

Definition 3. *An itemset is described by a set of items contained in an object.*

Definition 4. *An associative classification rule is of the form: $A_1, A_2, \ldots, A_n \Rightarrow c_i$ where the premise of the rule is an itemset and the conclusion is a class attribute.*

A classifier is a set of rules of the form $A_1, A_2, ..., A_n \Rightarrow c_i$ where A_i is an item and c_i is a class label. The classifier should be able to predict, as accurately as possible, the class of an unseen object belonging to the test data set. In fact, it should maximise the equality between the predicted class and the hidden actual class. Hence, the AC achieves higher classification accuracy than do traditional classification approaches [1,2].

2.2 Related Work

Associative classification approaches can be categorized into two groups according to the way of the classification rules extraction:

1. **Two-stages algorithms:**
   In the first stage, a set of associative classification rules is produced. Then, this latter is pruned and placed into a classifier. Examples of such approaches are CBA [6], CMAR [7], ARC-AC and ARC-BC [8,9].

   CBA [6] was one of the first algorithms to use association rule approach for the classification task. This approach, firstly, generates all the association rules with certain support and confidence thresholds as candidate rules by implementing the Apriori algorithm [10]. Then, it selects a small set from them by evaluating all the generated rules against the training data set. When predicting the class label for an example, the highest confidence rule, whose the body is satisfied by the example, is chosen for prediction.

   CMAR [7] generates rules in a similar way as do CBA with the exception that CMAR introduces a CR-tree structure to handle the set of generated rules and uses a set of them to make a prediction using a weighted $\chi 2$ metric [7]. The latter metric evaluates the correlation between the rules.

   ARC-AC and ARC-BC have been introduced in [8,9] in the aim of text categorization. They generate rules similar to the Apriori algorithm and rank them in the same way as do the CBA rules ranking method. Both ARC-AC and ARC-BC compute the average confidence of each set of rules grouped by class label in the conclusion part and select the class label of the group with the highest confidence average.

2. **Integrated algorithms:**
   The classifier is produced in a single processing step, *e.g.*, CPAR [2] and Harmony [11]. The CPAR [2] algorithm adopts FOIL [12] strategy in generating rules from data sets. It seeks for the best rule itemset that brings the highest gain value among the available ones in the data set. Once the itemset is identified, the examples satisfying it will be deleted until all the examples of the data set are covered. The searching process for the best rule itemset is a time consuming process, since the gain for every possible item needs to be calculated in order to determine the best item gain. During rule generation step, CPAR derives not only the best itemset but all close similar ones. It has been claimed that CPAR improves the classification accuracy whenever compared to popular associative methods like CBA and CMAR [2].

Another AC approach called Harmony was proposed in [11]. Harmony uses an instance-centric rule generation to discover the highest confidence discovering rules. Then, Harmony groups the set of rules into k groups according to their rules class labels, where k is the total number of distinct class labels in the training set. Within the same group of rules, Harmony sorts the rules in the same order as do CBA. To classify a new test instance, Harmony computes a score for each group of rules and assigns the class label with the highest score or a set of class labels if the underlying classification is a multi-class problem. It has been claimed that Harmony improves the efficiency of the rule generation process and the classification accuracy if compared to CPAR [2].

It is noteworthy that all the approaches, except CPAR, sort the classification rules using support and confidence measures in order to build the classifier. However, the support confidence measures could be misleading while classifying new objects. Moreover, all the approaches manipulate the totality number of the classification rules. Thus, regardless of the approach used to generate the classifier, there are an overwhelming number of rules manipulated during the learning stage, which is the main problem with AC approaches. In order to overcome this drawback, our proposed approach tries to gouge this fact by the use of generic bases of association rules in the classification framework. In the following, we begin by recalling some key notions about the Formal Concept Analysis (FCA), that are for need for the derivation of generic bases of association rules.

3 Generic Bases of Association Rules

The problem of the relevance and usefulness of extracted association rules is of primary importance. Indeed, in most real life databases, thousands and even millions of highly confident rules are generated among which many are redundant. In the following, we are interested in the lossless information reduction of association rules, which is based on the extraction of a generic subset of all association rules, called *generic basis* from which the remaining (redundant) association rules may be derived. In the following, we will present the generic basis of Bastide *et al.* [13,4] and $\mathcal{IGB}$ [5] after a brief description of FCA mathematical background necessary for the derivation of generic bases of association rules.

3.1 Mathematical Background

Interested reader for key results from the Galois lattice-based paradigm in FCA is referred to [14].

Formal context. A formal context is a triplet $\mathcal{K} = (\mathcal{O}, \mathcal{I}, \mathcal{R})$, where $\mathcal{O}$ represents a finite set of transactions, $\mathcal{I}$ is a finite set of items and $\mathcal{R}$ is a binary (incidence) relation (*i.e.*, $\mathcal{R} \subseteq \mathcal{O} \times \mathcal{I}$). Each couple $(o, i) \in \mathcal{R}$ expresses that the transaction $o \in \mathcal{O}$ contains the item $i \in \mathcal{I}$.

We define two functions summarizing links between subsets of objects and subsets of attributes induced by $\mathcal{R}$, that maps sets of objects to sets of attributes and *vice versa.* Thus, for a set $O \in \mathcal{O}$, we define $\phi(O) = \{i \in I | \forall o, o \in \mathcal{O} \Rightarrow (o, i) \in \mathcal{R}\}$ and for $i \in \mathcal{I}$, $\psi(I) = \{o \in \mathcal{O} | \forall i, i \in \mathcal{I} \Rightarrow (o, i) \in \mathcal{R}\}$.

Both operators $\phi(O)$ and $\psi(I)$, form a *Galois connection* between the sets $\mathcal{P}(\mathcal{I})$ and $\mathcal{P}(\mathcal{O})$ [14]. Consequently, both compound operators ϕ and ψ are closure operators in particular $\omega = \phi \circ \psi$ is a closure operator.

Frequent closed itemset. An itemset $I \subseteq \mathcal{I}$ is said to be *closed* if $\omega(I) = I$[1] [15]. I is said to be *frequent* if its *relative support*, Support$(I) = \frac{|\psi(I)|}{|\mathcal{O}|}$, exceeds a user-defined minimum threshold, denoted *minsup.*

Minimal generator. [4] An itemset $g \subseteq \mathcal{I}$ is said to be *minimal generator* of a closed itemset f, if and only if $\omega(g) = f$ and it does not exist $g_1 \subset g$ such that $\omega(g_1) = f$. The set $\mathcal{G}_f$ of the minimal generators of f is: $\mathcal{G}_f = \{g \subseteq \mathcal{I} \mid \omega(g) = f \wedge \nexists\, g_1 \subset g$ such as $\omega(g_1) = f\}$.

3.2 The Generic Basis for Exact Association Rules ($\mathcal{GBE}$) and the Informative Basis for Approximate Association Rules ($\mathcal{GBA}$)

Since the apparition of the approach based on the extraction of the frequent closed itemsets [15], several generic bases have been introduced among which those of Bastide *et al.* [4].

Exact association rules, of the form R: $X \stackrel{c}{\Rightarrow} Y$, are implications between two itemsets X and XY whose closures are identical, *i.e.*, $\omega(X) = \omega(XY)$. Thus, support(X)=support(XY), *i.e.*, confidence(R)=1. The *generic basis for exact association rules* is defined as follows:

Definition 5. *Let $\mathcal{FCI}_\mathcal{K}$ be the set of frequent closed itemsets extracted from an extraction context $\mathcal{K}$. For each frequent closed itemset $f \in \mathcal{FCI}_\mathcal{K}$, let $\mathcal{G}_f$ be the set of its minimal generators. The generic basis of exact association rules $\mathcal{GBE}$ is given by: $\mathcal{GBE} = \{R\colon g \Rightarrow (f - g) \mid f \in \mathcal{FCI}_\mathcal{K}$ and $g \in \mathcal{G}_f$ and $g \neq f$[2]$\}$.*

Bastide *et al.* also characterized the informative basis for approximate association rules, defined as follows [4]:

Definition 6. *Let $\mathcal{FCI}_\mathcal{K}$ be the set of frequent closed itemsets extracted from an extraction context $\mathcal{K}$. The $\mathcal{GBA}$ basis is defined as follows [4]:*
$\mathcal{GBA} = \{R \mid R\colon g \Rightarrow (f_1 - g) \mid f, f_1 \in \mathcal{FCI}_\mathcal{K}$ and $\omega(g) = f$ and $f \preceq f_1$ and Confidence$(R) \geq minconf\}$.

The pair ($\mathcal{GBE}$, $\mathcal{GBA}$) is informative, sound and lossless [4] and rules belonging to this pair are referred as *informative association rules.* In the following, we will present $\mathcal{IGB}$ basis.

3.3 Informative Generic Basis ($\mathcal{IGB}$)

The $\mathcal{IGB}$ basis has been shown to be informative and more compact than the generic basis of Bastide *et al.* [5]. The $\mathcal{IGB}$ basis is defined as follows:

---

1 The closure operator is indicated by ω.

2 The condition $g \neq f$ ensures discarding non-informative rules of the form $g \Rightarrow \emptyset$.

Definition 7. *Let $\mathcal{FCI}_{\mathcal{K}}$ be the set of frequent closed itemsets and $\mathcal{G}_f$ be the set of minimal generators of all the frequent itemsets included or equal to a closed frequent itemset f.*

$\mathcal{IGB} = \{R\text{: } g_s \Rightarrow (f_1 \text{ - } g_s) \mid f, f_1 \in \mathcal{FCI}_{\mathcal{K}}$ *and* $(f \text{ - } g_s) \neq \emptyset$ *and* $g_s \in \mathcal{G}_f$ *and* $f_1 \preceq f$ *and* $confidence(R) \geq minconf$ *and* $\nexists\, g' \subset g_s$ *such that* $confidence(g' \Rightarrow f_1\text{-}g') \geq minconf\}$.

Generic rules of $\mathcal{IGB}$ present implications between the smallest premises and the largest conclusions among possible ones. This feature is interesting, especially for our approach detailed hereafter.

4 GARC: Generic Associative Rules Based Classifier

In this section, we propose a new AC method GARC[3] that extracts the generic classification rules directly from a generic basis of association rules in order to overcome the drawback of the current AC approaches, *e.g.*, the generation of a large number of associative classification rules. Continuous attributes take values that are real numbers within some range defined by minimum and maximum limits. In such cases, the given range is split into a number of sub-ranges and a unique number is allocated to each sub-range. In the following, we will present and explain in details the GARC approach.

4.1 Rule Generation

In this step, GARC extracts the generic basis of association rules. Once obtained, generic rules are filtered out to retain only rules whose conclusions include a class label. Then, by applying the decomposition axiom[16], we obtain new rules of the form $A_1, A_2, ..., A_n \Rightarrow c_i$. Even though, the obtained rules are redundant, their generation is mandatory to guarantee a maximal cover of the necessary rules.

The $\mathcal{IGB}$ basis is composed of rules with a small premise which is an advantage for the classification framework when the rules imply the same class. For example, let us consider two rules R_1: `A B C D` $\Rightarrow$`cl1` and R_2: `B C` $\Rightarrow$`cl1`. R_1 and R_2 have the same attribute conclusion. R_2 is considered to be more interesting than R_1, since it is needless to satisfy the properties `A D` to choose the class `cl1`. Hence, R_2 implies less constraints and can match more objects from a given population than R_1.

Let us consider a new object O_x: `B C D`. If R_1 is the unique rule in the classifier, we wont be able to classify O_x, because the item A does not permit the matching. However, the rule R_2, which has a smaller premise than R_1, can classify O_x. This example shows the importance of the generic rules and, especially, the use of the $\mathcal{IGB}$ basis to extract the generic classification rules. In fact, such set of rules is smaller than the number of all the classification rules and their use is beneficial for classifying new objects.

[3] The acronym GARC stands for: Generic Association Rules based Classifier.

4.2 Classifier Builder

Once the generic classification rules obtained, a total order on rules is set as follows. Given two rules R_1 and R_2, R_1 is said to precede R_2, denoted $R_1 > R_2$ if the following conditions are fulfilled:

- $confidence(R_1) > confidence(R_2)$ or
- $confidence(R_1) = confidence(R_2)$ and $support(R_1) > support(R_2)$ or
- $confidence(R_1) = confidence(R_2)$ and $support(R_1) = support(R_2)$ and R_1 is generated before R_2.

The data set coverage is similar to that in CBA. In fact, a data object of the training set is removed after it is covered by a selected generic rule.

The major difference with current AC approaches [6,7,8,9,11] is that we use generic ACR directly deduced from generic bases of association rules to learn the classifier as shown by algorithm 1.

```
Data: D: Training data, GR: a set of generic classification rules
Results: C: Classifier
Begin
  GR=sort(GR) in a descending order according to confidence and support values;
  Foreach rule r ∈ GR do
    Foreach object d ∈ D do
      If d matches r.premise then
        remove d from D and mark r if it correctly classifies d;
    If r is marked then
      insert r at the end of C;
  select the major class from D as a default class;
  return Classifier C ;
End
```

Algorithm 1. GARC: *selected generic rules based on database coverage*

4.3 New Instance Classification

After a set of rules is selected for classification, GARC is ready to classify new objects. Some methods such as those described in [6,8,9,11] are based on the support-confidence order to classify a new object. However, the confidence measure selection could be misleading, since it may identify a rule $A \Rightarrow B$ as an interesting one even though, the occurrence of A does not imply the occurrence of B [17]. In fact, the confidence can be deceiving since it is only an estimate of the conditional probability of itemset B given an itemset A and does not measure the actual strength of the implication between A and B. Let us consider the example shown in Table 1 which shows the association between an item A and a class attribute B. A and $\overline{A}$ represent respectively the presence and absence of item A, B represents a class attribute and $\overline{B}$ the complement of B. We consider the associative classification $A \Rightarrow B$. The confidence of this rule is given by *confidence*($A \Rightarrow B$)=$\frac{support(AB)}{support(A)} = \frac{201}{250} = 80.4\%$. Hence, this rule has a high confidence value.

In the following, we will introduce interestingness measures of association rules and give a semantic interpretation for each of them.

Table 1. Example

	B	$\overline{B}$	Total
A	201	49	250
$\overline{A}$	699	51	750
Total	900	100	1000

Lift or Interest. The lift metric [17] computes the correlation between A and B. For the example depicted by Table 1, the lift of the rule $A \Rightarrow B$ is given by: $lift(A \Rightarrow B) = \frac{support(AB)}{support(A) \times support(B)} = \frac{0.201}{0.250 \times 0.900} = 0.893$.

The fact that this quantity is less than 1 indicates negative correlation between A and B.

If the resulting value is greater than 1, then A and B are said to be positively correlated. If the resulting value is equal to 1, then A and B are independent and there is no correlation between them.

Least Confidence (or Surprise). The least confidence (or surprise) [18] metric is computed as follows:

Surprise ($A \Rightarrow B$) = (support (AB) - support ($A\overline{B}$))/ support (B)
logical rule: surprise ($A \Rightarrow B$) = P (A)/ P (B)
A and B independent: surprise ($A \Rightarrow B$) = 2 P (A) - (P (A)/ P (B))
A and B incompatible: surprise ($A \Rightarrow B$) = - P (A)/ P (B)

The surprise metric selects rules, even with small support value, having the premise A always with the conclusion B and nowhere else.

Loevinger. Loevinger metric [18] is computed as follows:

loevinger($A \Rightarrow B$) = (P(B/A)-P(B))/P($\overline{B}$)

Unlike confidence metric, Loevinger metric does not suffer form the problem of producing misleading rules.

Score metric. To avoid the lacuna of using only confidence metric, we define a new lift based score formula as follows:

$$Score = \frac{1}{|\mathtt{Premise}|} \times lift^{\frac{|\mathtt{Premise}|}{numberofitems}}$$

$$= \frac{1}{|\mathtt{Premise}|} \times \left(\frac{support(\mathtt{Rule})}{support(\mathtt{Premise}) \times support(\mathtt{Conclusion})}\right)^{\frac{|\mathtt{Premise}|}{numberofitems}}$$

The introduced score includes the lift metric. In fact, the lift finds interesting relationships between A and B. It computes the correlation between the occurrence of A and B by measuring the actual strength of the implication between them which is interesting for the classification framework. Moreover, the lift is divided by the cardinality of the rule premise part in order to give a preference to rules with small premises.

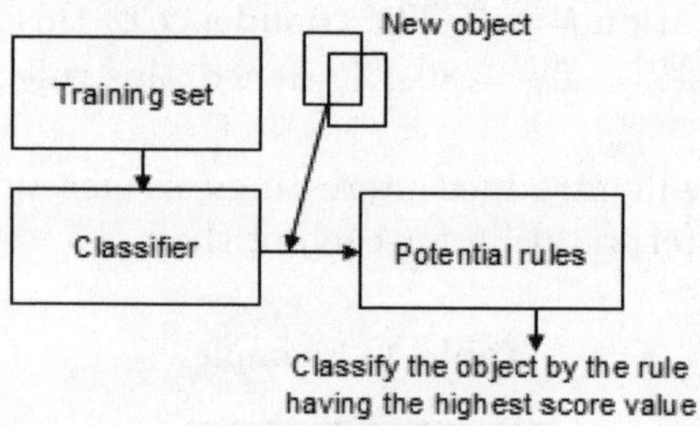

Fig. 1. GARC classifying instances strategy 1

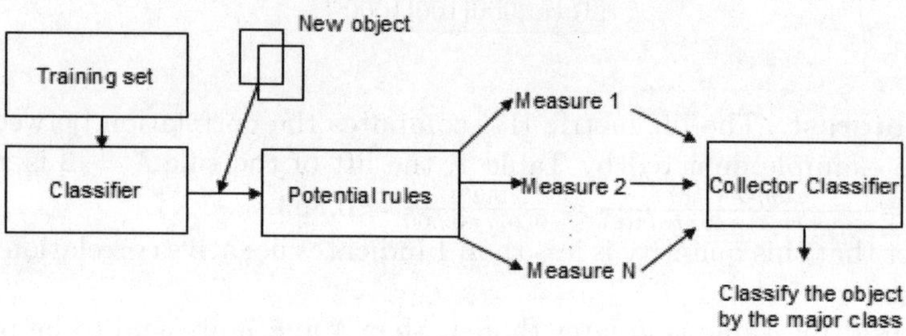

Fig. 2. GARC classifying instances strategy 2

In the following, we have considered two different strategies using the introduced measures in order to choose the best rules for classifying a new object.

1. **Strategy 1.** In the first strategy [19], illustrated by Figure 1, GARC collects the subset of rules matching the new object attributes from the classifier. Trivially, if all the rules matching it have the same class, GARC just assigns that class to the new object. If the rules do not imply the same class attribute, the score firing is computed for each rule. The rule with the highest score value is selected to classify the new object.
2. **Strategy 2.** The intuition behind this second strategy, illustrated by Figure 2, is that we cannot expect that a single measure can perfectly predict the class of an unseen object. That's why, GARC collects the subset of rules matching the new object attributes from the classifier. Then, for each measure presented earlier, GARC looks for the rule with the highest value among the set of the potential rules. The rules obtained from the different measures represent the collector classifier. From this latter set of rules, GARC assigns the major class to the new object.

Example 1. The training data set $\mathcal{D}$ shown by Figure 3 **(a)** is composed of twelve objects. Each object is described by six categorical attributes and belongs to a class. We have set the *minsup* and *minconf* values to 1% and 80%, respectively. We extract the generic basis $\mathcal{IGB}$ according to definition 7. Then, we generate generic ACR by applying the decomposition axiom to obtain rules of the form

	A_1	A_2	A_3	A_4	A_5	A_6	$Class$
$O1$	11	21	31	42	52	62	$Cl1$
$O2$	11	21	31	42	53	61	$Cl1$
$O3$	12	23	32	43	51	61	$Cl1$
$O4$	12	23	32	43	51	62	$Cl1$
$O5$	12	23	32	43	52	61	$Cl2$
$O6$	12	23	32	43	52	62	$Cl2$
$O7$	13	21	31	41	51	62	$Cl1$
$O8$	13	21	31	41	52	61	$Cl2$
$O9$	13	21	31	41	52	62	$Cl2$
$O10$	13	21	31	42	51	61	$Cl1$
$O11$	13	21	31	42	52	62	$Cl2$
$O12$	13	21	31	43	51	61	$Cl1$

(a)

R_1: $(A5 = 51) and (A6 = 62) \Rightarrow Cl1$
R_2: $(A5 = 52) \Rightarrow Cl2$
R_3: $(A1 = 11) \Rightarrow Cl1$
R_4: $(A3 = 32) and (A5 = 51) \Rightarrow Cl1$
R_5: $(A4 = 41) and (A6 = 61) \Rightarrow Cl1$

(b)

Fig. 3. (a): $\mathcal{D}$: Training data **(b)**: GARC Classifier for *minsup*=1% and *minconf*=80%

$A_1, A_2, \ldots, A_6 \Rightarrow c_i$ with $c_i \in \{Cl1, Cl2\}$. Once the generic ACR obtained, they are sorted on a descending order according to the support and confidence values as defined in section 4.2. Then, we apply the cover algorithm where each object of the training data $\mathcal{D}$ has to be satisfied (covered) by one rule before it is no longer considered in the classifier generation process and removed from the training data. The resulting classifier is given by Figure 3 **(b)**.

5 Experimental Study

We have conducted experiments to evaluate the accuracy of our proposed approach GARC, developed in C++, and compared it to the well known classifiers CBA, ID3, C4.5 and Harmony. Experiments were conducted using 12 data sets taken from UCI Machine Learning Repository[(4)]. The chosen data sets were discretized using the LUCS-KDD [(5)] software.

The features of these data sets are summarized in Table 2. All the experiments were performed on a 2.4 GHz Pentium IV PC under Redhat Linux.

The classification accuracy can be used to evaluate the performance of classification methods. It is the percentage of correctly classified examples in the test set and can be measured by splitting the data sets into a training set and a test set.

During experiments, we have used available test sets for data sets Monks1, Monks2 and Monks3 and we applied the 10 cross-validation for the rest of data sets, in which a data set is divided into 10 subsets; each subset is in turn used as testing data while the remaining data is used as the training data set; then the average accuracy across all 10 trials is reported.

---

4 *Available at* http://www.ics.uci.edu/~mlearn/MLRepository.html

5 *Available at* http://www.csc.liv.ac.uk/~frans/KDD/Software/LUCS-KDD-DN/lucs-kdd DN.html

Table 2. Data set description

Data set	# attributes	# transactions	# classes
MONKS1	6	124	2
MONKS2	6	169	2
MONKS3	6	122	2
PIMA	38	768	2
TICTACTOE	29	958	2
ZOO	42	101	7
IRIS	19	150	3
WINE	68	178	3
GLASS	48	214	7
FLARE	39	1389	9
PAGEBLOCKS	46	5473	5
NURSERY	32	12960	5

In order to extract generic association rules, we used the PRINCE algorithm [20] to generate both the pair ($\mathcal{GBE}$, $\mathcal{GBA}$) and $\mathcal{IGB}$ bases. To evaluate C4.5 and ID3, we used the WEKA[(6)] software and the Harmony prototype was kindly provided by its authors. We have implemented the CBA algorithm in C++ under Linux.

During these experiments, we evaluated the introduced new strategy 2 for classifying new instances *vs* the strategy 1 adopted initially by GARC. Then, we compared the effectiveness of using different interestingness measures of association rules for the classification framework with reference to accuracy. After that, we compared the effectiveness of the use of generic bases of the pair ($\mathcal{GBE}$, $\mathcal{GBA}$) and $\mathcal{IGB}$ for the classification framework. For this, we conducted experiments with reference to accuracy in order to compare the classifiers GARC_B and GARC_I issued respectively from the generic bases of the pair ($\mathcal{GBE}$, $\mathcal{GBA}$) and $\mathcal{IGB}$ without using the score firing. Moreover, to show the impact of the score firing on the quality of the produced classifiers, we report the accuracy results of GARCS_B and GARC deduced respectively from the generic bases of the pair ($\mathcal{GBE}$, $\mathcal{GBA}$) and $\mathcal{IGB}$ using the score firing.

In the following, we evaluate the introduced new strategy 2 for classifying new instances *vs* the strategy 1 adopted initially by GARC. For this, we conducted experiments with reference to accuracy in order to compare the measures impact while classifying new instances.

5.1 Comparison of Garc Classifying Instances Strategies

Table 3 shows that strategy 1 gives better accuracy for three data sets when compared to strategy 2. This is explained by the fact that the score used in strategy 1 gives the best prediction accuracy for these data sets. In fact, the use of the other measures in strategy 2 jeopardizes the classification accuracy. Thus, strategy 1 will be adopted by GARC in the following.

[6] *Available at* http://www.cs.waikato.ac.nz/ml/Weka

Table 3. Accuracy comparison of GARC classifying instances strategies for *minsup*=10% and *minconf*=80%

	GARC Accuracy	
Data set	Strategy 1	Strategy 2
MONKS1	**92.0**	88.9
MONKS2	56.0	**71.0**
MONKS3	**96.2**	95.0
PIMA	73.0	73.0
TICTACTOE	65.0	**67.5**
ZOO	90.0	90.0
IRIS	95.4	95.4
WINE	89.8	89.8
GLASS	**64.0**	58.8
FLARE	85.0	85.0
PAGEBLOCKS	89.7	89.7
NURSERY	66.2	66.2

5.2 Evaluating Measures Impact

Table 4 represents a comparison between the accuracy given by the measures used by GARC while classifying new instances [19].

Table 4 points out that the use of the score firing permits to achieve the best accuracy for eight data sets among eleven. The use of the surprise measure permits to achieve the best accuracy for three data sets. We can conclude that a multi-parameterizable tool will be efficient for users in order to choose the best measure suitable for the studied data set.

In the following, we introduce the experimental results showing the impact of the score firing on the quality of the produced classifiers, we report the accuracy results of $\text{GARCS}_{\mathcal{B}}$ and GARC deduced respectively from the generic bases of the pair ($\mathcal{GBE}$, $\mathcal{GBA}$) and $\mathcal{IGB}$ using the score firing.

Table 4. Evaluating measures *vs* accuracy

Data set	Surprise	Loevinger	Lift	Score
MONKS1	42.6	62.5	59.2	**92.0**
MONKS2	**67.1**	59.0	49.3	56.0
MONKS3	**97.2**	92.8	56.7	96.3
PIMA	72.9	72.9	72.9	**73.0**
TICTACTOE	63.0	63.0	63.0	**65.0**
ZOO	83.0	83.0	67.2	**90.0**
IRIS	94.0	89.3	95.3	**95.4**
WINE	**92.8**	81.1	88.3	89.8
GLASS	52.0	52.0	52.0	**64.0**
FLARE	84.7	84.6	84.7	**85.0**
PAGEBLOCKS	89.7	89.7	89.7	**89.8**
NURSERY	66,2	66,2	66,2	66,2

5.3 The Score Firing Impact

Table 5 represents a comparison between the classifiers deduced from the generic bases of the pair ($\mathcal{GBE}$, $\mathcal{GBA}$) and $\mathcal{IGB}$ when using or not the score firing.

Table 5 points out that the use of the score firing increases the accuracy performance for the classifiers deduced from the pair ($\mathcal{GBE}$, $\mathcal{GBA}$). In fact, GARCS$_B$ has a better average accuracy than GARC$_B$. Moreover, for the classifiers deduced from $\mathcal{IGB}$, the use of the score firing improves the accuracy for four data sets. In fact, GARC outperforms GARC$_I$ on Zoo, Iris, Wine and Glass data sets. Thus, the best average accuracy, highlighted in bold print, is given by GARC. Furthermore, as shown in Table 6, the number of rules generated by GARC is less than that generated by the approaches deduced from the pair ($\mathcal{GBE}$, $\mathcal{GBA}$), *i.e.*, GARC$_B$ and GARCS$_B$.

In the following, we put the focus on comparing GARC accuracy by using the score firing versus that of the well known classifiers ID3, C4.5, CBA and Harmony.

5.4 Generic Classification Rules Impact

Table 7 represents the accuracy of the classification systems generated by ID3, C4.5, CBA, Harmony and GARC on the twelve benchmark data sets. The best accuracy values obtained for each of data sets is highlighted in bold print. Table 7 shows that GARC outperforms the traditional classification approaches, *i.e.*, ID3 and C4.5 on six data sets and the associative classification approaches on nine data sets. Statistics depicted by Table 7 confirm the fruitful impact of the use of the generic rules. The main reason for this is that GARC classifier contains generic rules with small premises. In fact, this kind of rule allows to classify more objects than those with large premises.

Table 5. Accuracy comparison of GARC$_B$, GARC$_I$, GARCS$_B$ and GARC algorithms for *minsup*=10% and *minconf*=80%

	Without using the score		Using the score	
Data set	GARC$_B$	GARC$_I$	GARCS$_B$	GARC
MONKS1	92.0	92.0	92.0	92.0
MONKS2	56.0	56.0	56.0	56.0
MONKS3	96.3	96.3	96.3	96.3
PIMA	73.0	73.0	73.0	73.0
TICTACTOE	65.0	67.4	65.0	65.0
ZOO	89.0	89.0	89.0	90.0
IRIS	95.0	94.7	95.6	95.4
WINE	89.2	89.4	90.0	89.8
GLASS	58.0	59.3	58.0	64.0
FLARE	85.0	85.0	85.0	85.0
PAGEBLOCKS	92.0	89.8	92.0	89.8
NURSERY	66,2	66,2	66,2	66,2
Average accuracy	79.7	80.0	79.9	**80.4**

Table 6. Number of associative classification rules for *minsup*=10% and *minconf*=80%

Data set	# generic ACR deduced from $\mathcal{IGB}$	# generic ACR deduced from ($\mathcal{GBE}$, $\mathcal{GBA}$)
MONKS1	31	12
MONKS2	4	4
MONKS3	25	20
PIMA	20	20
TICTACTOE	15	15
ZOO	832	1071
IRIS	22	24
WINE	329	471
GLASS	31	36
FLARE	237	561
PAGEBLOCKS	128	128
NURSERY	12	12

Table 7. Accuracy comparison of ID3, C4.5, CBA, Harmony and GARC algorithms for *minsup*=1% and *minconf*=80%

Data set	ID3	C4.5	CBA	Harmony	GARC
MONKS1	77.0	75.0	**91.6**	83.0	**91.6**
MONKS2	64.0	**65.0**	56.0	48.0	**73.8**
MONKS3	94.0	**97.0**	95.1	82.0	95.1
PIMA	71.3	72.9	**73.0**	**73.0**	**73.0**
TICTACTOE	83.5	**85.6**	63.1	81.0	78.6
ZOO	**98.0**	92.0	82.2	90.0	95.1
IRIS	94.0	94.0	95.3	94.7	**95.4**
WINE	84.8	87.0	89.5	63.0	**94.4**
GLASS	64.0	69.1	52.0	**81.5**	65.9
FLARE	80.1	84.7	**85.0**	83.0	**85.0**
PAGEBLOCKS	92.3	**92.4**	89.0	91.0	91.0
NURSERY	95,0	**95,4**	88,8	90,3	88,8

6 Conclusion and Future Work

In this paper, we reported a synthetic discussion about AC related works and divided them into two groups according to the way of the classification rules extraction. We also introduced a new classification approach called GARC that aims to prune the set of classification rules without jeopardizing the accuracy and even ameliorates the predictive power. To this end, GARC uses generic bases of association rules to drastically reduce the number of associative classification rules. We presented the results given by the use of two different generic basis, *i.e.*, the pair ($\mathcal{GBE}$, $\mathcal{GBA}$) and $\mathcal{IGB}$ in order to build the classifier. Moreover, GARC proposes two strategies to classify new instances based in the use of interestingness measures in order to ameliorate the rules selection for unseen objects.

Carried out experiments outlined that GARC is highly competitive in terms of accuracy in comparison with popular classification methods.

Thus, associative classification is becoming a common approach in classification since it extracts very competitive classifiers in terms of accuracy if compared with rule induction, probabilistic, covering and decision tree approaches.

However, challenges such as efficiency of rule discovery methods, the exponential growth of rules, rule ranking and new metrics investigation need more consideration. Furthermore, another associative classification avenues for future work address the following issues:

- **Missing values in test data.** The problem of dealing with missing values in test data sets has not yet been explored well in AC approaches. In fact, most of the existing AC approaches assume that data set objects are complete and there is no missing values by building classifiers in training data sets without considering the missing values problem. In fact, it will be more interesting if we treat in a particular way missing values that could provide a good deal of information.
- **Incremental learning.** AC approaches generate classifiers by considering the hole training data set. However, in the real world data, modification could occur on the training data set. For instance, in applications like medical diagnosis, adding or editing information is an operation which could occur in any time. Thus, the necessity of updating the classifier which needs to scan one more time the training set. The repetitive scan has an expensive cost in terms of computational time. That's why, it will be interesting to consider incremental AC as future research work.

Acknowledgments

This work is partially supported by French-Tunisian project PAI CMCU 05G1412.

References

1. Zaïane, O.R., Antonie, M.-L.: On Pruning and Tuning Rules for Associative Classifiers. In: Khosla, R., Howlett, R.J., Jain, L.C. (eds.) KES 2005. LNCS (LNAI), vol. 3683, pp. 966–973. Springer, Heidelberg (2005)
2. Xiaoxin Yin, J.H.: CPAR: Classification based on Predictive Association Rules. In: Proceedings of the SDM, San Francisco, CA, pp. 369–376 (2003)
3. Quinlan, J.R.: C4.5: Programs for Machine Learning (1993)
4. Stumme, G., Pasquier, N., Bastide, Y., Taouil, R., Lakhal, L.: Mining Minimal Non-redundant Association Rules Using Frequent Closed Itemsets. In: Palamidessi, C., Moniz Pereira, L., Lloyd, J.W., Dahl, V., Furbach, U., Kerber, M., Lau, K.-K., Sagiv, Y., Stuckey, P.J. (eds.) CL 2000. LNCS (LNAI), vol. 1861, Springer, Heidelberg (2000)
5. Gasmi, G., BenYahia, S., Nguifo, E.M., Slimani, Y.: $\mathcal{IGB}$: A new informative generic base of association rules. In: Ho, T.-B., Cheung, D., Liu, H. (eds.) PAKDD 2005. LNCS (LNAI), vol. 3518, pp. 81–90. Springer, Heidelberg (2005)

6. Liu, B., Hsu, W., Ma, Y.: Integrating classification and association rule mining. In: Knowledge Discovery and Data Mining, pp. 80–86 (1998)
7. Li, W., Han, J., Pei, J.: CMAR: Accurate and efficient classification based on multiple class-association rules. In: Proceedings of IEEE International Conference on Data Mining (ICDM 2001), San Jose, CA, pp. 369–376. IEEE Computer Society, Los Alamitos (2001)
8. Antonie, M., Zaiane, O.: Text Document Categorization by Term Association. In: Proceedings of the IEEE International Conference on Data Mining (ICDM 2002), Maebashi City, Japan, pp. 19–26 (2002)
9. Antonie, M., Zaiane, O.: Classifying Text Documents by Associating Terms with Text Categories. In: Proceedings of the Thirteenth Austral-Asian Database Conference (ADC 2002), Melbourne, Australia (2002)
10. Agrawal, R., Srikant, R.: Fast algorithms for mining association rules. In: Bocca, J.B., Jarke, M., Zaniolo, C. (eds.) Proceedings of the 20th Intl. Conference on Very Large Databases, Santiago, Chile, pp. 478–499 (1994)
11. Wang, J., Karypis, G.: HARMONY: Efficiently mining the best rules for classification. In: Proceedings of the International Conference of Data Mining, pp. 205–216 (2005)
12. Quinlan, J., Cameron-Jones, R.: FOIL: A midterm report. In: Proceedings of European Conference on Machine Learning, Vienna, Austria, pp. 3–20 (1993)
13. Bastide, Y.: Data mining: algorithmes par niveau, techniques d'implantation et applications. Phd thesis, Ecole Doctorale Sciences pour l'Ingénieur de Clermont-Ferrand, Université Blaise Pascal, France (2000)
14. Ganter, B., Wille, R.: Formal Concept Analysis. Springer, Heidelberg (1999)
15. Pasquier, N., Bastide, Y., Taouil, R., Lakhal, L.: Efficient Mining of Association Rules Using Closed Itemset Lattices. Journal of Information Systems 24, 25–46 (1999)
16. BenYahia, S., Nguifo, E.M.: Revisiting generic bases of association rules. In: Kambayashi, Y., Mohania, M., Wöß, W. (eds.) DaWaK 2004. LNCS, vol. 3181, pp. 58–67. Springer, Heidelberg (2004)
17. Han, J., Kamber, M.: Data Mining: Concepts and Techniques. Morgan Kaufmann, San Francisco (2001)
18. Lallich, S., Teytaud, O.: Evaluation et validation de l'intérêt des règles d'association. In: RNTI-E, pp. 193–217 (2004)
19. Bouzouita, I., Elloumi, S.: GARC-M: Generic association rules based classifier multi-parameterizable. In: Proceedings of 4th International Conference of the Concept Lattices and their Applications (CLA 2006), Hammamet, Tunisia (2006)
20. Hamrouni, T., BenYahia, S., Slimani, Y.: Prince: An algorithm for generating rule bases without closure computations. In: Tjoa, A.M., Trujillo, J. (eds.) DaWaK 2005. LNCS, vol. 3589, pp. 346–355. Springer, Heidelberg (2005)

Galois Lattices and Bases for M_{GK}-Valid Association Rules

Jean Diatta[1], Daniel R. Feno[1,2], and A. Totohasina[2]

[1] IREMIA, Université de La Réunion, 15, Avenue René Cassin - B.P. 7151
97715, Saint-Denis. Messag Cedex 9, France
{jdiatta,drfeno}@univ-reunion.fr
[2] ENSET, Université d'Antsiranana - B.P. 0 - Antsiranana 201 Madagascar
{fenodaniel2,totohasina}@yahoo.fr

Abstract. We review the main properties of the quality measure M_{GK}, which has been shown to be the normalized quality measure associated to most of the quality measures used in the data mining literature, and which enables to handle negative association rules. On the other hand, we characterize bases for M_{GK}-valid association rules in terms of a closure operator induced by a Galois connection. Thus, these bases can be derived from a Galois lattice, as do well known bases for Confidence-valid association rules.

Keywords. Closure operator, Basis, Galois connection, Negative association rule, Quality measure.

1 Introduction

Association rules reveal attributes (or attribute values) that occur together frequently in a data set. Their relevance is commonly assessed by means of quality measures. Several quality measures have been proposed in the literature [1], the most popular of them being the well-known Support and Confidence [2]. A major problem faced in association rule mining is the large number of valid rules, *i.e.*, rules that meet specific constraints relative to a given (set of) quality measure(s). Such a situation is generally due to the presence of many redundant and/or trivial rules in the set of valid ones. A way to cope with these redundant and trivial rules is to generate a basis, *i.e.*, a minimal set of rules from which all the valid rules can be derived, using some inference axioms.

In this paper, we consider the quality measure M_{GK} independently introduced in [3] and in [4], and which has been shown to be the normalized quality measure associated to most of the quality measures used in the data mining literature [5]. On the one hand, we review its main properties. On the other hand, we characterize bases for M_{GK}-valid association rules in terms of a closure operator induced by a Galois connection [6]. This result shows that these bases can be derived from a Galois lattice, as do well known bases for Confidence-valid association rules [7,8]. The rest of the paper is organized as follows.

Basic concepts relative to association rules and Galois lattices, and the main properties of the quality measure M_{GK} are presented in Section 2. Section 3 is

S. Ben Yahia et al. (Eds.): CLA 2006, LNAI 4923, pp. 186–197, 2008.

devoted to two known bases for Confidence-valid association rules, whereas the bases we propose for M_{GK}-valid rules are dealt with in Section 4. Finally, a short conclusion is included in the end of the paper.

2 Association Rules, Quality Measures, Galois Lattices

2.1 Association Rules

In this paper, we place ourselves in the framework of a binary context $(\mathcal{E}, \mathcal{V})$, where $\mathcal{E}$ is a finite entity set and $\mathcal{V}$ a finite set of boolean variables (or items) defined on $\mathcal{E}$. The subsets of $\mathcal{V}$ will be called *itemsets*, and an entity "e" will be said to contain an item "x" if $x(e) = 1$.

Definition 1. *An association rule of $(\mathcal{E}, \mathcal{V})$ is an ordered pair (X, Y) of itemsets, denoted by $X \rightarrow Y$, where Y is required to be nonempty. The itemsets X and Y are respectively called the "*premise*" and the "*consequent*" of the association rule $X \rightarrow Y$.*

Given an itemset X,

- X' will denote the set of entities containing all the items of X, *i.e.*,

$$X' = \{e \in \mathcal{E} : \forall x \in X[x(e) = 1]\}, and$$

- $\overline{X}$ will denote the negation of X, *i.e.*, $\overline{X}(e) = 1$ if and only if there exists $x \in X$ such that $x(e) = 0$; it may be noticed that $\overline{X}' = \mathcal{E} \setminus X'$.

Table 1 presents a binary context $\mathbb{K} = (\mathcal{E}, \mathcal{V})$, where $\mathcal{E} = \{e_1, e_2, e_3, e_4, e_5\}$ and $\mathcal{V} = \{A, B, C, D, E\}$. If we let $X = \{B, C\}$ then $X' = \{e_2, e_3, e_5\}$ and $\overline{X}' = \{e_1, e_4\}$.

Table 1. A binary context

	A	B	C	D	E
e_1	1	0	1	1	0
e_2	0	1	1	0	1
e_3	1	1	1	0	1
e_4	0	1	0	0	1
e_5	1	1	1	0	1

According to the definition above, the binary context $(\mathcal{E}, \mathcal{V})$ contains $2^{|\mathcal{V}|}$ $(2^{|\mathcal{V}|} - 1)$ association rules among which several are certainly irrelevant. To cope with this, quality measures, also called interestingness measures, are used to capture only those association rules meeting some given constraints [1]. In the sequel, $\mathcal{E}$ will denote a finite entity set, $\mathcal{V}$ a finite set of boolean variables defined on $\mathcal{E}$, and $\mathbb{K}$ the binary context $(\mathcal{E}, \mathcal{V})$.

2.2 Quality Measures for Association Rules

Let Σ denote the set of association rules of the binary context $\mathbb{K}$.

Definition 2. *A* quality measure *for the association rules of* $\mathbb{K}$ *is a real-valued map* μ *defined on* Σ.

There are several quality measures introduced in the literature, the most popular of them being Support and Confidence [2].

The support of an itemset X, denoted by $\mathrm{Supp}(X)$, is the proportion of entities in $\mathcal{E}$ containing all the items belonging to X; it is defined by $\mathrm{Supp}(X) = \frac{|X'|}{|\mathcal{E}|}$, where, for any finite set W, $|W|$ denotes the number of its elements. Denoting by p the intuitive probability measure defined on $(\mathcal{E}, \mathcal{P}(\mathcal{E}))$ by $p(E) = \frac{|E|}{|\mathcal{E}|}$ for $E \subseteq \mathcal{E}$, the support of X can be written in terms of p as $\mathrm{Supp}(X) = p(X')$.

The support of an association rule $X \rightarrow Y$ is defined by:

$$\mathrm{Supp}(X \rightarrow Y) = \mathrm{Supp}(X \cup Y) = p((X \cup Y)') = p(X' \cap Y').$$

The confidence of $X \rightarrow Y$, denoted by $\mathrm{Conf}(X \rightarrow Y)$, is the proportion of entities containing all the items belonging to Y, among those entities containing all the items belonging to X; it is defined by:

$$\mathrm{Conf}(X \rightarrow Y) = \frac{\mathrm{Supp}(X \rightarrow Y)}{\mathrm{Supp}(X)} = \frac{p(X' \cap Y')}{p(X')} = p(Y'|X'),$$

where $p(Y'|X')$ is the conditional probability of Y' given X'.

The two following straightforward inequalities involving conditional probabilities may help to understand the definition of the quality measure $\mathrm{M_{GK}}$ below.

(i) if $p(Y'|X') \geq p(Y')$, then $0 \leq p(Y'|X') - p(Y') \leq 1 - p(Y')$;
(ii) if $p(Y'|X') \leq p(Y')$, then $-p(Y') \leq p(Y'|X') - p(Y') \leq 0$.

The quality measure $\mathrm{M_{GK}}$ independently introduced in [3] and in [4], is defined by:

$$\mathrm{M_{GK}}(X \rightarrow Y) = \begin{cases} \frac{p(Y'|X') - p(Y')}{1 - p(Y')} & \text{if } p(Y'|X') \geq p(Y'); \\ \frac{p(Y'|X') - p(Y')}{p(Y')} & \text{if } p(Y'|X') \leq p(Y'). \end{cases}$$

In this paper, we will be mainly concerned with the quality measures Confidence and $\mathrm{M_{GK}}$. The quality measure Confidence is clearly a probability measure and its properties are more or less well known. For instance, $\mathrm{Conf}(X \rightarrow Y) = 0$ if and only if X and Y are incompatible. Moreover, the Confidence measure is not symmetric (*i.e.* $\mathrm{Conf}(X \rightarrow Y)$ is not always equal to $\mathrm{Conf}(Y \rightarrow X)$), and $\mathrm{Conf}(X \rightarrow Y) = 1$ if and only if $X' \subseteq Y'$, *i.e.*, if X logically implies Y. However, the Confidence measure does not reflect the independence between the premise and the consequent of an association rule. Indeed, in case of independence between X and Y, $p(Y'|X') = p(Y')$ and, equivalently, $p(X'|Y') = p(X')$. Furthermore, as quoted in [9], Confidence does not satisfy the logical principle of contraposition, *i.e.*, $\mathrm{Conf}(\overline{Y} \rightarrow \overline{X})$ is not always equal to $\mathrm{Conf}(X \rightarrow Y)$.

On the other hand, it can be easily checked that $\mathrm{M_{GK}}$ satisfies the five following properties:

1. $\mathrm{M_{GK}}(X \to Y) = -1$ if and only if X and Y are incompatible, *i.e.*, if $p(X' \cap Y') = 0$;
2. $-1 \leq \mathrm{M_{GK}}(X \to Y) < 0$ if and only if X disfavors Y (or X and Y are negatively dependent), *i.e.*, if $p(Y'|X') < p(Y')$;
3. $\mathrm{M_{GK}}(X \to Y) = 0$ if and only if X and Y are independent, *i.e.*, if $p(X' \cap Y') = p(X')p(Y')$;
4. $0 < \mathrm{M_{GK}}(X \to Y) \leq 1$ if and only if X favors Y (or X and Y are positively dependent), *i.e.*, if $p(Y'|X') > p(Y')$;
5. $\mathrm{M_{GK}}(X \to Y) = 1$ if and only if X logically implies Y, *i.e.*, if $p(Y'|X') = 1$.

This shows that the values of $\mathrm{M_{GK}}$ lie into the interval $[-1, +1]$ as well as they reflect references situations such as incompatibility, negative dependence, independence, positive dependence, and logical implication between the premise and the consequent. Thus, according to [5], $\mathrm{M_{GK}}$ is a normalized quality measure. Moreover, it has been shown in [5] that $\mathrm{M_{GK}}$ is the normalized quality measure associated to most of the quality measures proposed in the literature, including Support and Confidence [2], ϕ- coefficient [10], Laplace, Rule interest, Cosine and Kappa (cf. [11]), and Lift [12]. That is, if we normalize such a quality measure by transforming its expression in order to make its values both lie into the interval $[-1, +1]$ and reflect the five reference situations mentioned above, then we obtain the quality measure $\mathrm{M_{GK}}$. In other words, all these quality measures can be written as affine functions of $\mathrm{M_{GK}}$, with coefficients depending on the support of the premise and/or the support of the consequent. Furthermore, unlike several other quality measures, $\mathrm{M_{GK}}$ satisfies the logical principle of contraposition in case of positive dependence, *i.e.*, $\mathrm{M_{GK}}(\overline{Y} \to \overline{X}) = \mathrm{M_{GK}}(X \to Y)$ when X favors Y [13]. In addition, the greater the absolute value of $\mathrm{M_{GK}}(X \to Y)$, the stronger the (positive or negative) dependence between X and Y.

The following result provides us with relationships between positive dependence and negative dependence.

Proposition 1. *Let X and Y be two itemsets. Then the three following conditions are equivalent.*

(1) X disfavors Y.
(2) X favors $\overline{Y}$.
(3) $\overline{X}$ favors Y.

This result shows that the so-called right-hand side negative (RHSN) rule $X \to \overline{Y}$ and/or the so-called left-hand side negative (LHSN) rule ($\overline{X} \to Y$) can be of interest when X disfavors Y. This is an additional motivation for our choice of $\mathrm{M_{GK}}$ because $\mathrm{M_{GK}}$ enables to handle negative rules as well as positive ones, *i.e.*, those which do not involve negation of itemsets.

It should be noticed that only a rule whose premise favors its consequent is interesting, regardless if it is a positive or a negative rule. Thus, let $\mathrm{M_{GK}}^f(X \to Y)$ denote the value of $\mathrm{M_{GK}}(X \to Y)$ when X favors Y, *i.e.*,

$\mathrm{M_{GK}}^f(X \rightarrow Y) = \frac{p(Y'|X')-p(Y')}{1-p(Y')}$, and let $\mathrm{M_{GK}}^{df}(X \rightarrow Y)$ denote the value of $\mathrm{M_{GK}}(X \rightarrow Y)$ when X disfavors Y, *i.e.*, $\mathrm{M_{GK}}^{df}(X \rightarrow Y) = \frac{p(Y'|X')-p(Y')}{p(Y')}$. The next result shows that the value of $\mathrm{M_{GK}}$ for a RHSN rule is equal to that of $\mathrm{M_{GK}}$ for the corresponding positive rule, on the one hand, and, on the other hand, this value both determines and can be determined from that for the corresponding LHSN rule.

Proposition 2. *Let X and Y be two itemsets. Then the two following properties hold.*

(1) $\mathrm{M_{GK}}^f(X \rightarrow \overline{Y}) = -\mathrm{M_{GK}}^{df}(X \rightarrow Y)$.
(2) $\mathrm{M_{GK}}^f(\overline{X} \rightarrow Y) = \frac{P(X')}{1-P(X')}\frac{P(Y')}{1-P(Y')}\mathrm{M_{GK}}^f(X \rightarrow \overline{Y})$.

Definition 3. *Let μ be a quality measure, and let $\alpha > 0$ be a positive real number. Let $X \rightarrow Y$ be a positive or a (right-hand side, left-hand side or both side) negative association rule. Then $X \rightarrow Y$ is said to be* valid *w.r.t. α in the sense of μ or, simply, (μ, α)-valid if $\mu(X \rightarrow Y) \geq \alpha$. When the meaning is clear from the context, we omit the validity threshold α and/or the quality measure μ, and talk about μ-valid or, simply, valid association rules.*

In the sequel, α will denote a minimum validity threshold belonging to the interval $]0, 1[$. As a consequence of Proposition 2 above, LHSN $\mathrm{M_{GK}}$-valid rules can be obtained from RHSN ones, as stated in the next corollary.

Corollary 1. *If X and Y are two itemsets such that X disfavors Y, then* $\mathrm{M_{GK}}^f(\overline{X} \rightarrow Y) \geq \alpha$ *if and only if* $\mathrm{M_{GK}}^f(X \rightarrow \overline{Y}) \geq \alpha(\frac{1}{\mathrm{Supp}(X)} - 1)(\frac{1}{\mathrm{Supp}(Y)} - 1)$.

To summarize, we need to consider negative rules as well as positive ones. However, LHSN $\mathrm{M_{GK}}$-valid rules can be derived from RHSN ones w.r.t. a corresponding validity threshold, so that we can restrict ourselves to generate only RHSN $\mathrm{M_{GK}}$-valid rules. Moreover, as $\mathrm{M_{GK}}$ satisfies the logical principle of contraposition when the premise favors the consequent, the both side negative $\mathrm{M_{GK}}$-valid rules can also be derived from their corresponding positive $\mathrm{M_{GK}}$-valid ones w.r.t. the same validity threshold. Therefore, we will consider only positive rules and RHSN rules in the sequel. Hence, we will simply use the term negative rule to mean RHSN rule.

One of the major problems faced in association rule mining is the huge number of generated rules. Indeed, despite the fact that a (set of) quality measure(s) is used in order to capture only those rules meeting some given constraints, the set of generated rules can still be of a very large size, due to the presence of redundant and/or trivial rules. Indeed, for a given quality measure μ, the set of μ-valid association rules often contains many rules that are redundant in the sense that they can be derived from other μ-valid rules. For instance, if $\mathrm{Conf}(X \rightarrow Y) = 1$ and $\mathrm{Conf}(Y \rightarrow Z) = 1$, then $\mathrm{Conf}(X \rightarrow Z) = 1$. Thus, if we look for Confidence-exact association rules, *i.e.* rules whose confidence is equal to 1, then the rule $X \rightarrow Z$ is redundant when the rules $X \rightarrow Y$ and $Y \rightarrow Z$ are given, since it can be derived from these ones.

On the other hand, some rules are valid whatever the validity threshold is, and thus, are not informative at all. For instance, for any itemsets X and Y with $Y \subseteq X$, the rule $X \rightarrow Y$ is Confidence-exact. Therefore, if we are interested in informative Confidence-exact association rules, then the rules of the form $X \rightarrow Y$ with $Y \subseteq X$ are not worth generating.

A way to cope with redundant or non informative association rules without loss of information is to generate a basis for the set of valid rules. Indeed, a basis is a set of rules from which any valid rule can be derived using given inference axioms, and which is minimal (w.r.t. set inclusion) among the rule sets having this property. In this paper, we characterize bases for M$_{GK}$-valid association rules of a binary context, in terms of the closure operator induced by a Galois connection. Thus, these bases can be derived from a Galois lattice, as do bases for positive Confidence-valid rules.

2.3 The Galois Lattices of a Binary Context

The binary context $\mathbb{K}$ induces a Galois connection between the partially ordered sets $(\mathcal{P}(\mathcal{E}), \subseteq)$ and $(\mathcal{P}(\mathcal{V}), \subseteq)$ by means of the maps

$$f : X \mapsto \bigcap_{x \in X} \{v \in \mathcal{V} : v(x) = 1\} = X'$$

and

$$g : Y \mapsto \bigcap_{v \in Y} \{x \in \mathcal{E} : v(x) = 1\},$$

for $X \subseteq \mathcal{E}$ and $Y \subseteq \mathcal{V}$ [14]. Moreover, the Galois connection (f, g) induces, in turn, a closure operator $\varphi := f \circ g$ on $(\mathcal{P}(\mathcal{V}), \subseteq)$ [6]. That is, for $X, Y \subseteq \mathcal{V}$:

(C1) $X \subseteq \varphi(X)$ (extensivity);
(C2) $X \subseteq Y$ implies $\varphi(X) \subseteq \varphi(Y)$ (isotony);
(C3) $\varphi(\varphi(X)) = \varphi(X)$ (idempotence).

Let $G(\mathbb{K})$ denote the set of all pairs $(X, Y) \in \mathcal{P}(\mathcal{E}) \times \mathcal{P}(\mathcal{V})$ such that $\varphi(Y) = Y$ and $g(Y) = X$. Then $G(\mathbb{K})$, endowed with the order defined by $(X_1, Y_1) \leq (X_2, Y_2)$ if and only if $X_1 \subseteq X_2$ (or, equivalently $Y_2 \subseteq Y_1$), is a complete lattice called the *Galois lattice* of the binary context $\mathbb{K}$ [14], also known as the concept lattice of the formal context $(\mathcal{E}, \mathcal{V}, \mathcal{R})$, where $\mathcal{R}$ is the binary relation from $\mathcal{E}$ to $\mathcal{V}$ defined by $x\mathcal{R}v$ if and only if $v(x) = 1$ [15].

Example 1. *Consider the binary context $\mathbb{K}$ given in Table 1. Then, the pair $(\{e_2, e_3, e_5\}, \{B, C\})$ is a member of $G(\mathbb{K})$. But though $\varphi(\{B, C\}) = \{B, C\}$, $(\{e_2, e_3\}, \{B, C\})$ does not belong to $G(\mathbb{K})$ since $g(\{B, C\}) \neq \{e_2, e_3\}$.*

3 Bases for Confidence-Valid Association Rules

This section is intended to remind two known bases for positive Confidence-valid association rules, namely, the Luxenburger basis for approximate rules and the Guigues-Duquenne basis for exact ones.

The set of positive Confidence-exact association rules is a full implicational system, *i.e.*, it satisfies the following Armstrong's inference axioms for all itemsets X, Y, Z [16]:

(PE1) $Y \subseteq X$ implies $X \rightarrow Y$;
(PE2) $X \rightarrow Y$ and $Y \rightarrow Z$ imply $X \rightarrow Z$;
(PE3) $X \rightarrow Y$ and $Z \rightarrow T$ imply $X \cup Z \rightarrow Y \cup T$.

Thus, the Guigues-Duquenne basis [7] for full implicational systems is by the way a basis for positive Confidence-exact rules. To define this basis, we need to recall the notion of a critical set of a closure operator.

Consider the closure operator φ, induced on $\mathcal{P}(\mathcal{V})$ by the Galois connection (f, g) defined above. An itemset X is said to be *φ-closed* if $\varphi(X) = X$; it is said to be *φ-quasi-closed* if it is not φ-closed and for all $Y \subset X$, either $\varphi(Y) \subset X$ or $X \subset \varphi(Y)$ [17]; it is said to be *φ-critical* if it is minimal among the φ-quasi-closed itemsets Y such that $\varphi(Y) = \varphi(X)$ [18]. A definition of quasi-closed sets in terms of Moore families can be found in [19,20,21], as well as other characterizations of φ–critical sets.

The Guigues-Duquenne basis [7] for positive Confidence-exact association rules is the set BPE defined by

$$\text{BPE} = \{X \rightarrow \varphi(X) \setminus X : X \text{ is } \varphi\text{-critical}\}.$$

This basis has been adapted to Support-and-Confindence-exact association rules by [22] and [23], who placed association rule mining problem within the theoretic framework of Galois lattices.

Example 2. *The rules $B \rightarrow E$ and $D \rightarrow AC$ are two rules belonging to* BPE, *from which many other positive* M_{GK}*-exact rules such as, for instance, $BD \rightarrow ACE$, $AB \rightarrow E$ and $AD \rightarrow ACE$ can be derived, using (PE1), (PE2) and (PE3).*

The Luxenburger basis [8] for Confidence-approximate association rules is the set LB defined by

$$\text{LB} = \{X \rightarrow Y : X = \varphi(X), Y = \varphi(Y), X \prec Y \text{ and } \text{Conf}(X \rightarrow Y) \geq \alpha\},$$

where $X \prec Y$ means that $X \subset Y$ and there is no φ-closed set Z such that $X \subset Z \subset Y$.

4 Bases for M_{GK}-Valid Association Rules

In this section, we characterize a basis for $(\text{M}_{\text{GK}}, \alpha)$-valid association rules. This basis is in fact the union of four bases: a basis for positive exact rules (*i.e.* the rules $X \rightarrow Y$ such that $\text{M}_{\text{GK}}(X \rightarrow Y) = 1$), a basis for negative exact rules (*i.e.* the rules $X \rightarrow \overline{Y}$ such that $\text{M}_{\text{GK}}(X \rightarrow \overline{Y}) = 1$), a basis for positive approximate rules (*i.e.* the rules $X \rightarrow Y$ such that $\alpha \leq \text{M}_{\text{GK}}(X \rightarrow Y) < 1$), and a basis for negative approximate rules (*i.e.* the rules $X \rightarrow \overline{Y}$ such that $\alpha \leq \text{M}_{\text{GK}}(X \rightarrow \overline{Y}) < 1$).

4.1 Basis for Positive M_{GK}-Exact Association Rules

The set of positive M_{GK}-exact rules coincides with that of positive Confidence-exact ones. Thus, the basis BPE for Confidence-exact association rules is by the way a basis for positive M_{GK}-exact rules.

4.2 Basis for Negative M_{GK}-Exact Association Rules

Recall that negative association rules are rules of the form $X \rightarrow \overline{Y}$, where X and Y are itemsets. The following straightforward but instrumental properties define their support and confidence.

Proposition 3. *Let X and Y be two itemsets. Then the three following conditions hold.*

(1) $\text{Supp}(\overline{X}) = 1 - \text{Supp}(X)$.
(2) $\text{Supp}(X \rightarrow \overline{Y}) = \text{Supp}(X) - \text{Supp}(X \rightarrow Y)$.
(3) $\text{Conf}(X \rightarrow \overline{Y}) = 1 - \text{Conf}(X \rightarrow Y)$.

Negative M_{GK}-exact association rules are those negative rules $X \rightarrow \overline{Y}$ such that $M_{GK}(X \rightarrow \overline{Y}) = 1$. The next easily-checked result characterizes them in terms of the support or the confidence of their corresponding positive rules.

Proposition 4. *Let X and Y be two itemsets such that* $\text{Supp}(X) \neq 0$ *and* $\text{Supp}(Y) \neq 0$. *Then the following conditions are equivalent:*

(1) $M_{GK}(X \rightarrow \overline{Y}) = 1$. *(2)* $M_{GK}(X \rightarrow Y) = -1$.
(3) $\text{Conf}(X \rightarrow Y) = 0$. *(4)* $\text{Supp}(X \rightarrow Y) = 0$.

In the sequel, for $x \in \mathcal{V}$ and $X, Y \subseteq \mathcal{V}$, we will sometimes denote $\{x\}$ by x, $X \cup Y$ by XY and $\{x\} \cup X$ by $x + X$. Proposition 4 leads us to consider the following inference axioms for any itemsets X, Y, Z:

(NE1) $X \rightarrow \overline{Y}$ and $\text{Supp}(YZ) > 0$ imply $X \rightarrow \overline{YZ}$;
(NE2) $X \rightarrow \overline{Y}$, $Z \subset X$ and $\text{Supp}(ZY) = 0$ imply $Z \rightarrow \overline{Y}$.

The next result shows that every association rule derived from negative M_{GK}-exact ones using (NE1) and (NE2) is also negative M_{GK}-exact.

Proposition 5. *The inference axioms (NE1) and (NE2) are sound for negative M_{GK}-exact association rules.*

Proposition 4 also leads us to consider the positive border of the set of itemsets having a null support [24], *i.e.*, the set

$$\text{Bd}^{+}(0) = \{X \subseteq \mathcal{V} : \text{Supp}(X) > 0 \text{and for all } x \notin X [\text{Supp}(x + X) = 0]\}$$

consisting of maximal itemsets (w.r.t. set inclusion) having a non null support.

Example 3. *For the context given in Table 1,* $\mathrm{Bd}^+(0) = \{ACD, BCE, ABCE\}$.

We now go on to characterize the basis we propose for the set of negative $\mathrm{M_{GK}}$-exact association rules.

Theorem 1. *The set* BNE *defined by*

$$\mathrm{BNE} = \{X \to \overline{x} : X \in Bd^+(0) \text{ and } x \notin X\}$$

is a basis for negative $\mathrm{M_{GK}}$*-exact association rules w.r.t. the inference axioms (NE1) and (NE2).*

Example 4. *For the context given in Table 1,* $\mathrm{BNE} = \{ABCE \to \overline{D}, ACD \to \overline{B}, ACD \to \overline{E}, BCE \to \overline{A}, BCE \to \overline{D}\}$. *Moreover, the eleven rules* $ABCE \to \overline{D}$, $ABCE \to \overline{AD}$, $ABCE \to \overline{CD}$, $ABE \to \overline{ACD}$, $BE \to \overline{AD}$, $E \to \overline{AD}$, $B \to \overline{AD}$, $E \to \overline{CD}$, $B \to \overline{CD}$, $E \to \overline{ACD}$, $B \to \overline{ACD}$ *can be derived from the rule* $ABCE \to \overline{D}$, *using (NE1) and (NE2).*

It may be noticed that the positive border $\mathrm{Bd}^+(0)$ is nothing else than the set of maximal φ-closed itemsets having a strictly positive support. Thus, the basis BNE is clearly characterized in terms of the closure operator φ. It may also be noticed that $Y \to \overline{X}$ is a negative $\mathrm{M_{GK}}$-exact rule whenever $X \to \overline{Y}$ is. However these two rules are not always equally informative. Indeed, if, for instance, $|X_1| > |X_2| > |Y_1| > |Y_2|$, then the rule $X_2 \to \overline{Y_2}$ is more informative than any other negative rule involving the itemsets X_1, X_2, Y_1, Y_2.

4.3 Basis for Positive $\mathrm{M_{GK}}$-Approximate Association Rules

Positive $(\mathrm{M_{GK}}, \alpha)$-approximate association rules are those positive rules $X \to Y$ such that $\alpha \leq \mathrm{M_{GK}}(X \to Y) < 1$. The following straightforward result characterizes them in terms of their confidence.

Proposition 6. *Let X and Y be two itemsets such that X favors Y. Then* $\alpha \leq \mathrm{M_{GK}}(X \to Y) < 1$ *if and only if* $\mathrm{Supp}(Y)(1-\alpha) + \alpha \leq \mathrm{Conf}(X \to Y) < 1$.

This result leads us to consider the following inference axiom for any itemsets X, Y, Z, T:

(PA) $X \to Y$, $\varphi(X) = \varphi(Z)$ and $\varphi(Y) = \varphi(T)$ imply $Z \to T$.

The two following technical lemmas will be helpful for proving the soundness of the axiom (PA). The first lemma shows that every itemset has the same support as its φ-closure [25].

Lemma 1. *For any itemset X,* $\mathrm{Supp}(\varphi(X)) = \mathrm{Supp}(X)$.

The second lemma is a characterization of closure operators by means of path independence property [26,21].

Lemma 2. *An extensive function ϕ on a finite powerset, say $\mathcal{P}$, is a closure operator on $\mathcal{P}$ if and only if it satisfies the path independence property:* $\phi(X \cup Y) = \phi(\phi(X) \cup \phi(Y))$, *for any* $X, Y \in \mathcal{P}$.

The next proposition shows that every association rule derived from a positive M_{GK}-approximate one, using the inference axiom (PA), is also positive M_{GK}-approximate.

Proposition 7. *The inference axiom (PA) is sound for positive* (M_{GK}, α)*-approximate association rules.*

We now go on to characterize the basis we propose for the set of positive M_{GK}-approximate association rules.

Theorem 2. *The set* $\mathrm{BPA}(\alpha)$ *defined by*

$$\mathrm{BPA}(\alpha) = \{X \to Y : \varphi(X) = X, \varphi(Y) = Y, \mathrm{Supp}(Y)(1-\alpha)+\alpha \leq \mathrm{Conf}(X \to Y) < 1\}$$

is a basis for positive (M_{GK}, α)*-approximate association rules w.r.t. the inference axiom (PA).*

Example 5. *Consider the context given in Table 1 and let the minimum validity threshold* α *be set to* $\frac{1}{10}$*. Then, the rule* $AC \to BCE$ *is a member of* $\mathrm{BPA}(\alpha)$ *from which can be derived the five rules* $A \to BC$*,* $A \to CE$*,* $A \to BCE$*,* $AC \to BC$ *and* $AC \to CE$*, using the inference axiom (PA).*

4.4 Basis for Negative M_{GK}-Approximate Association Rules

Negative (M_{GK}, α)-approximate association rules are those negative rules $X \to \overline{Y}$ such that $\alpha \leq M_{GK}(X \to \overline{Y}) < 1$. The next straightforward result characterizes them in terms of the confidence of their corresponding positive rules.

Proposition 8. *Let* X *and* Y *be two itemsets such that* X *disfavors* Y*. Then* $\alpha \leq M_{GK}(X \to \overline{Y}) < 1$ *if and only if* $0 < \mathrm{Conf}(X \to Y) \leq \mathrm{Supp}(Y)(1-\alpha)$.

This result leads us to consider the following inference axiom for any itemsets X, Y, Z, T:

(NA) $X \to \overline{Y}$, $\varphi(X) = \varphi(Z)$ and $\varphi(Y) = \varphi(T)$ imply $Z \to \overline{T}$.

The next result shows the soundness of the inference axiom (NA).

Proposition 9. *The inference axiom (NA) is sound for negative* (M_{GK}, α)*-approximate association rules.*

Theorem 3 below characterizes the basis we propose for the set of negative M_{GK}-approximate association rules.

Theorem 3. *The set* $\mathrm{BNA}(\alpha)$ *defined by*

$$\mathrm{BNA}(\alpha) = \{X \to \overline{Y} : \varphi(X) = X, \varphi(Y) = Y, 0 < \mathrm{Conf}(X \to Y) \leq \mathrm{Supp}(Y)(1-\alpha)\}$$

is a basis for negative (M_{GK}, α)*-approximate association rules w.r.t. the inference axiom (NA).*

Example 6. *Consider the context given in Table 1 and let the minimum validity threshold* α *be set to* $\frac{1}{10}$*. Then, the rule* $AC \to \overline{BE}$ *is a member of* $\mathrm{BNA}(\alpha)$ *from which can be derived the five rules* $A \to \overline{B}$*,* $A \to \overline{E}$*,* $A \to \overline{BE}$*,* $AC \to \overline{B}$ *and* $AC \to \overline{E}$*, using the inference axiom (NA).*

5 Conclusion

We reviewed the main properties of the quality measure for association rules, M_{GK}, independently introduced in [3] and in [4], and which has been shown to be the normalized quality measure associated to most of the quality measures proposed in the data mining literature [5]. On the other hand, we characterized bases for M_{GK}-valid association rules in terms of a closure operator induced by a Galois connection [6]: two bases for positive rules (exact and approximate) and two bases for negative rules (exact and approximate). Thus, these bases can be derived from a Galois lattice, as do well known bases for Confidence-valid association rules [7,8].

References

1. Hilderman, R.J., Hamilton, H.J.: Knowledge discovery and interestingness measures: A survey. Technical Report CS 99-04, Department of Computer Science, University of Regina (1999)
2. Agrawal, R., Imielinski, T., Swami, A.: Mining association rules between sets of items in large databases. In: Buneman, P., Jajodia, S. (eds.) Proc. of the ACM SIGMOD International Conference on Management of Data, Washington, vol. 22, pp. 207–216. ACM Press, New York (1993)
3. Guillaume, S.: Traitement des données volumineuses. Mesures et algorithmes d'extraction des règles d'association et règles ordinales. PhD thesis, Université de Nantes, France (2000)
4. Wu, X., Zhang, C., Zhang, S.: Mining both positive and negative rules. ACM J. Information Systems 22, 381–405 (2004)
5. Feno, D., Diatta, J., Totohasina, A.: Normalisée d'une mesure probabiliste de qualité des règles d'association: étude de cas. In: Actes du 2nd Atelier Qualité des Données et des Connaissances, Lille, France, pp. 25–30 (2006)
6. Birkhoff, G.: Lattice theory. 3rd edn., Coll. Publ., XXV. American Mathematical Society, Providence, RI (1967)
7. Guigues, J.L., Duquenne, V.: Famille non redondante d'implications informatives résultant d'un tableau de données binaires. Mathématiques et Sciences humaines 95, 5–18 (1986)
8. Luxemburger, M.: Implications partielles dans un contexte. Math. Inf. Sci. hum. 113, 35–55 (1991)
9. Brin, S., Motwani, R., Ullman, J.D., Tsur, S.: Dynamic itemset counting and implication rules for market basket data. In: Proc. of the ACM SIGMOD Conference, pp. 255–264 (1997)
10. Lerman, I., Gras, R., Rostam, H.: Elaboration et évaluation d'un indice d'implication pour des données binaires. Math Sc. Hum. 74, 5–35 (1981)
11. Huynh, X., Guillet, F., Briand, H.: Une plateforme exploratoire pour la qualité des règles d'association: Apport pour l'analyse implicative. In: Troisièmes Rencontres Internationales A.S.I., pp. 339–349 (2005)
12. Brin, S., Motwani, R., Silverstein, C.: Beyond market baskets: Generalizing association rules to correlation. In: Proc. of the ACM SIGMOD Conference, pp. 265–276 (1997)
13. Totohasina, A., Ralambondrainy, H.: Lon: A pertinent new measure for mining information from many types of data. In: IEEE SITIS 2005, pp. 202–207 (2005)

14. Barbut, M., Monjardet, B.: Ordre et classification. In: Hachette, Paris (1970)
15. Wille, R.: Restructuring lattice theory: An approach based on hierarchies of concepts. In: Rival, I. (ed.) Ordered sets, pp. 445–470. Ridel, Dordrecht-Boston (1982)
16. Armstrong, W.W.: Dependency structures of data base relationships. Information Processing 74, 580–583 (1974)
17. Diatta, J.: Charactérisation des ensembles critiques d'une famille de Moore finie. In: Rencontres de la Société Francophone de Classification, Montréal, Canada, pp. 126–129 (2005)
18. Day, A.: The lattice theory of functional dependencies and normal decompositions. Internat. J. Algebra Comput. 2, 409–431 (1992)
19. Caspard, N.: A characterization theorem for the canonical basis of a closure operator. Order 16, 227–230 (1999)
20. Caspard, N., Monjardet, B.: The lattices of closure systems, closure operators, and implicational systems on a finite set: a survey. Discrete Applied Mathematics 127, 241–269 (2003)
21. Domenach, F., Leclerc, B.: Closure systems, implicational systems, overhanging relations and the case of hierarchical classification. Mathematical Social Sciences 47, 349–366 (2004)
22. Zaki, M.J., Ogihara, M.: Theoretical Foundations of Association Rules. In: 3rd SIGMOD 1998 Workshop on Research Issues in Data Mining and Knowledge Discovery (DMKD), pp. 1–8 (1998)
23. Pasquier, N., Bastide, Y., Taouil, R., Lakhal, L.: Closed set based discovery of small covers for association rules. In: Proc. 15emes Journees Bases de Donnees Avancees, BDA, pp. 361–381 (1999)
24. Mannila, H., Toivonen, H.: Levelwise search and borders of theories in knowledge discovery. Data Mining Knowledge Discovery 1, 241–258 (1997)
25. Pasquier, N., Bastide, Y., Taouil, R., Lakhal, L.: Efficient mining of association rules using closed itemset lattices. Information Systems 24, 25–46 (1999)
26. Plott, C.R.: Path independence, rationality and social choice. Econometrica 41, 1075–1091 (1973)

Generic Association Rule Bases: Are They so Succinct?

Tarek Hamrouni[1,2], Sadok Ben Yahia[1], and Engelbert Mephu Nguifo[2]

[1] Department of Computer Science, Faculty of Sciences of Tunis, Tunis, Tunisia
{tarek.hamrouni,sadok.benyahia}@fst.rnu.tn
[2] CRIL-CNRS, IUT de Lens, Lens, France
{hamrouni,mephu}@cril.univ-artois.fr

Abstract. In knowledge mining, current trend is witnessing the emergence of a growing number of works towards defining "concise and lossless" representations. One main motivation behind is: tagging a unified framework for drastically reducing large sized sets of association rules. In this context, generic bases of association rules – whose backbone is the conjunction of the concepts of minimal generator (MG) and closed itemset (CI) – constituted so far irreducible compact nuclei of association rules. However, the inherent absence of a unique MG associated to a given CI offers an "ideal" gap towards a tougher redundancy removal even from generic bases of association rules. In this paper, we adopt the succinct system of minimal generators (SSMG), as newly redefined in [1], to be an *exact* representation of the MG set. Then, we incorporate the SSMG into the framework of generic bases to only maintain the *succinct* generic association rules. After that, we give a thorough formal study of the related inference mechanisms allowing to derive *all redundant* association rules starting from succinct ones. Finally, an experimental study shows that our approach makes it possible to eliminate without information loss an important number of *redundant* generic association rules and thus, to only present *succinct* and *informative* ones to users.

1 Introduction

As an important topic in data mining, association rule mining research [2] has progressed in various directions. Unfortunately, one problem with the current trend is that it mainly favoured the efficient extraction of interesting itemsets regardless the effectiveness of the mined knowledge. Indeed, by laying stress on the "algorithmic" improvement of the *frequent* (closed) itemset extraction step, the current trend neglects user's needs: "concise with add-value knowledge". Hence, the number of association rules, which can be extracted even from small datasets, is always a real hampering towards their effective exploitation by the users. Indeed, at the end of the extraction process, the user is faced to an overwhelming quantity of association rules among which a large number is *redundant*, what badly affects the quality of their interpretability. Nevertheless, some approaches have been devoted to the reduction of the number of association rules such as generic bases [3,4,5,6,7,8], concise representations of *frequent* itemsets [9,10,11], quality measures [12], user-defined templates or constraints [13,14]. Among them, generic bases constitute an interesting starting point to reduce without loss of information the size of the association rule set. Indeed, using the mathematical settings of the Formal Concept Analysis (FCA) [15], generic bases were flagged as irreducible

S. Ben Yahia et al. (Eds.): CLA 2006, LNAI 4923, pp. 198–213, 2008.

nuclei of association rules from which *redundant* ones can be derived without any loss of information [3]. In this context, different works have shown that generic bases, containing association rules whose implications are between minimal generators (MGs) [3] and closed itemsets (CIs) [9], convey the maximum of information since they are of minimal premises and of maximal conclusions [3,16]. For these reasons, such association rules are considered as the most informative ones [3].

Nevertheless, a recent study, proposed by Dong *et al.*, showed that the MG set still present a kind of redundancy [17]. Indeed, they consider the set of MGs associated to a given CI by distinguishing two distinct types: *succinct* MGs and *redundant* ones. Thus, Dong *et al.* introduce the succinct system of minimal generators (SSMG) as a concise representation of the MG set. They state that *redundant* MGs can be withdrawn from the MG set since they can straightforwardly be inferred, without loss of information, using the knowledge gleaned from the *succinct* ones [17]. However, in [1], we showed that the *succinct* MGs, as defined by Dong *et al.*, prove not to be an *exact* representation (no loss of information *w.r.t. redundant* MGs) in contrary to authors' claims. We also presented new definitions allowing to overcome the limitations of their work and, hence, to make of the SSMG really an *exact* representation.

In this paper, we propose to incorporate the SSMG, as redefined in [1], into the framework of generic bases to reduce as far as possible the redundancy within generic association rules. Thus, after a study of the best known generic bases of association rules, we apply the SSMG to the couple proposed by Bastide *et al.* [3]. This couple presents at least two complementary advantages. On the one hand, association rules composing it are of minimum premises and of maximal conclusions, and, hence, convey the maximum of information [3,16]. On the other hand, this couple gathers the *ideal* properties of an association rule representation since it is lossless, sound and informative [5]. We then study the obtained generic bases - once the SSMG is applied - to check whether they are extracted without loss of information. Finally, an experimental evaluation illustrates the potential of our approach towards offering to users a redundancy-free set of generic association rules. Please note that it is out of the scope of this paper to discuss how the *succinct* generic association rules are efficiently discovered.

The organization of the paper is as follows: Section 2 recalls some preliminary notions that will be used in the remainder of the paper. We devote Section 3 to the presentation of the main definition of the SSMG proposed in [1]. Section 4 is dedicated to the presentation of the *succinct* generic bases of association rules. In order to derive *all redundant* association rules that can be extracted from a context, an axiomatic system and a study of its main properties are also provided. In Section 5, several experiments illustrate the utility of our approach followed by a summary of our contributions and avenues for future work in Section 6.

2 Preliminary Notions

In this section, we present some notions that will be used in the following.

Definition 1. (EXTRACTION CONTEXT) *An extraction context is a triplet* $\mathcal{K}=(\mathcal{O},\mathcal{I},\mathcal{R})$, *where* $\mathcal{O}$ *represents a finite set of objects,* $\mathcal{I}$ *is a finite set of items and* $\mathcal{R}$ *is a binary* *(incidence) relation (i.e., $\mathcal{R} \subseteq \mathcal{O} \times \mathcal{I}$). Each couple $(o, i) \in \mathcal{R}$ indicates that the object $o \in \mathcal{O}$ has the item $i \in \mathcal{I}$.*

Example 1. Consider the extraction context in Table 1 where $\mathcal{O} = \{1, 2, 3, 4\}$ and $\mathcal{I} = \{a, b, c, d, e, f, g\}$. The couple $(4, b) \in \mathcal{R}$ since it is crossed in the matrix[1].

Table 1. An extraction context $\mathcal{K}$

	a	b	c	d	e	f	g
1			×	×	×	×	×
2	×	×	×	×	×		
3	×	×	×			×	×
4	×	×	×	×	×		×

For arbitrary sets $I \subseteq \mathcal{I}$ and $O \subseteq \mathcal{O}$, the following derivation operators are defined [18]: $I' = \{o \in \mathcal{O} \mid \forall\, i \in I,\, (o, i) \in \mathcal{R}\}$, and, $O' = \{i \in \mathcal{I} \mid \forall\, o \in O,\, (o, i) \in \mathcal{R}\}$. The composite operators $''$ define closures on $(2^{\mathcal{I}}, \subseteq)$ and $(2^{\mathcal{O}}, \subseteq)$. A pair (I, O), of mutually corresponding subsets, *i.e.*, $I = O'$ and $O = I'$, is called a (formal) concept [18], where I is the intent and O is the extent (*e.g.*, $(abcde, 24)$[2] is a concept from Table 1). Once applied, the corresponding operator $''$ induces an equivalence relation on the power set of items $2^{\mathcal{I}}$ partitioning it into distinct subsets called *equivalence classes* [19], which will further be denoted *γ-equivalence classes*. In each class, all itemsets appear in the same set of objects and, hence, have the same closure. The largest element (*w.r.t.* set inclusion) is called a *closed itemset* (CI) – the intent part of a formal concept – while the minimal incomparable ones are called *minimal generators* (MGs). These notions are defined as follows:

Definition 2. (CLOSED ITEMSET)*[9] An itemset $f \subseteq \mathcal{I}$ is said to be closed if and only if $f'' = f$.*

Example 2. Given the extraction context depicted by Table 1, the itemset "*cdeg*" is a closed itemset since it is the maximal set of items common to the set of objects $\{1, 4\}$. The itemset "*cdg*" is not a closed itemset since all objects containing the itemset "*cdg*" also contain the item "*e*".

Definition 3. (MINIMAL GENERATOR)*[3] An itemset $g \subseteq \mathcal{I}$ is said to be a minimal generator of a closed itemset f if and only if $g'' = f$ and $\nexists\, g_1 \subset g$ s.t. $g_1'' = f$.*

The set MG_f of the MGs associated to an CI f is hence $\text{MG}_f = \{g \subseteq \mathcal{I} \mid g'' = f \wedge \nexists\, g_1 \subset g \text{ s.t. } g_1'' = f\}$.

Example 3. Consider the CI "*cdeg*" described by the previous example. "*cdeg*" has "*dg*" as an MG. Indeed, $(dg)'' = cdeg$ and the closure of every subset of "*dg*" is different from "*cdeg*". Indeed, $(\emptyset)'' = c$, $(d)'' = cde$ and $(g)'' = cg$. The CI "*cdeg*" has also another MG which is "*eg*". Hence, $\text{MG}_{cdeg} = \{dg, eg\}$. "*cdeg*" is then the largest element of its γ-equivalence class, whereas "*dg*" and "*eg*" are the minimal incomparable ones. All these itemsets share the set of objects $\{1, 4\}$.

[1] For the sake of homogeneity, we borrowed our running context from [1].

[2] We use a separator-free form for the sets, *e.g.*, the set *abcde* stands for $\{a, b, c, d, e\}$.

Since in practice, we are mainly interested in itemsets that occur at least in a given number of objects, we introduce the notion of support.

Definition 4. (SUPPORT) *The support of an itemset $I \subseteq \mathcal{I}$, denoted by Supp(I), is equal to the number of objects in $\mathcal{K}$ that have all items from I. I is said to be frequent in $\mathcal{K}$ if Supp(I) is greater than or equal to a minimum support threshold, denoted minsupp.*

Example 4. Consider the itemset "*cde*" of the extraction context depicted by Table 1. The objects 1, 2 and 4 contain the itemset "*cde*". Hence, Supp(*cde*) = **3**. If *minsupp* = **2**, then "*cde*" is *frequent* in $\mathcal{K}$ since Supp(*cde*) = $\mathbf{3} \geq \mathbf{2}$.

The *frequent* CIs can be structured as follows:

Definition 5. (ICEBERG LATTICE) *Let $\mathcal{FCI}_{\mathcal{K}}$ be the set of frequent* CI*s extracted from a context $\mathcal{K}$. When the set $\mathcal{FCI}_{\mathcal{K}}$ is partially ordered with set inclusion, the resulting structure only preserves the Join operator [18]. This structure is called a join semi-lattice or an upper semi-lattice [20], and is hereafter referred to as "Iceberg lattice" [21].*

Example 5. An example of an Iceberg lattice is shown in Figure 1.

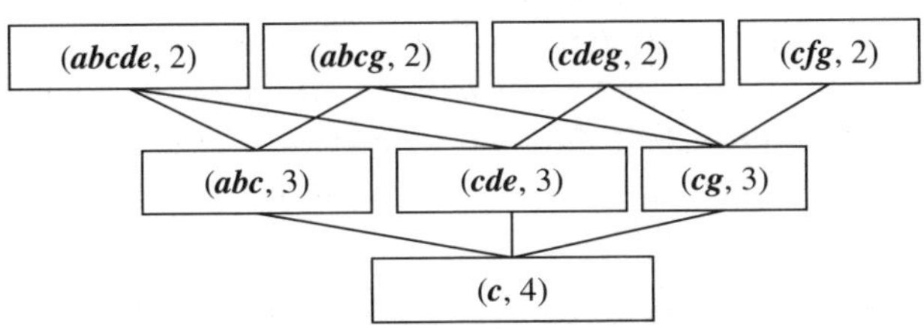

Fig. 1. For *minsupp* = **2**, the Iceberg lattice associated to the extraction context $\mathcal{K}$ given by Table 1. Each one of its γ-equivalence classes contains a *frequent* CI f with its support

Each node (or equivalently, a *frequent* CI) in the Iceberg lattice has a set of nodes that immediately cover it. This set is called *upper cover* and is formally defined as follows:

Definition 6. (UPPER COVER) *The upper cover of a frequent* CI f *(denoted $Cov^u(f)$) consists of the frequent* CI*s that immediately cover f in the Iceberg lattice. The set $Cov^u(f)$ is given as follows: $Cov^u(f) = \{f_1 \in \mathcal{FCI}_{\mathcal{K}} \mid f \subset f_1 \wedge \nexists\, f_2 \in \mathcal{FCI}_{\mathcal{K}}$ s.t. $f \subset f_2 \subset f_1\}$.*

Example 6. Let us consider the *frequent* CI "*c*" of the Iceberg lattice depicted by Figure 1. Cov^u(c) = $\{abc, cde, cg\}$.

3 Succinct System of Minimal Generators

In this section, we briefly describe the main structural properties of the succinct system of minimal generators (SSMG) newly redefined in [1] to make of it an *exact* representation of the minimal generator (MG) set.

The set MG_f of the MGs associated to a given closed itemset (CI) f can be divided into different equivalence classes thanks to a substitution process. To avoid confusion with the γ-equivalence classes induced by the closure operator $''$, the substitution-based ones will be denoted *σ-equivalence classes*. The substitution process uses an operator denoted *Subst*. This substitution operator is a partial one allowing to substitute a subset of an itemset X, say Y, by another itemset Z belonging to the same γ-equivalence class of Y (*i.e.*, $Y'' = Z''$). This operator is then defined as follows:

Definition 7. (SUBSTITUTION OPERATOR) *Let X, Y and Z be three itemsets s.t. $Y \subset X$ and $Y'' = Z''$. The substitution operator Subst, w.r.t. X, Y and Z, is defined as follows: Subst$(X, Y, Z) = (X \backslash Y) \bigcup Z$.*

It is shown in [1] that X and *Subst*(X, Y, Z) have the same closure.

For each γ-equivalence class $\mathcal{C}$ (or equivalently, for each CI f), the substitution operator induces an equivalence relation on the set MG_f of the MGs of f portioning it into distinct σ-equivalence classes. The definition of a σ-equivalence class requires that we define the notion of *redundant* MG under the substitution process point of view as follows:

Definition 8. (MINIMAL GENERATORS' REDUNDANCY) *Let g and g_1 be two* MGs *belonging to the same γ-equivalence class.*

• *g is said to be a* ***direct redundant*** *(resp. derivable) with respect to (resp. from) g_1, denoted $g_1 \vdash g$, if* Subst$(g_1, g_2, g_3) = g$ *where $g_2 \subset g_1$ and $g_3 \in \mathcal{MG}_\mathcal{K}$ s.t. $g_3'' = g_2''$.*

• *g is said to be a* ***transitive redundant*** *with respect to g_1, denoted $g_1 \vdash^+ g$, if it exists a sequence of n* MGs *($n \geq 2$), gen_1, gen_2, ..., gen_n, s.t. $gen_i \vdash gen_{i+1}$ ($i \in$ [1..(n-1)]) where $gen_1 = g_1$ and $gen_n = g$.*

Proposition 1. *The substitution relations $\vdash$ and $\vdash^+$ have the following properties:*

• *The substitution relation $\vdash$ is reflexive, symmetric but not necessarily transitive.*

• *The substitution relation $\vdash^+$ is reflexive, symmetric and transitive.*

The definition of a *succinct* minimal generator that we give hereafter requires that we adopt a total order relation among itemsets defined as follows:

Definition 9. (TOTAL ORDER RELATION) *Let $\preceq$ be a total order relation among item literals, i.e., $\forall\ i_1, i_2 \in \mathcal{I}$, we have either $i_1 \preceq i_2$ or $i_2 \preceq i_1$. This relation is extended to also cope with itemsets of different sizes by first considering their cardinality. This is done as follows: Let X and Y be two itemsets and let Card(X) and Card(Y) be their respective cardinalities. We then have:*

- *If Card(X) < Card(Y), then $X \prec Y$.*
- *If Card(X) = Card(Y), then X and Y are compared using their lexicographic order. Hence, $X \prec Y$ if and only if $X \preceq Y$ and $X \neq Y$.*

Example 7. Consider the alphabetic order on items as the basis for the total order relation $\preceq$ on itemsets[3]:

[3] In the remainder of the paper, we will only mention the criterion used to order items (*e.g.*, alphabetic order, ascending/descending support order, etc). The latter is then extended to be the total order relation on itemsets, as shown in Definition 9.

- Since Card(d) < Card(be), then $d \prec be$.
- Since Card(abd) = Card(abe), then abd and abe are compared using their lexicographic order. We then have $abd \prec abe$ since $abd \preceq abe$ and $abd \neq abe$.

The formal definition of a σ-equivalence class is then as follows:

Definition 10. (σ-EQUIVALENCE CLASS) *The operator $\vdash^{+}$ induces an equivalence relation on the set MG_f, of the MGs associated to an CI f, portioning it into distinct subsets called σ-equivalence classes. If $g \in \mathrm{MG}_f$, then the σ-equivalence class of g, denoted by $[g]$, is the subset of MG_f consisting of all elements that are transitively redundant w.r.t. g. In other words, we have: $[g] = \{g_1 \in \mathrm{MG}_f \mid g \vdash^{+} g_1\}$.*

*The smallest MG in each σ-equivalence class, w.r.t. the total order relation $\preceq$, will be considered as its **succinct** MG. While, the other MGs will be qualified as **redundant** MGs.*

Example 8. Let us consider the extraction context $\mathcal{K}$ depicted by Table 1. The total order relation $\preceq$ is set to the alphabetic order. Table 2 shows, for each CI, the following information: its MGs, its *succinct* MGs and its support. The MG "*adg*" is a *succinct* one, since it is the smallest MG, *w.r.t.* $\preceq$, among those of "*abcdeg*". Indeed, when extracting the first σ-equivalence class associated to "*abcdeg*", the whole MG set associated to "*abcdeg*" is considered. We then have: $adg \preceq aeg$, $adg \preceq bdg$ and $adg \preceq beg$. The MG "*aeg*" is a redundant one since $Subst(adg, ad, ae) = aeg \in \mathrm{MG}_{abcdeg}$ ($adg \vdash aeg$ and, hence, $adg \vdash^{+} aeg$). It is the same for the MGs "*bdg*" and "*beg*" since $adg \vdash^{+} bdg$ and $adg \vdash^{+} beg$.

Table 2. The CIs extracted from $\mathcal{K}$ and for each one, the corresponding MGs, *succinct* MGs and support

#	CI	MGs	Succinct MGs	Support
1	*c*	$\emptyset$	$\emptyset$	4
2	*abc*	*a, b*	*a, b*	3
3	*cde*	*d, e*	*d, e*	3
4	*cg*	*g*	*g*	3
5	*cfg*	*f*	*f*	2
6	*abcde*	*ad, ae, bd, be*	*ad*	2
7	*abcg*	*ag, bg*	*ag*	2
8	*abcfg*	*af, bf*	*af*	1
9	*cdeg*	*dg, eg*	*dg*	2
10	*cdefg*	*df, ef*	*df*	1
11	*abcdeg*	*adg, aeg, bdg, beg*	*adg*	1

The succinct system of minimal generators (SSMG) is then defined as follows:

Definition 11. *[1]* (SUCCINCT SYSTEM OF MINIMAL GENERATORS) *A succinct system of minimal generators* (SSMG) *is a system where only succinct* MGs *are retained among all* MGs *associated to each* CI.

Proposition 2. *[1] The* SSMG *is an exact representation of the* MG *set.*

In the remainder, the set of *succinct* (*resp. redundant*) *frequent* MGs that can be extracted from a context $\mathcal{K}$ will be denoted $\mathcal{FMG}\mathtt{suc}_{\mathcal{K}}$ (*resp.* $\mathcal{FMG}\mathtt{red}_{\mathcal{K}}$).

4 Succinct and Informative Association Rules

We now put the focus on integrating the concept of succinct system of minimal generators (SSMG) within the generic association rule framework. Our purpose is to obtain, without information loss, a more compact set of all association rules, from which the remaining *redundant* ones can be generated if desired.

4.1 Association Rules: Some Basic Notations

The formalization of the association rule extraction problem was introduced by Agrawal *et al.* [22]. The derivation of association rules is achieved starting from a set of *frequent* itemsets [23] extracted from a context $\mathcal{K}$ (denoted $\mathcal{FI}_{\mathcal{K}}$), for a minimal support threshold *minsupp*. An association rule R is a relation between itemsets and is of the form R: $X \Rightarrow (Y \backslash X)$, such that X and Y are *frequent* itemsets, and $X \subset Y$. The itemsets X and $(Y \backslash X)$ are, respectively, called the *premise* and the *conclusion* of the association rule R (also called *antecedent* and *consequent* of R [3], and *condition* and *consequence* of R [16]). The support of R, *Supp*(R), is equal to *Supp*(Y). R is said to be *valid* (or *strong*) if its confidence measure, $Conf(R) = \frac{Supp(Y)}{Supp(X)}$, is greater than or equal to a minimal threshold of confidence denoted *minconf*. If $Conf(R) = \mathbf{1}$, then R is called *exact association rule*, otherwise it is called *approximate association rule*. Please note that the confidence of R is always greater than or equal to its frequency (*i.e.*, $Conf(R) \geq \frac{Supp(R)}{|\mathcal{O}|}$).

4.2 Extraction of Succinct and Informative Association Rules

The problem of the relevance and the usefulness of association rules is of paramount importance. Indeed, an overwhelming quantity of association rules can be extracted even from small real-life datasets, among which a large number is *redundant* (*i.e.*, conveying the same information) [4,6]. This fact boosted the interest in novel approaches aiming to reduce this large association rule list, while preserving the most interesting rules. These approaches are mainly based on the battery of results provided by the Formal Concept Analysis (FCA) mathematical settings [15]. Thus, they focused on extracting irreducible nuclei of all association rules, commonly referred to as "*generic bases*", from which the remaining *redundant* association rules can be derived. Definition 12 describes the properties that characterize a generic basis once it is extracted without loss of information.

Definition 12. *A generic basis $\mathcal{B}$, associated with an appropriate inference mechanism, is said to fulfill the ideal properties of an association rule representation if it is [5]:*

1. ***lossless***: *$\mathcal{B}$ must enable the derivation of all valid association rules,*
2. ***sound***: *$\mathcal{B}$ must forbid the derivation of association rules that are not valid, and,*
3. ***informative***: *$\mathcal{B}$ must allow to exactly retrieve the support and confidence values of each derived association rule.*

The generic basis $\mathcal{B}$ is said to verify the property of derivability if it is lossless and sound.

The majority of the generic bases that were proposed in the literature convey association rules presenting implications between minimal generators (MGs) and closed itemsets (CIs) [3,5,7]. Indeed, it was proven that such association rules, with minimal premises and maximal conclusions, convey the maximum of information [3,16] and are hence qualified as the most informative association rules [3]. Furthermore, *succinct* MGs are very well suited for such association rules, since they offer the minimal possible premises. Indeed, they are the smallest ones in their respective σ-equivalence classes. They are also the most interesting ones since correlations in each *succinct* MG can not be predicted given correlations of its subsets and those of the other (redundant) MGs.

Hence, in order to extract much more compact sets of association rules, we propose to integrate the concept of the succinct system of minimal generators (SSMG) within the framework of generic bases. Although, our approach can be applied to different generic bases, we concentrate our presentation on the couple ($\mathcal{GB}$, $\mathcal{RI}$) proposed by Bastide *et al.* [3]. Indeed, in addition to the quality of the conveyed knowledge, the selected couple has the advantage to fulfill the *ideal* association rule representation's properties (summarized by Definition 12) in comparison to other generic bases (like the couple ($\mathcal{GDB}$, $\mathcal{LB}$) [4], $\mathcal{RR}$ [5], $\mathcal{NRR}$ [6], etc[4]) [5]. Moreover, as this will be shown in the remainder, these properties are still maintained after the application of the SSMG which ensures the derivation of *all redundant* association rules *without loss of information*. Unfortunately, this is not the case for the informative generic basis $\mathcal{IGB}$ [7]. Indeed, even if it was proven in [7] that $\mathcal{IGB}$ also fulfills the ideal properties of an association rule representation, the obtained generic basis, once the SSMG is applied to $\mathcal{IGB}$, is with information loss because some *succinct* MGs can sometimes be missing (*w.r.t.* the definition of $\mathcal{IGB}$, see [7]). Finally, the couple ($\mathcal{GB}$, $\mathcal{RI}$) offers quite interesting compactness rates (*vs.* the whole set of association rules) when compared to the remaining representations of association rules.

The couple ($\mathcal{SGB}$, $\mathcal{SRI}$) of *succinct* generic bases of association rules is defined as follows[5]:

Definition 13. (THE SUCCINCT GENERIC BASIS ($\mathcal{SGB}$) FOR EXACT ASSOCIATION RULES) *Let $\mathcal{FCI}_{\mathcal{K}}$ be the set of frequent* CI*s extracted from a context $\mathcal{K}$. For each entry f in $\mathcal{FCI}_{\mathcal{K}}$, let* FMG$suc_f$ *be the set of its succinct frequent* MG*s. The succinct generic basis for exact association rules $\mathcal{SGB}$ is given by: $\mathcal{SGB} = \{R\text{: } g \Rightarrow (f \backslash g) \mid f \in \mathcal{FCI}_{\mathcal{K}} \wedge g \in \text{FMG}suc_f \wedge g \neq f$ [6]$\}$.*

[4] $\mathcal{GDB}$ (*resp.* $\mathcal{LB}$, $\mathcal{RR}$, and $\mathcal{NRR}$) stands for Guigues-Duquenne Basis [4] (*resp.* Luxenburger Basis [4], Representative Rules [5], and Non-Redundant Rules [6]).

[5] The definition of the couple ($\mathcal{GB}$, $\mathcal{RI}$) can be derived from that of the couple ($\mathcal{SGB}$, $\mathcal{SRI}$) by considering *all* MGs instead of only *succinct* ones.

[6] The condition $g \neq f$ ensures discarding non-informative association rules of the form $g \Rightarrow \emptyset$.

Definition 14. (THE SUCCINCT TRANSITIVE REDUCTION ($\mathcal{SRI}$) FOR APPROXIMATE ASSOCIATION RULES) *Let $\mathcal{FMG}suc_{\mathcal{K}}$ be the set of the succinct frequent* MG*s extracted from a context $\mathcal{K}$. The succinct transitive reduction $\mathcal{SRI}$ is given by: $\mathcal{SRI}$ = $\{R$: $g \Rightarrow (f \backslash g) \mid f \in \mathcal{FCI}_{\mathcal{K}} \wedge g \in \mathcal{FMG}suc_{\mathcal{K}} \wedge f \in Cov^u(f_1)$ where $f_1 = g'' \wedge$ $Conf(R) = \frac{Supp(f)}{Supp(g)} \geq minconf\}$.*

Example 9. Consider the extraction context $\mathcal{K}$ given by Table 1 for a *minsupp* value equal to 1. The alphabetic order relation is used as a total one. The associated Iceberg concept lattice is depicted by Figure 2 (Left). A *succinct* exact generic rule is an "intra-node" association, with a confidence value equal to 1, within a γ-equivalence class of the Iceberg concept lattice. The use of the SSMG allows, for example, to only extract the *succinct* exact generic association rule $adg \Rightarrow bce$ from the γ-equivalence class having "*abcdeg*" for *frequent* CI, instead of four if *redundant frequent* MGs were of use (as indicated by the last entry in Table 2). A *succinct* approximate generic rule represents an "inter-node" association, assorted with a confidence measure, between a γ-equivalence class and another one belonging to its upper cover. For example, for *minconf* = 0.40, only the association rule $ad \overset{0.50}{\Rightarrow} bceg$ is extracted from both γ-equivalence classes having, respectively, "*abcde*" and "*abcdeg*" for *frequent* CI instead of four if *redundant frequent* MGs were of use (as indicated by the seventh entry in Table 2). The complete set of *succinct* generic association rules, extracted from $\mathcal{K}$, is reported in Figure 2 (Right). The cardinality of $\mathcal{SGB}$ (*resp.* $\mathcal{GB}$) is equal to $\mathbf{13}$ (*resp.* $\mathbf{23}$), while that of $\mathcal{SRI}$ (*resp.* $\mathcal{RI}$) is equal to $\mathbf{21}$ (*resp.* $\mathbf{28}$). Hence, thanks to the SSMG, we are able to discard $\mathbf{43.48\%}$ (*resp.* $\mathbf{25.00\%}$) of the exact (*resp.* approximate) generic association rules since they are *redundant*. It is important to mention that the total number of association rules, which can be retrieved from $\mathcal{K}$, is equal to $\mathbf{943}$.

Remark 1. It is worth noting that in [24], the authors baptized *succinct association rules* those obtained using a pruning strategy based on a model called *maximal potentially useful* (MaxPUF) association rules. However, such reduction is done with information loss since the regeneration of the whole set of valid association rules is not ensured. It is important to mention that this approach and ours can be easily combined towards a more reduced set of association rules.

4.3 Derivation of Redundant Association Rules

In the following, we study the structural properties of the new generic bases introduced in the previous subsection. The study requires checking the *ideal* properties of an association rule representation (see Definition 12). Since, it was shown in [5] that the couple ($\mathcal{GB}$, $\mathcal{RI}$) is extracted without loss of information, it is sufficient to show that it is possible to derive without loss of information *all* association rules that belong to the couple ($\mathcal{GB}$, $\mathcal{RI}$) starting from the couple ($\mathcal{SGB}$, $\mathcal{SRI}$). If so, *all redundant* association rules can be derived from ($\mathcal{SGB}$, $\mathcal{SRI}$).

Association rules belonging to the couple ($\mathcal{SGB}$, $\mathcal{SRI}$) are implications between *succinct frequent* minimal generators (MGs) and *frequent* closed itemsets (CIs).

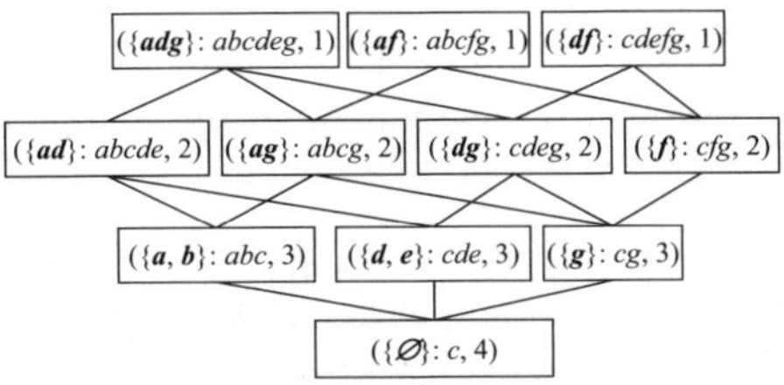

$\mathcal{SGB}$		
$R_1: \emptyset \Rightarrow c$	$R_2: a \Rightarrow bc$	$R_3: b \Rightarrow ac$
$R_4: d \Rightarrow ce$	$R_5: e \Rightarrow cd$	$R_6: g \Rightarrow c$
$R_7: ad \Rightarrow bce$	$R_8: ag \Rightarrow bc$	$R_9: dg \Rightarrow ce$
$R_{10}: f \Rightarrow cg$	$R_{11}: adg \Rightarrow bce$	$R_{12}: af \Rightarrow bcg$
$R_{13}: df \Rightarrow ceg$		

$\mathcal{SRI}$		
$R_1: \emptyset \overset{0.75}{\Rightarrow} abc$	$R_2: \emptyset \overset{0.75}{\Rightarrow} cde$	$R_3: \emptyset \overset{0.75}{\Rightarrow} cg$
$R_4: a \overset{0.67}{\Rightarrow} bcg$	$R_5: b \overset{0.67}{\Rightarrow} acg$	$R_6: a \overset{0.67}{\Rightarrow} bcde$
$R_7: b \overset{0.67}{\Rightarrow} acde$	$R_8: d \overset{0.67}{\Rightarrow} abce$	$R_9: e \overset{0.67}{\Rightarrow} abcd$
$R_{10}: d \overset{0.67}{\Rightarrow} ceg$	$R_{11}: e \overset{0.67}{\Rightarrow} cdg$	$R_{12}: g \overset{0.67}{\Rightarrow} abc$
$R_{13}: g \overset{0.67}{\Rightarrow} cde$	$R_{14}: g \overset{0.67}{\Rightarrow} cf$	$R_{15}: ad \overset{0.50}{\Rightarrow} bceg$
$R_{16}: ag \overset{0.50}{\Rightarrow} bcde$	$R_{17}: ag \overset{0.50}{\Rightarrow} bcf$	$R_{18}: dg \overset{0.50}{\Rightarrow} abce$
$R_{19}: dg \overset{0.50}{\Rightarrow} cef$	$R_{20}: f \overset{0.50}{\Rightarrow} abcg$	$R_{21}: f \overset{0.50}{\Rightarrow} cdeg$

Fig. 2. (**Left**) For *minsupp* = **1**, the Iceberg concept lattice associated to the extraction context $\mathcal{K}$ of Table 1. Each one of its γ-equivalence classes contains a *frequent* CI f accompanied by the set of its *succinct frequent* MGs FMGsuc$_f$ and its support, in the form (FMGsuc$_f$: f, Supp(f)). (**Right**) The complete set of *succinct* generic association rules extracted from $\mathcal{K}$.

Hence, to derive the couple ($\mathcal{GB}$, $\mathcal{RI}$), *redundant frequent* MGs need to be deduced since they form the premises of *redundant* generic association rules, *i.e.*, association rules belonging to ($\mathcal{GB}$, $\mathcal{RI}$) and discarded from ($\mathcal{SGB}$, $\mathcal{SRI}$). In order to derive *all* association rules belonging to ($\mathcal{GB}$, $\mathcal{RI}$), we propose a new axiom called the *substitution axiom*. Thus, from each association rule R: $X \Rightarrow (Y \backslash X)$ of ($\mathcal{SGB}$, $\mathcal{SRI}$) where $X \in \mathcal{FMG}\mathtt{suc}_{\mathcal{K}}$ and $Y \in \mathcal{FCI}_{\mathcal{K}}$, we propose to derive, using the substitution axiom, the set of *redundant* generic association rules given by: *Red_Gen_Assoc_Rules*$_{R:\ X \Rightarrow (Y \backslash X)}$ = $\{R'$: $Z \Rightarrow (Y \backslash Z) \mid Z \in \mathcal{FMG}\mathtt{red}_{\mathcal{K}}$ *s.t.* $X \vdash^{+} Z\}$. The substitution axiom proceeds according to the following steps:

Step 1: The set $\mathcal{GB}$ (*resp.* $\mathcal{RI}$) is firstly initialized to $\mathcal{SGB}$ (*resp.* $\mathcal{SRI}$).

Step 2: Association rules belonging to ($\mathcal{GB}$, $\mathcal{RI}$) are processed in an ascending order of their respective sizes [7], *i.e.*, that for an association rule R: $X \Rightarrow (Y \backslash X) \in$ ($\mathcal{GB}$, $\mathcal{RI}$) where $X \in \mathcal{FMG}\mathtt{suc}_{\mathcal{K}}$ and $Y \in \mathcal{FCI}_{\mathcal{K}}$, the set of *redundant* generic association rules associated to each association rule R_1: $X_1 \Rightarrow (Y_1 \backslash X_1)$, *s.t.* $X_1 \subset X$ and $Y_1 \subset Y$, were already derived.

Step 2.1: For each association rule R: $X \Rightarrow (Y \backslash X) \in \mathcal{GB}$, derive the set of *redundant* generic association rules *Red_Gen_Assoc_Rules*$_R$ = $\{R'$: $Z \Rightarrow (Y \backslash Z) \mid$ Z is the result of the substitution of a subset of X, say V, by T *s.t.* (R_1: $V \Rightarrow (I \backslash V)$, R_2: $T \Rightarrow (I \backslash T)) \in \mathcal{GB}$ where $I \in \mathcal{FCI}_{\mathcal{K}}$ and $\nexists\ Z_1 \subseteq Z$ *s.t.* $Z_1 \Rightarrow (Y \backslash Z_1) \in \mathcal{GB}\}$.

[7] The size of an association rule $X \Rightarrow Y$ is equal to the cardinality of $X \bigcup Y$.

Step 2.2: For each association rule R: $X \Rightarrow (Y\backslash X) \in \mathcal{RI}$, derive the set of *redundant* generic association rules *Red_Gen_Assoc_Rules*$_R$ = $\{R'$: $Z \Rightarrow (Y\backslash Z)$ | Z is the result of the substitution of a subset of X, say V, by T *s.t.* (R_1: $V \Rightarrow (I\backslash V)$, R_2: $T \Rightarrow (I\backslash T)) \in \mathcal{GB}$ where $I \in \mathcal{FCI}_\mathcal{K}$ and $\nexists$ $Z_1 \subseteq Z$ *s.t.* $Z_1 \Rightarrow (Y\backslash Z_1) \in \mathcal{RI}\}$. ♦

It is worth noting that comparing Z to Z_1 ensures discarding the case where a substitution leads to an already existing association rule or to a one having a *non-minimal* generator as a premise.

Example 10. From the association rule R: $adg \Rightarrow bce$ belonging to $\mathcal{SGB}$ (*cf.* Figure 2 (Right)), we will show how to derive association rules belonging to $\mathcal{GB}$ which are *redundant w.r.t.* R. Before that R is processed, *all* association rules whose respective sizes are lower than that of R (*i.e.*, lower than **6**) were handled and *redundant* generic association rules were derived from such association rules. Among the handled association rules, we find those having for premises the **2**-subsets of "*adg*", *i.e.*, $ad \Rightarrow bce$, $ag \Rightarrow bc$ and $dg \Rightarrow ce$. To derive the *redundant* generic association rules associated to R, the first **2**-subset of "*adg*", *i.e.*, "*ad*", is replaced by the *frequent* MGs having its closure, *i.e.*, the *redundant frequent* MGs "*ae*", "*bd*" and "*be*". Indeed, generic association rules using these latter as premises were already derived as *redundant w.r.t.* $ad \Rightarrow bce$. Hence, we augment $\mathcal{GB}$ by the following association rules: $aeg \Rightarrow bcd$, $bdg \Rightarrow ace$ and $beg \Rightarrow acd$. The same process is applied to the second subset of "*adg*", *i.e.*, "*ag*". Nevertheless, the obtained association rule, namely $bdg \Rightarrow ace$, will not be added to $\mathcal{GB}$. Indeed, it already exists an association rule in $\mathcal{GB}$ such that $Z_1 \Rightarrow (abcdeg\backslash Z_1)$ and $Z_1 \subseteq abg$ (Z_1 being itself equal to "*abg*"). It is the same for the derived association rule using the third subset "*dg*", *i.e.*, $aeg \Rightarrow bcd$ (Z_1 being equal to "*aeg*").

Now, we prove that the substitution axiom allows the couple ($\mathcal{SGB}$, $\mathcal{SRI}$) to be *lossless* and *sound*. Then, we show that this couple is also informative.

Proposition 3. *The couple* ($\mathcal{SGB}$, $\mathcal{SRI}$) *of generic bases is lossless:* $\forall$ R: $X \Rightarrow (Y\backslash X) \in (\mathcal{SGB}, \mathcal{SRI})$, *the set* Red_Gen_Assoc_Rules$_R = \{R'$: $Z \Rightarrow (Y\backslash Z) \mid Z \in \mathcal{FMG}\mathtt{red}_\mathcal{K}$ *s.t.* $X \vdash^+ Z\}$ *of the redundant generic association rules with respect to R, is completely derived thanks to the proposed substitution axiom.*

Proof. The sorting imposed in Step 2 ensures that, before R is processed, all association rules whose respective sizes are lower than that of R were handled, and redundant generic association rules were then derived from such association rules. Hence, all information required to derive association rules belonging to Red_Gen_Assoc_Rules$_R$ *are gathered thanks to the different sets* Red_Gen_Assoc_Rules$_{R_1: X_1 \Rightarrow (Y_1\backslash X_1)}$ *such that* $X_1 \in \mathcal{FMG}\mathtt{suc}_\mathcal{K}$, $Y_1 \in \mathcal{FCI}_\mathcal{K}$ *and* $Y_1 \subset Y$. *Indeed, using these sets, all redundant frequent* MG*s, with respect to* X, *are straightforwardly derivable since, for each subset* X_1 *of* X, *the different frequent* MG*s belonging to its* γ*-equivalence class are already known as they are the premises of association rules belonging to the sets* Red_Gen_Assoc_Rules$_{R_1}$ *defined above. Hence, all association rules belonging to* ($\mathcal{GB}$, $\mathcal{RI}$) *can be deduced from* ($\mathcal{SGB}$, $\mathcal{SRI}$) *using the substitution axiom. Therefore, the couple* ($\mathcal{SGB}$, $\mathcal{SRI}$) *is lossless.* ♦

Proposition 4. *The couple* ($\mathcal{SGB}$, $\mathcal{SRI}$) *of generic bases is sound:* $\forall$ R': $Z \Rightarrow (Y \backslash Z)$ $\in$ *Red_Gen_-*Assoc*_Rules*$_{R:\ X \Rightarrow (Y \backslash X)}$, *Supp*($R'$)=*Supp*($R$) *and Conf*($R'$)= *Conf*($R$).

Proof. On the one hand, Supp(R) *is equal to Supp*(Y). *It is the same for Supp*(R'). *Hence, Supp*(R') = *Supp*(R). *On the other hand,* X *and* Z *are two frequent* MG*s belonging to the same* γ*-equivalence class. Hence, Supp*(X) *is equal to Supp*(Z). *Thus, Conf*(R') = $\frac{Supp(Y)}{Supp(Z)} = \frac{Supp(Y)}{Supp(X)}$ = *Conf*(R). *Therefore, the couple* ($\mathcal{SGB}$, $\mathcal{SRI}$) *is sound.* ♦

The property of derivability is fulfilled by the couple ($\mathcal{SGB}$, $\mathcal{SRI}$) of generic bases since it is lossless and sound. Now, we show that this couple allows the retrieval of the exact values of the support and the confidence associated to each derived association rule.

Proposition 5. *The couple* ($\mathcal{SGB}$, $\mathcal{SRI}$) *of generic bases is informative: the support and the confidence of all derived association rules can exactly be retrieved from* ($\mathcal{SGB}$, $\mathcal{SRI}$).

Proof. Association rules belonging to the couple ($\mathcal{SGB}$, $\mathcal{SRI}$) *are of the following form:* $g \Rightarrow (f \backslash g)$ *where* $g \in \mathcal{FMG}suc_{\mathcal{K}}$ *and* $f \in \mathcal{FCI}_{\mathcal{K}}$. *Therefore, we are able to reconstitute all necessary frequent* CI*s by concatenation of the premise and the conclusion parts of the generic association rules belonging to* ($\mathcal{SGB}$, $\mathcal{SRI}$). *Since the support of a frequent itemset* I *is equal to the support of the smallest frequent* CI *containing it [9], then the support of* I *and its closure can be straightforwardly derived from* ($\mathcal{SGB}$, $\mathcal{SRI}$). *Hence, the support and the confidence values of all redundant association rules can exactly be retrieved. Thus, the couple* ($\mathcal{SGB}$, $\mathcal{SRI}$) *is informative.* ♦

The substitution axiom is proved to be lossless, sound and informative; allowing to derive *all* association rules forming ($\mathcal{GB}$, $\mathcal{RI}$) as well as their *exact* support and confidence values. Since the couple ($\mathcal{GB}$, $\mathcal{RI}$) is shown to be extracted without loss of information [5], we can deduce that the couple ($\mathcal{SGB}$, $\mathcal{SRI}$) is also extracted without information loss. In order to find the complete set of *valid redundant* association rules that can be extracted from a context $\mathcal{K}$, the axiom of transitivity proposed by Luxenburger [25] should be applied to the $\mathcal{RI}$ basis to derive association rules forming the informative basis $\mathcal{IB}$ for the approximate association rules [3]. Then, the cover operator proposed by Kryszkiewicz [5] or the lossless and sound axiomatic system proposed by Ben Yahia and Mephu Nguifo [26] makes it possible to derive *all valid redundant* association rules starting from the couple ($\mathcal{GB}$, $\mathcal{IB}$). The complete process allowing to derive *all valid* (*redundant*) association rules (denoted $\mathcal{AR}$), starting from the couple ($\mathcal{SGB}$, $\mathcal{SRI}$), is hence as follows:

$$(\mathcal{SGB}, \mathcal{SRI}) \overset{substitution\ axiom}{- - - - - \longrightarrow} (\mathcal{GB}, \mathcal{RI}) \overset{transitivity\ axiom}{- - - - - \longrightarrow} (\mathcal{GB}, \mathcal{IB}) \overset{cover\ operator\ or\ Ben\ Yahia\ and\ Mephu\ Nguifo\ axiomatic\ system}{- - - - - - - - - - - - - - - - - - \longrightarrow} \mathcal{AR}$$

5 Experimental Study

We carried out experimentations on benchmark datasets[8] in order to evaluate the number of (*succinct*) generic association rules. Characteristics of these datasets are summarized by Table 3. Hereafter, we use a logarithmically scaled ordinate axis in all figures.

Table 3. Dataset characteristics

Dataset	Number of items	Number of objects	Average object size	*minsupp* interval (%)
PUMSB	7, 117	49, 046	74.00	90 - 60
MUSHROOM	119	8, 124	23.00	1 - 0.01
CONNECT	129	67, 557	43.00	90 - 50
T40I10D100K	1, 000	100, 000	39.61	10 - 1

We compared both couples ($\mathcal{SGB}$, $\mathcal{SRI}$) and ($\mathcal{GB}$, $\mathcal{RI}$) using the couple size as evaluation criterion, for a fixed *minsupp* value. Indeed, this was carried out for the PUMSB (*resp.* CONNECT, MUSHROOM and T40I10D100K) dataset for a *minsupp* value equal to **70%** (*resp.* **50%**, **0.01%** and **1%**). Obtained results are graphically sketched by Figure 3. For each dataset, the *minconf* value varies between the aforementioned *minsupp* value and **100%**.

Figure 3 points out that removing redundancy within the *frequent* MG set[9] offers an interesting lossless reduction of the number of the extracted generic association rules. Indeed, the use of the SSMG allows to remove in average **63.03%** (*resp.* **49.46%**) of the *redundant* generic association rules extracted from the PUMSB (*resp.* MUSHROOM) dataset. The maximum rate of redundancy reaches **68.11%** (*resp.* **53.84%**) for the PUMSB (*resp.* MUSHROOM) dataset, for a *minconf* value equal to **100%** (*resp.* **20%**). For the CONNECT and T40I10D100K datasets, the respective curves representing the size of the couple ($\mathcal{SGB}$, $\mathcal{SRI}$) and those representing the size of the couple ($\mathcal{GB}$, $\mathcal{RI}$) are strictly overlapping. Indeed, these two datasets do not generate *redundant frequent* MGs and, hence, there are no *redundant* generic association rules. Furthermore, for the T40I10D100K dataset, none *exact* association rule is generated since each *frequent* MG is equal to its closure.

We also set the *minconf* value to **0%** to evaluate the reduction rate within *valid exact* generic association rules (*i.e.*, the generic basis $\mathcal{GB}$) compared to that within *approximate* ones (*i.e.*, the $\mathcal{RI}$ basis). In this context, Figure 4 shows that, for the PUMSB dataset, in average **62.46%** (*resp.* **49.11%**) of the exact (*resp.* approximate) generic association rules are *redundant*, and the maximum rate of redundancy reaches **68.46%** (*resp.* **62.65%**) for a *minsupp* value equal to **65%** (*resp.* **65%**). For the MUSHROOM dataset, in average **50.55%** (*resp.* **52.65%**) of the exact (*resp.* approximate) generic association rules are *redundant*, and the maximum rate of redundancy reaches **53.23%** (*resp.* **57.86%**) for a *minsupp* value equal to **0.20%** (*resp.* **0.10%**).

[8] These benchmark datasets are downloadable from: *http://fimi.cs.helsinki.fi/data.*

[9] Interested readers are referred to [1] for more details.

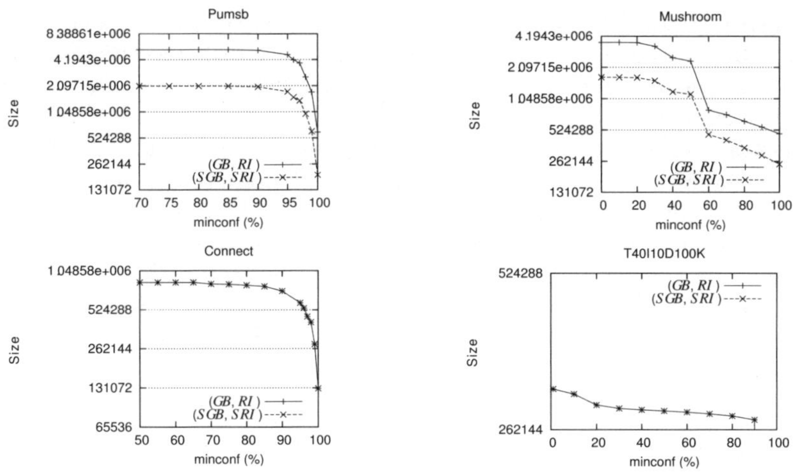

Fig. 3. For a fixed *minsupp* value, the size of the couple ($\mathcal{GB}$, $\mathcal{RI}$) of generic bases compared to that of the couple ($\mathcal{SGB}$, $\mathcal{SRI}$) of *succinct* generic bases

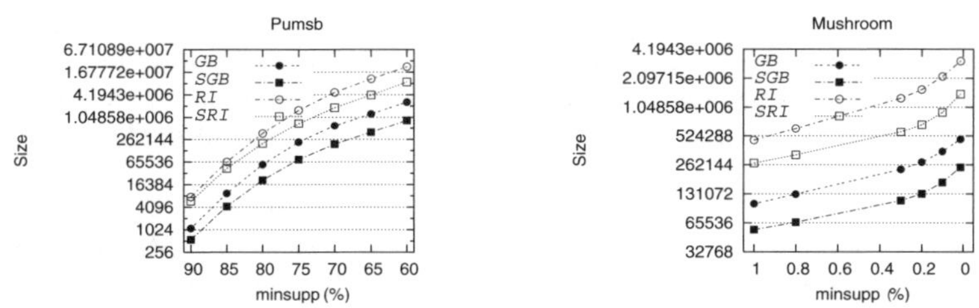

Fig. 4. For a fixed *minconf* value, the size of the generic basis $\mathcal{GB}$ (*resp.* $\mathcal{RI}$) compared to that of the *succinct* generic basis $\mathcal{SGB}$ (*resp.* $\mathcal{SRI}$)

These experiments clearly indicate that our approach can be advantageously used to eliminate, without loss of information, a large number of *redundant* generic association rules.

6 Conclusion and Future Work

In this paper, we briefly described the principal structural properties of the succinct system of minimal generators (SSMG) redefined in [1]. We then incorporated it into the framework of generic bases to tackle the problem of succinctness within generic association rule sets. Thus, we introduced two new *succinct* generic bases of association rules, namely the couple ($\mathcal{SGB}$, $\mathcal{SRI}$). We also showed that, starting from this couple, it is possible to derive without loss of information *all valid* association rules belonging to the couple ($\mathcal{GB}$, $\mathcal{RI}$) thanks to the application of a new substitution process. Consequently, any *valid redundant* association rule, which can be extracted from a context, can be inferred starting from the couple ($\mathcal{SGB}$, $\mathcal{SRI}$). Finally, carried out experiments confirmed that the application of the SSMG makes it possible to eliminate, as much as possible, *redundant* generic association rules and, hence, to only offer succinct and informative ones to users.

In the near future, from the viewpoint of the presentation and quality of knowledge, we plan to set up an association rule visualization platform based on *succinct* generic bases, which, in our opinion, will constitute a helpful tool for the users. In this setting, integrating quality measures and user-defined constraints will be interesting for more association rule pruning. In addition, we think that a careful study of the effect of the total order relation choice, on the quality of the extracted *succinct* association rules according to the data under consideration, presents an interesting issue towards increasing the knowledge usefulness.

Acknowledgements

We are grateful to the anonymous reviewers for their helpful comments. This work is partly supported by the French-Tunisian project PAI CMCU 05G1412.

References

1. Hamrouni, T., Ben Yahia, S., Mephu Nguifo, E.: Succinct system of minimal generators: A thorough study, limitations and new definitions. In: Ben Yahia, S., Mephu Nguifo, E., Behlohlavek, R. (eds.) CLA 2006. LNCS (LNAI), vol. 4923, pp. 80–95. Springer, Heidelberg (2008) (this volume)
2. Ceglar, A., Roddick, J.F.: Association mining. ACM Computing Surveys 38(2) (July 2006)
3. Bastide, Y., Pasquier, N., Taouil, R., Stumme, G., Lakhal, L.: Mining Minimal Non-redundant Association Rules Using Frequent Closed Itemsets. In: Palamidessi, C., Moniz Pereira, L., Lloyd, J.W., Dahl, V., Furbach, U., Kerber, M., Lau, K.-K., Sagiv, Y., Stuckey, P.J. (eds.) CL 2000. LNCS (LNAI), vol. 1861, pp. 972–986. Springer, Heidelberg (2000)
4. Stumme, G., Pasquier, N., Bastide, Y., Taouil, R., Lakhal, L.: Intelligent Structuring and Reducing of Association Rules with Formal Concept Analysis. In: Baader, F., Brewka, G., Eiter, T. (eds.) KI 2001. LNCS (LNAI), vol. 2174, pp. 335–350. Springer, Heidelberg (2001)
5. Kryszkiewicz, M.: Concise representation of frequent patterns and association rules. In: Habilitation dissertation, Institute of Computer Science, Warsaw University of Technology, Warsaw, Poland (August 2002)
6. Zaki, M.J.: Mining non-redundant association rules. In: Data Mining and Knowledge Discovery (DMKD), vol. 9(3), pp. 223–248. Springer, Heidelberg (2004)
7. Gasmi, G., Ben Yahia, S., Mephu Nguifo, E., Slimani, Y.: $\mathcal{IGB}$: A new informative generic base of association rules. In: Ho, T.-B., Cheung, D., Liu, H. (eds.) PAKDD 2005. LNCS (LNAI), vol. 3518, pp. 81–90. Springer, Heidelberg (2005)
8. Li, J.: On optimal rule discovery. IEEE Transactions on Knowledge and Data Engineering (TKDE) 18(4), 460–471 (2006)
9. Pasquier, N., Bastide, Y., Taouil, R., Stumme, G., Lakhal, L.: Generating a condensed representation for association rules. Journal of Intelligent Information Systems 24(1), 25–60. Kluwer Academic Publisher, Dordrecht (2005)
10. Boulicaut, J.F., Bykowski, A., Rigotti, C.: Free-sets: A condensed representation of Boolean data for the approximation of frequency queries. In: Data Mining and Knowledge Discovery (DMKD), vol. 7(1), pp. 5–22. Springer, Heidelberg (2003)

11. Calders, T., Goethals, B.: Non-derivable itemset mining. In: Data Mining and Knowledge Discovery (DMKD), vol. 14(1), pp. 171–206. Springer, Heidelberg (2007)
12. Ohsaki, M., Kitaguchi, S., Okamoto, K., Yokoi, H., Yamaguchi, T.: Evaluation of rule interestingness measures with a clinical dataset on Hepatitis. In: Boulicaut, J.-F., Esposito, F., Giannotti, F., Pedreschi, D. (eds.) PKDD 2004. LNCS (LNAI), vol. 3202, pp. 362–373. Springer, Heidelberg (2004)
13. Srikant, R., Vu, Q., Agrawal, R.: Mining association rules with item constraints. In: Proceedings of the 3rd International Conference on Knowledge Discovery in Databases and Data Mining (KDD 1997), Newport Beach, California, USA, pp. 67–73 (August 1997)
14. Bonchi, F., Lucchese, C.: On condensed representations of constrained frequent patterns. In: Journal of Knowledge and Information Systems, pp. 1–22 (April 2005)
15. Wille, R.: Restructuring lattices theory: An approach based on hierarchies of concepts. In: Ordered Sets, pp. 445–470. Reidel, Dordrecht-Boston (1982)
16. Kryszkiewicz, M.: Representative association rules and minimum condition maximum consequence association rules. In: Żytkow, J.M. (ed.) PKDD 1998. LNCS, vol. 1510, pp. 361–369. Springer, Heidelberg (1998)
17. Dong, G., Jiang, C., Pei, J., Li, J., Wong, L.: Mining succinct systems of minimal generators of formal concepts. In: Zhou, L.-z., Ooi, B.-C., Meng, X. (eds.) DASFAA 2005. LNCS, vol. 3453, pp. 175–187. Springer, Heidelberg (2005)
18. Ganter, B., Wille, R.: Formal Concept Analysis. Springer, Heidelberg (1999)
19. Bastide, Y., Taouil, R., Pasquier, N., Stumme, G., Lakhal, L.: Mining frequent patterns with counting inference. ACM-SIGKDD Explorations 2(2), 66–75 (2000)
20. Mephu Nguifo, E.: Galois lattice: A framework for concept learning, design, evaluation and refinement. In: Proceedings of the IEEE International Conference on Tools with Artificial Intelligence (ICTAI 1994), New-Orleans, USA, pp. 461–467 (November 1994)
21. Stumme, G., Taouil, R., Bastide, Y., Pasquier, N., Lakhal, L.: Computing Iceberg concept lattices with TITANIC. Journal on Knowledge and Data Engineering (KDE) 2(42), 189–222 (2002)
22. Agrawal, R., Imielinski, T., Swami, A.: Mining association rules between sets of items in large databases. In: Proceedings of the ACM-SIGMOD International Conference on Management of Data, Washington D. C., USA, pp. 207–216 (May 1993)
23. Agrawal, R., Mannila, H., Srikant, R., Toivonen, H., Verkamo, A.I.: Fast discovery of association rules. In: Proceedings of Advances in Knowledge Discovery and Data Mining, pp. 307–328. AAAI Press, Menlo Park (1996)
24. Deogun, J.S., Jiang, L.: SARM - succinct association rule mining: An approach to enhance association mining. In: Hacid, M.-S., Murray, N.V., Raś, Z.W., Tsumoto, S. (eds.) ISMIS 2005. LNCS (LNAI), vol. 3488, pp. 121–130. Springer, Heidelberg (2005)
25. Luxenburger, M.: Implications partielles dans un contexte. Mathématiques, Informatique et Sciences Humaines 29(113), 35–55 (1991)
26. Ben Yahia, S., Mephu Nguifo, E.: Revisiting generic bases of association rules. In: Kambayashi, Y., Mohania, M., Wöß, W. (eds.) DaWaK 2004. LNCS, vol. 3181, pp. 58–67. Springer, Heidelberg (2004)

Concept Lattice Representations of Annotated Taxonomies

Tim B. Kaiser[1], Stefan E. Schmidt[2], and Cliff A. Joslyn[3]

[1] Department of Mathematics, Darmstadt University of Technology
[2] Institute of Algebra, Technische Universität Dresden
[3] Computer Science, Los Alamos National Laboratory
tkaiser@mathematik.tu-darmstadt.de, midt1@msn.com, joslyn@lanl.gov

Abstract. We show how the concept of an *annotated ordered set* can be used to model large taxonomically structured ontologies such as the Gene Ontology. By constructing a formal context consistent with a given annotated ordered set, their concept lattice representations are derived. We develop the fundamental mathematical relations present in this formulation, in particular deriving a conceptual pre-ordering of the taxonomy, and constructing a correspondence between the annotations of an ordered set and the closure systems of its filter lattice. We study an example from the Gene Ontology to demonstrate how the introduced technique can be utilized for ontology review.

1 Introduction

Ontologies, taxonomies, and other semantic hierarchies are increasingly necessary for organizing large quantities of data, and recent years have seen the emergence of new large taxonomically structured ontologies such as the Gene Ontology (GO) [AsMBaC00][1], the UMLS Meta-Thesaurus [BoOMiJ02], object-oriented typing hierarchies [KnTReJ00], and verb typing hierarchies in computational linguistics [DaA00a]. Cast as Directed Acyclic Graphs (DAGs), these all entail canonical mathematical representations as *annotated ordered sets* (previously called "poset ontologies" [JM04]).

The size and complexity of these modern taxonomic hierachies requires algorithmic treatement of tasks which could previously be done by hand or by inspection. These include reviewing the consistency and completeness of the underlying hierarchical structure, and the coherence of the *labeling* (the assignment of objects to ontological categories). The close similarity of the annotated ordered set representations of these taxonomies to concept lattices in Formal Concept Analysis (FCA) [GW99] suggests pursuing their representation within FCA, in order to gain a deeper understanding of their mathematical structure and optimize their management and analytical tractibility (see also [JoCGeD06]).

We begin this paper by defining annotated ordered sets, and demonstrate their appropriateness for representing the GO. Then, we define a formal context appropriate for annotated ordered sets, and thereby construct their concept

[1] http://www.geneontology.org

S. Ben Yahia et al. (Eds.): CLA 2006, LNAI 4923, pp. 214–225, 2008.

lattices. We analyze the relationship between an annotated ordered set and its concept lattice representation, which includes the formulation of a correspondence between the annotations of an ordered set and the closure systems of its filter lattice. Additionally, we study an example from the GO. The paper is concluded with a discussion of future applications and extensions of the outlined approach. Throughout, we assume that the reader is knowledgable of the theory of FCA [GW99].

2 Taxonomic Ontologies as Annotated Ordered Sets

We use the GO as our touchstone for the general concept of an annotated ordered set. Fig. 1 (from [AsMBaC00]) shows a sample portion of the GO. Nodes in black represent functional categories of biological processes, basically things that proteins "do". Nodes are connected by links indicating subsumptive, "is-a" relations between categories, so that, for example, "DNA ligation" is a kind of "DNA repair". Elsewhere in the GO, nodes can also be connected by compositional, "has-part" relations, but for our purposes, we will consider the GO as singly-typed.

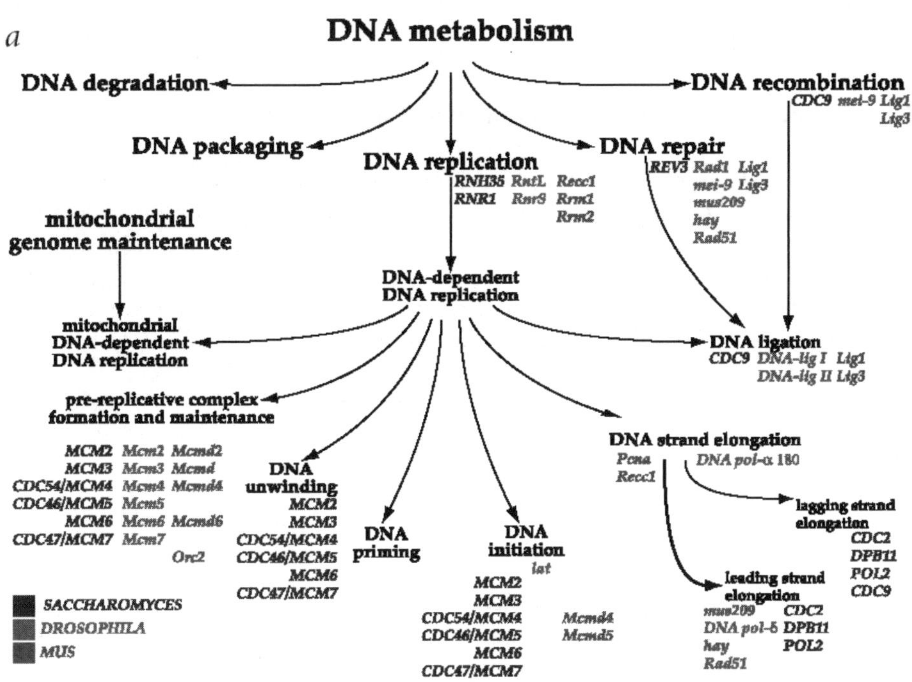

Fig. 1. A portion of the BP branch of the GO (used with permission from [AsMBaC00]). GO nodes in the hierarchy have genes from three species annotated below them.

Colored terms attached to each node indicate particular proteins in particular species which perform those functions. This assignment is called "annotation". Note that proteins can be annotated to multiple functions, for example yeast MCM2 does both "DNA initiation" and "DNA unwinding". Furthermore, an annotation to a node should be considered a simultaneous annotation to all ancestor nodes, so that yeast CDC9 does both "DNA ligation" and "DNA repair". So explicit such annotations, for example CDC9 annotation to both "DNA ligation" and "DNA recombination" in Fig. 1, are actually redundant. Finally, note the presence of multiple inheritance: "DNA ligation" is both "DNA repair" and "DNA recombination".

It is therefore appropriate to model structures such as the GO as structures called annotated ordered sets (previously referred to as poset ontologies [JM04]).

Definition 1 (Annotated Ordered Set). *Let $\mathcal{P} := (P, \leq_{\mathcal{P}})$ be a finite ordered set (poset), let X be a finite set of* labels, *and let $F : X \to 2^P$ be an* annotation function. *Then we call $\mathbb{O} := (\mathcal{P}, X, F)$ an* annotated ordered set *and refer to (X, F) as an* annotation *of $\mathcal{P}$. In case $\mathcal{P}$ is a (complete) lattice we call $\mathbb{O}$ an* annotated (complete) lattice *denoted $\mathbb{L}$. If $|F(x)| = 1$ for all $x \in X$, for convenience, we regard F as a map from X to P and say that $\mathbb{O}$ is* elementary.

It should be emphasized that Fig. 1 shows only a small fragment of the GO, which currently has on the order of 20,000 nodes in three disjoint taxonomies, annotated by hundreds of thousands of proteins from dozens of species.

3 Concept Lattice Representations

We are now prepared to construct concept lattice representations of annotated ordered sets by deriving the appropriate formal contexts. For an ordered set $\mathcal{P} := (P, \leq_{\mathcal{P}})$ and node $q \in P$ we denote by $\uparrow q := \{p \in P \mid q \leq_{\mathcal{P}} p\}$ the principal filter of q and dually by $\downarrow q$ the principal ideal. In general, for $Q \subseteq P$ we define $\uparrow Q := \{p \in P \mid \exists q \in Q : q \leq_{\mathcal{P}} p\}$ and dually $\downarrow Q$. Given an annotated ordered set $\mathbb{O} := (\mathcal{P}, X, F)$ we can construct a formal context $\mathbb{K}_{\mathbb{O}} := (X, P, I)$ where

$$xIp :\Longleftrightarrow \downarrow p \cap F(x) \neq \emptyset$$

for $x \in X, p \in P$. Note also that

$$xIp \quad \Longleftrightarrow \quad \exists q \leq_{\mathcal{P}} p : q \in F(x) \quad \Longleftrightarrow \quad p \in \bigcup_{q \in F(x)} \uparrow q.$$

The concept lattice of $\mathbb{K}_{\mathbb{O}}$ will be denoted by $\underline{\mathfrak{B}}_{\mathbb{O}} := (\mathfrak{B}_{\mathbb{O}}, \leq_{\underline{\mathfrak{B}}_{\mathbb{O}}})$, where $\mathfrak{B}_{\mathbb{O}} := \mathfrak{B}(\mathbb{K}_{\mathbb{O}})$ is the set of formal concepts of the formal context $\mathbb{K}_{\mathbb{O}}$ [GW99]. $\underline{\mathfrak{B}}_{\mathbb{O}}$ is called the *concept lattice representation* of the annotated ordered set $\mathbb{O}$.

In case $\mathbb{O}$ forms an annotated complete lattice and $(A, B) \in \mathfrak{B}_{\mathbb{O}}$ is a formal concept in $\mathfrak{B}_{\mathbb{O}}$, we observe that $A = B^I$ is the set of all $x \in X$ such that $\bigwedge B$

is an upper bound of $F(x)$. Also, for convenience, for a node $p \in P$ denote $p^I := \{p\}^I \subseteq X$.

We can define a new relation on P induced by the concept lattice $\underline{\mathfrak{B}}_{\mathbb{O}}$. We say p is *conceptually* less or equal than q if and only if $(p^I, p^{II}) \leq_{\underline{\mathfrak{B}}_{\mathbb{O}}} (q^I, q^{II})$, denoted by $p \leq_{\mathbb{O}} q$. We call $\leq_{\mathbb{O}}$ the *conceptual pre-order* of $\mathbb{O}$. In general, the relation $\leq_{\mathbb{O}}$ is not an order since for different $p, q \in P$ the corresponding attribute concepts (p^I, p^{II}) and (q^I, q^{II}) can match. Two annotations $(X, F_1), (X, F_2)$ of $\mathcal{P}$ are called *annotationally equivalent* if their conceptual pre-orders coincide. In Sections 4, 5, and 6 we will explore the relationship between the original ordered set $\mathcal{P}$ and the constructed conceptual pre-order $(P, \leq_{\mathbb{O}})$ and hint at potential applications arising from this comparison.

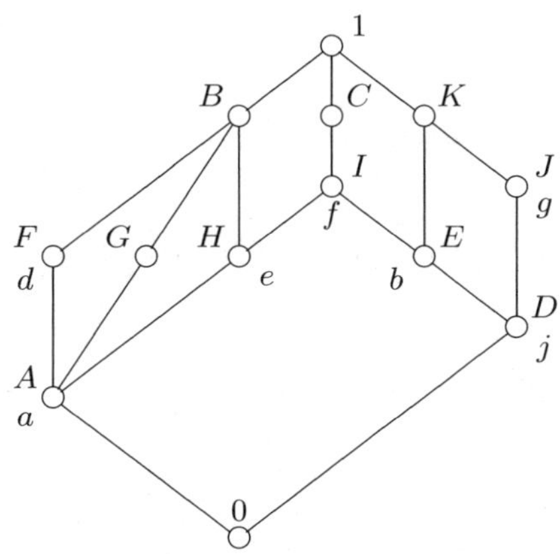

Fig. 2. Example of an annotated lattice

	0	A	B	C	D	E	F	G	H	I	J	K	1
a		×	×	×			×	×	×				×
b				×		×				×		×	×
d			×				×						×
e			×	×					×	×			×
f				×						×			×
g											×	×	×
j				×	×	×				×	×	×	×

Fig. 3. Context for the annotated lattice in Fig. 2

Example 1. An example for an elementary annotated lattice $\mathbb{L} := (\mathcal{P}, X, F)$ is given in Fig. 2 where $\mathcal{P} := (\{A, B, \ldots, K, 0, 1\}, \leq_{\mathcal{P}}), X = \{a, b, d, e, f, g, j\}$, and F and $\leq_{\mathcal{P}}$ are defined as illustrated. Fig. 3 shows the formal context $\mathbb{K}_{\mathbb{L},\cdot}$ and Fig. 4 the resulting concept lattice $\underline{\mathfrak{B}}_{\mathbb{O}}$.

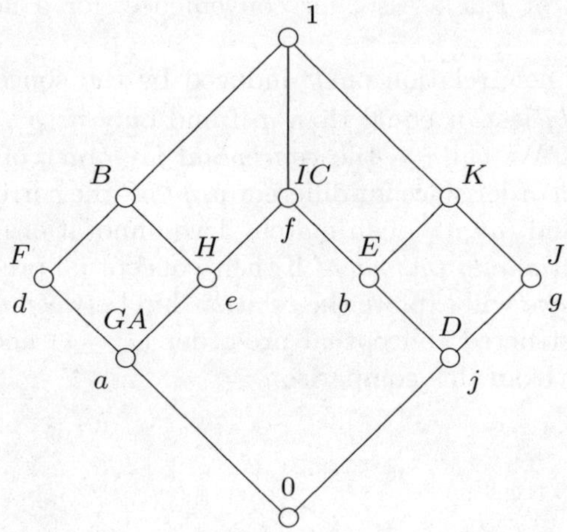

Fig. 4. Concept lattice representation of the annotated lattice in Fig 2

4 Mathematical Properties of Concept Lattice Representations

In the first part of this section we will analyze how the order of the annotated ordered set and the order of its concept lattice are related. In the second part we will investigate how the concept lattice representations, derived from a given ordered set using different annotations, can be classified.

4.1 Annotated Ordered Sets and Their Conceptual Pre-order

The following proposition connects an annotated ordered set with its concept lattice representation.

Proposition 1. *Let* $\mathbb{O} := (\mathcal{P}, X, F)$ *be an annotated ordered set and let* $\mu : P \to \mathfrak{B}_{\mathbb{O}}$ *be such that* $\mu(p) = (p^I, p^{II})$ *maps each poset node to its* attribute concept *in* $\mathfrak{B}_{\mathbb{O}}$. *Then* μ *constitutes an order-homomorphism between* $\mathcal{P}$ *and the concept lattice representation of* $\mathbb{O}$.

Proof. Let $p, q \in P$. *Then we have* $p \leq_{\mathcal{P}} q \Longleftrightarrow \downarrow p \subseteq \downarrow q$ *which implies*

$$p^I = \{x \in X \mid \downarrow p \cap F(x) \neq \emptyset\} \subseteq \{x \in X \mid \downarrow q \cap F(x) \neq \emptyset\} = q^I.$$

Since the last statement is equivalent to $\mu(p) \leq_{\mathfrak{B}_{\mathbb{O}}} \mu(q)$ *this asserts the proposition.* □

By definition of $\leq_{\mathbb{O}}$, it follows that $p \leq_{\mathcal{P}} q \implies p \leq_{\mathbb{O}} q$. Clearly, the converse is wrong, since in general $\leq_{\mathbb{O}}$ is only a pre-order. Even for the factor order associated with the pre-order, the converse implication does not hold as is verified by the example shown in Figure 5, where $c \leq_{\mathbb{O}} a$, but c and a are non-comparable in $\mathcal{P}$.

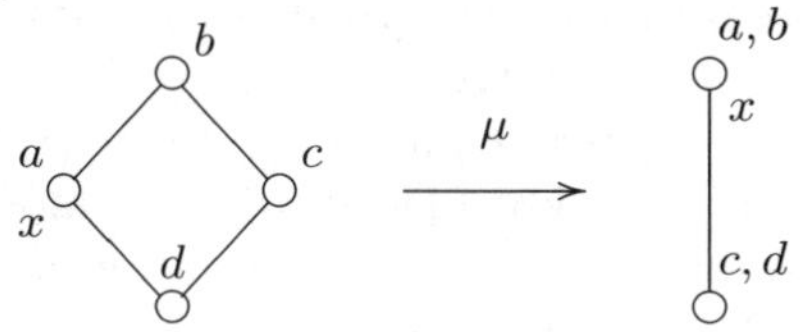

Fig. 5. Counter-example to μ inducing an order isomorphism

But for elementary annotated complete lattices we find the following connection which goes further than the results for annotated ordered sets.

Proposition 2. *Let $\mathbb{L} = (\mathcal{P}, X, F)$ be an elementary annotated complete lattice. The concept lattice of $\mathbb{L}$ is order-embedded into $\mathcal{P}$ via the map $\varphi : \mathfrak{B}_{\mathbb{L}} \to P$ where $(A, B) \mapsto \bigwedge B$.*

Proof. For all $\mathfrak{c}_1, \mathfrak{c}_2 \in \mathfrak{B}_{\mathbb{L}}$, we have to show that $\mathfrak{c}_1 \leq_{\underline{\mathfrak{B}}_{\mathbb{L}}} \mathfrak{c}_2$ holds if and only if $\varphi(\mathfrak{c}_1) \leq_{\mathcal{P}} \varphi(\mathfrak{c}_2)$. Let $\mathfrak{c}_1 = (A, B)$ and $\mathfrak{c}_2 = (C, D)$ be concepts in $\mathfrak{B}_{\mathbb{L}}$.
"$\Rightarrow$": Assume $(A, B) \leq_{\underline{\mathfrak{B}}_{\mathbb{L}}} (C, D)$. This is equivalent to $D \subseteq B$ which implies $\bigwedge B \leq_{\mathcal{P}} \bigwedge D$.
"$\Leftarrow$": Assume $\bigwedge B \leq_{\mathcal{P}} \bigwedge D$. Since $\mathbb{L}$ is elementary, B^I is the set of all labels $x \in X$ such that $\bigwedge B$ is an upper bound of $F(x)$ it follows that $B^I \subseteq D^I$ and therefore we have $(B^I, B) \leq_{\underline{\mathfrak{B}}_{\mathbb{L}}} (D^I, D)$ as required. □

For elementary annotated lattices the previously introduced mappings μ and φ combine in a surprising way.

Theorem 1. *Let $\mathbb{L} := (\mathcal{P}, X, F)$ be an elementary annotated complete lattice. Then (φ, μ) forms a residuated pair between the concept lattice representation of $\mathbb{L}$ and $\mathcal{P}$. In particular, φ is an injective $\bigvee$-morphism and μ is a surjective $\bigwedge$-morphism.*

Proof. Firstly, we deduce from Proposition 2 that φ is injective. For residuated pairs this implies the surjectivity of the second map. It remains to show that (φ, μ) forms a residuated pair.

Since $\mathbb{L}$ is elementary, F can be regarded as a map from X to P and then the incidence relation I of $\mathbb{K}_{\mathbb{L}}$ is defined via xIp if and only if $F(x) \leq_{\mathcal{P}} p$; therefore $x^I = \uparrow F(x)$ for all $x \in X$. In the following let (A, B) be an arbitrary concept in $\mathfrak{B}_{\mathbb{L}}$. We derive $B = A^I = \bigcap_{a \in A} x^I = \bigcap_{a \in A} \uparrow F(x) = \uparrow \bigvee F(A)$; hence, we receive $\varphi(A, B) = \bigwedge B = \bigvee F(A)$. We conclude the proof as follows:

$$\varphi(A, B) \leq_{\mathcal{P}} p \Longleftrightarrow \bigvee F(A) \leq_{\mathcal{P}} p \Longleftrightarrow p \in \uparrow \bigvee F(A) = A^I$$
$$\Longleftrightarrow \quad A \subseteq p^I \quad \Longleftrightarrow \quad (A, B) \leq_{\underline{\mathfrak{B}}_{\mathbb{L}}} \mu(p)$$

□

As a consequence of our theorem we know that φ embeds the concept lattice representation of an elementary annotated complete lattice into its underlying lattice as a kernel system. This fact applies to Example 1 and is visualized in Fig. 6.

Though, in general, the concept lattice representations of elementary annotated ordered sets cannot be embedded into their underlying ordered set, it is feasible to embed them into a well-known extension of the former. For a subset Q of an ordered set $\mathcal{P}$, we will use the notation $Q^{\downarrow}$ for the set of all lower bounds of Q in $\mathcal{P}$.

Theorem 2. *Let* $\mathbb{O} := (\mathcal{P}, X, F)$ *be an elementary annotated ordered set and let* $\mathbb{K} := (P, P, \leq_{\mathcal{P}})$*. Then the map* $\mathfrak{B}_{\mathbb{O}} \to \mathfrak{B}(\mathbb{K})$ *with* $(A, B) \mapsto (B^{\downarrow}, B)$ *forms a* $\bigvee$*-embedding of the concept lattice representation of* $\mathbb{O}$ *into the Dedekind-MacNeille completion of* $\mathcal{P}$*.*

Proof. Firstly, we refer to Theorem 4 in [GW99], p.48, for details regarding the Dedekind-MacNeille completion.

Since $\mathbb{O}$ is elementary, $x^I = \uparrow F(x)$ is an intent not only of $\mathbb{K}_{\mathbb{O}}$ but also of $\mathbb{K}$ for every $x \in X$; trivially, p^I is an extent of $\mathbb{K}_{\mathbb{O}}$ for every $p \in P$. By Definition 69 in [GW99], p. 185, this means that I is a bond from $\mathbb{K}_{\mathbb{O}}$ to $\mathbb{K}$. Now Corollary 112 in [GW99], p. 256, implies that the map φ_I from $\underline{\mathfrak{B}}_{\mathbb{O}}$ to $\underline{\mathfrak{B}}(\mathbb{K})$ with $\varphi_I(A, B) = (A^{I\downarrow}, A^I) = (B^{\downarrow}, B)$ is a $\bigvee$-morphism, which clearly is injective. □

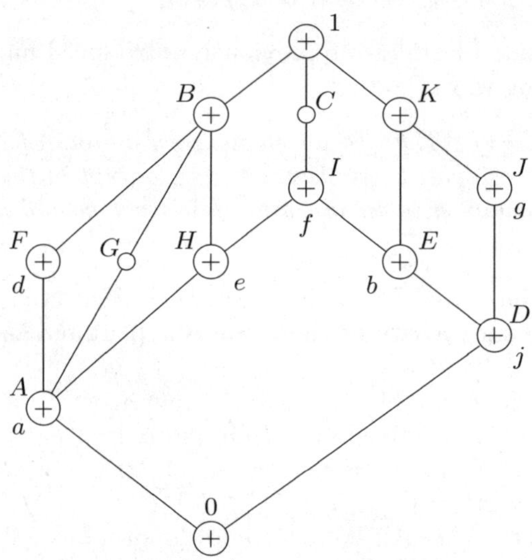

Fig. 6. The concept lattice representation from Fig. 4 embedded as kernel system in its annotated lattice from Fig. 2

4.2 Classifying the Annotations of an Ordered Set

We start with giving two rather extreme examples for different concept lattices derived from the same ordered set via different annotations. In the following, it is more convenient to regard an annotated ordered set $\mathbb{O} := (\mathcal{P}, X, F)$ as a formal context with an ordered set of attributes. Since the annotation function $F : X \to 2^P$ set-theoretically is a relation $F \subseteq X \times P$, the formal context (X, P, F) together with the ordered set $\mathcal{P} = (P, \leq_{\mathcal{P}})$ yields another way of looking at an annotated ordered set. It is obvious, that the formal context $\mathbb{K}_{\mathbb{O}}$ is equal to $(X, P, F \circ \leq_{\mathcal{P}})$, where $\circ$ denotes the relational product. We recall Theorem 4 from [GW99] which states that for an ordered set $\mathcal{P}$ its Dedekind-MacNeille completion is isomorphic to the concept lattice of the formal context $(P, P, \leq_{\mathcal{P}})$. Now it is easy to see, that the *identical* labelling function $F_{id} : P \to 2^P$ with $p \mapsto \{p\}$ yields an annotated ordered set $\mathbb{O}_{id} := (\mathcal{P}, P, F_{id})$ which is isomorphic to the Dedekind-MacNeille completion of $\mathcal{P}$, because $F_{id} \circ \leq_{\mathcal{P}} = \leq_{\mathcal{P}}$ which yields $\mathbb{K}_{\mathbb{O}} = (P, P, \leq_{\mathcal{P}})$. On the other hand – as complicated as $\mathcal{P}$ might be – if the labelling function is constant with $F_{\mathcal{P}}(x) = P$ for any label $x \in X$ we get a formal context with $I = X \times P$. That means, the concept lattice representation shrinks the ordered set into a single element.

To get a more comprehensive description of the interplay of the annotations of an ordered set $\mathcal{P}$ and its concept lattice representations we will use the *filter lattice* of an ordered set $\mathcal{P}$ – which is defined as $\mathcal{F}(\mathcal{P}) := (\{F \subseteq P \mid \uparrow F = F\}, \subseteq)$ – as a framing structure.

Theorem 3. *The annotations of an ordered set $\mathcal{P} = (P, \leq_{\mathcal{P}})$ are, up to annotational equivalence, in one-to-one correspondence to the closure systems in the filter lattice of $\mathcal{P}$.*

Proof. Let $x \in X$ be a label. The object intent $x^{F \circ \leq_{\mathcal{P}}}$ of x in $(X, P, F \circ \leq_{\mathcal{P}})$ is of the form $\{p \in P \mid \exists q \in x^F : q \leq_{\mathcal{P}} p\}$ which is equal to the filter $\uparrow x^F$ in $\mathcal{P}$. Since the intents of all concepts of a concept lattice are exactly the meets of the object intents, the intents of the concepts of $\mathfrak{B}(X, P, F \circ \leq_{\mathcal{P}})$ are exactly the meets of filters of the form $x^{F \circ \leq_{\mathcal{P}}}$ with $x \in X$, and therefore, form a closure system in the filter lattice of $\mathcal{P}$.

Let us assume that $\mathcal{X} \subseteq 2^P$ is a closure system in the filter lattice of $\mathcal{P}$. We consider the formal context $(\mathcal{X}, P, \ni)$. For $X \in \mathcal{X}$, we get $X^{\ni} = \{p \in P \mid p \in X\} = X$. Therefore the intents of the associated concept lattice constitute exactly the closure system $\mathcal{X}$. And since in our situation $\ni \circ \leq_{\mathcal{P}}$ is equal to $\ni$, an annotation $(\mathcal{X}, \ni)$ corresponding to $\mathcal{X}$ is found. □

The above theorems say that the cosmos of possible structures which can be produced via annotating an ordered set and forming its concept lattice are restricted to closure systems in the filter lattice of the original ordered set – and also exhaust them.

5 Application to the Gene Ontology

In this section we apply our proposed technique to the GO cutout depicted in Figure 1. The given diagram can be seen as an annotated ordered set where the underlying ordered set consists of the functional categories of biological processes (as e.g. *DNA replication*) and the order is given by the arrows. The set of labels consists of the proteins and the annotation function maps a protein to a function category if it is listed at the respective function category node. Clearly, this annotated ordered set can not be interpreted as an annotated lattice, since infima and suprema do not exist for any subset of nodes, e.g. the infimum over all function categories is not present. Also the annotation function attaches some proteins to several nodes as it is the case for *Lig1* and *Lig3* who are attached to the functional categories of *DNA ligation*, *DNA recombination*, and *DNA repair*. Figure 8 shows a diagram of the concept lattice representation of this annotated ordered set where we have omitted function categories where there is no protein attached to the nodes or to some subnode. Figure 7 shows the conceptual pre-ordering of the functions derived from the concept lattice representation (in this case it is an order).

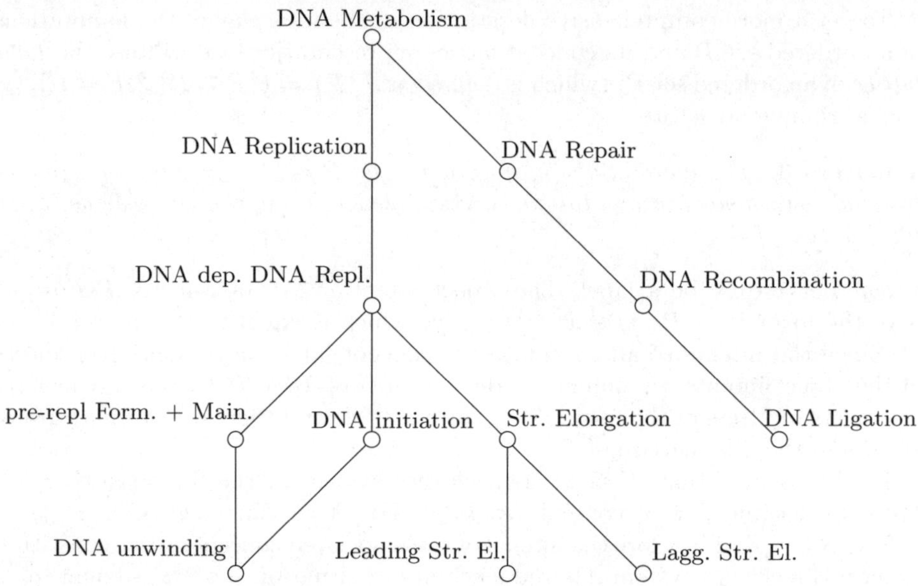

Fig. 7. Function categories ordered conceptually

We want to point out some interesting differences between the annotated ordered set and its concept lattice. Conceptually, the function category *DNA Recombination* is less than *DNA Repair* while in the GO the two nodes are not comparable. This change occurs because *DNA Repair* "inherits" the proteins from *DNA Ligation* which yields a superset of proteins annotated to *DNA Repair* compared to *DNA Recombination.* Since the design of the function category ordering of the GO differs from the conceptual pre-ordering the question arises if some proteins exists but are not present in the GO which justify the non-comparability or if the ordering should be redesigned.

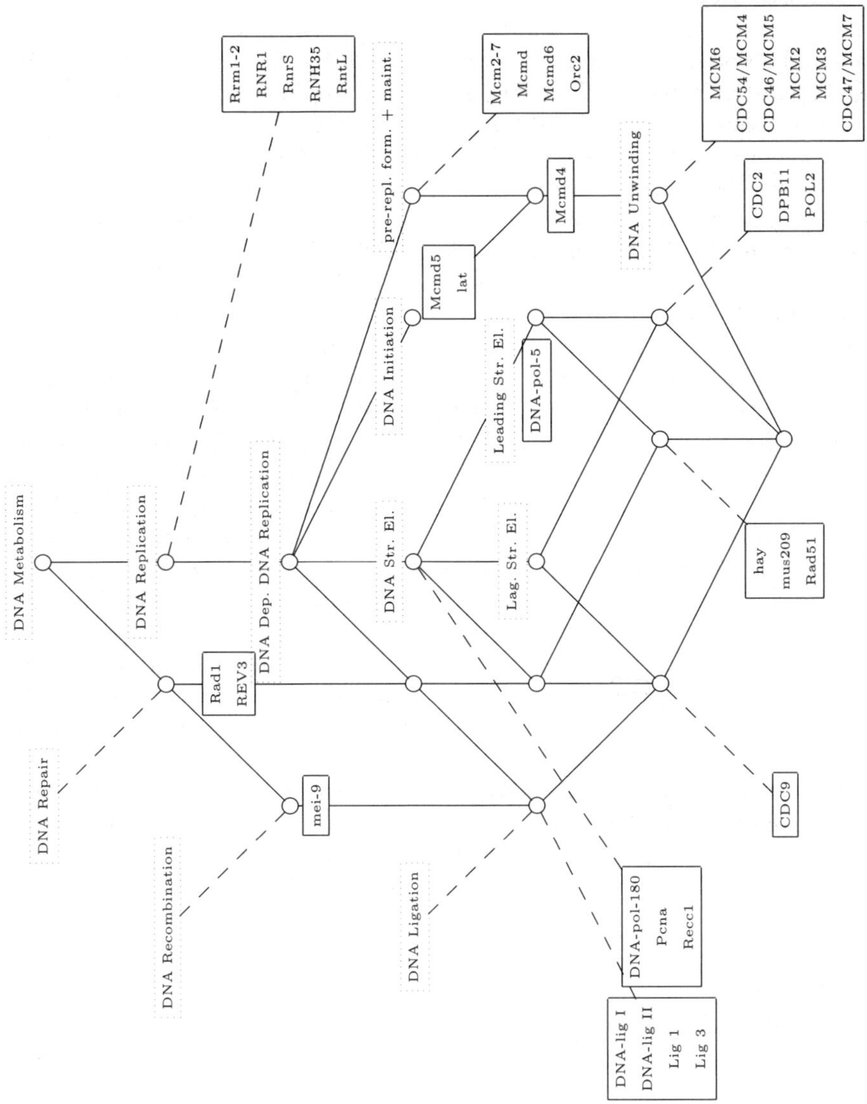

Fig. 8. Concept Lattice for the GO cutout depicted in Figure 1

If we focus our attention on the protein *CDC9* we see that it is annotated to two quite horizontally distinct nodes in the GO, *Lagging strand elongation* and *DNA ligation*. In the concept lattice representation, the new object concept node for *CDC9* thus ties together these two GO nodes through the intermediate concept shown there, the *CDC9* object concept atom on the left. Now, we could ask the question if there is a meaningful label for this node and if it should eventually be introduced in the GO.

6 Discussion

It should be noted that the formal properties of the GO are just now beginning to be explored. Joslyn *et al.* [JM04] have done preliminary measurements of its poset properties, including height, width, and ranks. And while we've noted that the GO is not specifically lower-bounded, if a lower bound is asserted, then it can be questioned how many pairs of nodes do not have unique meets and joins, and thus how close it comes to our idealized annotated lattice. This is something we have addressed specifically elsewhere [JoCGeD06], including proposing a method to measure this degree of lattice-ness based on the FCA reconstruction of the (un-annotated) GO.

As a main application area of our technique we see the task of ontology review or refinement as insinuated in the last section. We want to emphasize two aspects. Firstly, one can investigate all pairs of nodes which are not comparable in the ontology but become comparable in the conceptual pre-order. Secondly, one can consider concepts which are not attribute concepts in the concept lattice representation. Those concepts might be considered as proposals for new nodes in the ontology. This task could even be supported by software tools which could automatically extract the conceptual pre-order of the nodes of the ontology and compute all pairs of nodes which are not comparable in the ontology but become comparable in the conceptual pre-order. The number of those nodes could be interpreted as a degree of *conceptual soundness* of the ontology. Many additional measures are possible, e.g. counting the number of concepts which are not attribute concepts in the concept lattice representation, where this number could be interpreted as *conceptual completeness*. In both cases lower numbers would be considered as better results. We see future work in this line of research in evolving measures and tools to make the technique operable for large ontologies (see also [JoCGeD06]). This would involve the design of expert systems, which support a semi-automated ontology review or reengineering process.

References

[AsMBaC00] The Gene Ontology Consortium: Gene Ontology: Tool For the Unification of Biology. Nature Genetics v. 25(1), 25–29 (2000)

[BoOMiJ02] Bodenreider, O., Mitchell, J.A., McCray, A.T.: Evaluation of the UMLS As a Terminology and Knowledge Resource for Biomedical Informatics. In: AMIA 2002 Annual Symposium, pp. 61–65 (2002)

[DP90] Davey, B.A., Priestly, H.A.: Introduction to Lattices and Order. Cambridge University Press, Cambridge (1990)

[DaA00a] Davis, A.R.: Types and Constraints for Lexical Semantics and Linking, Cambridge UP (2000)

[GW99] Ganter, B., Wille, R.: Formal Concept Analysis, Mathematical Foundations. Springer, Berlin Heidelberg New York (1999)

[JM04] Joslyn, C.A., Mniszewski, S.M., Fulmer, A., Heaton, G.G.: The Gene Ontology Categorizer. Bioinformatics v. 20:s1, 169–177 (2004)

[JoCGeD06] Joslyn, C.A., Gessler, D.D.G., Schmidt, S.E., Verspoor, K.M.: Distributed Representations of Bio-Ontologies for Semantic Web Services (2006) (submitted to the 2006 Bio-Ontologies SIG 2006)

[KnTReJ00] Knoblock, B.T., Rehof, J.: Type Elaboration and Subtype Completion for Java Bytecode. In: Proc. 27th ACM SIGPLAN-SIGACT Symposium on Principles of Programming Languages (2000)

An Algorithm to Find Frequent Concepts of a Formal Context with Taxonomy

Peggy Cellier[1], Sébastien Ferré[1], Olivier Ridoux[1], and Mireille Ducassé[2]

[1] IRISA/University of Rennes 1
[2] IRISA/INSA,
Campus universitaire de Beaulieu, 35042 Rennes, France
firstname.lastname@irisa.fr
http://www.irisa.fr/LIS/

Abstract. Formal Concept Analysis (FCA) considers attributes as a non-ordered set. This is appropriate when the data set is not structured. When an attribute taxonomy exists, existing techniques produce a completed context with all attributes deduced from the taxonomy. Usual algorithms can then be applied on the completed context for finding frequent concepts, but the results systematically contain redundant information. This article describes an algorithm which allows the frequent concepts of a formal context with taxonomy to be computed. It works on a non-completed context and uses the taxonomy information when needed. The results avoid the redundancy problem with equivalent performance.

1 Introduction

Formal Concept Analysis (FCA) [GW99] finds interesting clusters, called *concepts*, in data sets. FCA is based on a formal *context*, i.e. a binary relation describing a set of objects by a set of properties (*attributes*). A *formal concept* is defined by a pair (*extent*, *intent*), where *extent* is the maximal set of objects that have in their description all attributes of *intent*, and *intent* is the maximal set of attributes common to the description of all objects of *extent*. Searching all concepts is, in general, costly and not always relevant. Thus some of these algorithms search for *frequent* concepts. A concept is called frequent, with respect to a threshold, if the cardinal of its extent is greater than the threshold. Algorithms have been designed in order to find frequent concepts ([STB+02]).

FCA considers attributes as a non-ordered set. There are, however, numerous cases where attribute taxonomies are genuinely available. For example, most corpus of knowledge in natural science are organized in rich taxonomies. *Conceptual Scaling* [GW99] can treat contexts with ordered attributes. A preprocessing step produces a completed context where new attributes deduced from the taxonomy are included. Namely, let o be an object with initial attribute a, if in the taxonomy a implies b, Conceptual Scaling adds attribute b to the description of o. After the transformation, usual data mining algorithms can be applied on the completed context for finding frequent concepts. However, the explicit links

S. Ben Yahia et al. (Eds.): CLA 2006, LNAI 4923, pp. 226–231, 2008.

between initial attributes and deduced attributes are lost. As a consequence, the resulting frequent concepts will systematically contain redundant information. This might be a problem. For example, it is not always relevant to recall that a nightingale is a Muscicapidae, order Passeriformes, class Aves, category Bird, phylum Chordata, kingdom Animalia.

In this paper, we propose an algorithm for finding frequent concepts in a context with taxonomy. The context needs not be completed, because the taxonomy is taken into account, when needed, during the computation. It is based on Bordat's algorithm which computes the concept lattice of a formal context [Bor86].

The algorithm is implemented into LISFS [PR03], a file system based on Logical Concept Analysis (LCA) [FR04], a version of FCA.

The contribution of this article is to describe, and experimentally validate, an algorithm which allows frequent concepts of a formal context with taxonomy to be computed. Thus, it is able to compute answers at the proper level of abstraction with respect to the taxonomy, without redundancy in the resulting frequent concepts.

In the following, Section 2 describes the algorithm. Section 3 gives experimental results. Section 4 concludes this paper.

2 Finding Frequent Concepts

Our algorithm is an adaptation of Bordat's algorithm [Bor86, KO02]. The differences are: firstly the strategy to explore the concept lattice; secondly the underlying data structures, and most importantly, the possibility to use a taxonomy to compute concepts.

The strategy of the method is top-down. The concept lattice is traversed by first exploring one non-explored concept whose extent has the greatest cardinal. The algorithm starts with the top concept. The taxonomy is taken into account, when needed, during the computation.

In the following, we first present the data structures used by the algorithm, then we give the details of the algorithm, its properties and we show the first 2 steps of computation on one example.

2.1 Data Structures

The algorithm manages 2 data structures: a set of computed frequent concepts with respect to a threshold min_sup, called SOLUTION, and a set of concepts to explore called EXPLORATION.

Notation: given a concept c, ext_c (resp. int_c) is the extent of c (resp. its intent). Given an intent i (resp. an extent e), $ext(i)$ is the extent of i (resp. the intent of e).

Appart from the top concept, each concept s is computed from a concept $g(s)$, which we call the *generator* of s. That generator is such that there exists a set of attributes, X, such that $ext_s = ext_{g(s)} \cap ext(X)$. We call X an *increment* of

Algorithm 1. Frequent_concepts

Require: $\mathcal{K}$, a context with taxonomy; and *min_sup*, a minimal support
Ensure: SOLUTION, a set of all concepts of $\mathcal{K}$ that are frequent with respect to *min_sup*

SOLUTION $:= \emptyset$
EXPLORATION.add$((\mathcal{O} \to \{root_{tax}\}, \emptyset, \emptyset)$
while EXPLORATION $\neq \emptyset$ **do**
  let $(ext_s \to X, int_{g(s)}, incr_{g(s)}) = max_{ext}($EXPLORATION$)$ **in**
  $\text{int}_s := (\text{int}_{g(s)} \cup_{tax} X) \cup_{tax} \{y \in succ^{+}_{tax}(X) \mid ext_s \subseteq ext(\{y\})\}$
  $\text{incr}_s := \{(c \to X) \mid \exists c' : (c' \to X) \in incr_{g(s)} \wedge c = ext_s \cap c' \wedge \|c\| \geq min\_sup\}$
  for all $y \in succ_{tax}(X)$ **do**
    let $c = ext_s \cap ext(\{y\})$ **in**
    if $\|c\| \geq sup\_min$ **then**
      $incr_s := incr_s[c \to (incr_s(c) \cup \{y\})]$
    end if
  end for
  for all $(ext \to Y)$ in $incr_s$ **do**
    EXPLORATION.add$(ext \to Y, \text{int}_s, \text{incr}_s)$
  end for
  SOLUTION.add$(\text{ext}_s, \text{int}_s)$
end while

$g(s)$. A concept c may have several increments, but we are only interested in increments that lead to different frequent immediate subconcepts of c. This is approximated by a data structure $incr_c$ which contains at least all frequent immediate subconcepts of c. In this data structure, every subconcept is associated with its increment with respect to c. Thus, $incr_c$ is a mapping from subconcepts to increments, and we write $incr_c[s \to X]$ to express that the mapping is modified so that c maps to X.

An invariant for the correction of the algorithm is that

$$incr_c \subseteq \{(s \to X) \mid ext_s = ext_c \cap ext(X) \wedge \|ext_s\| \geq min\_sup\} .$$

All elements of $incr_c$ are frequent subconcepts of c.

An invariant for completness is that

$$\begin{aligned} & ext_c \supset ext_s \wedge \|ext_s\| \geq min\_sup \wedge \neg\exists X : (s \to X) \in incr_c \\ \Longrightarrow\ & \exists s' : ext_c \supset ext_{s'} \supset ext_s \wedge \exists X : (s' \to X) \in incr_c . \end{aligned}$$

All frequent subconcepts of c that are not in $incr_c$ are subconcepts of a subconcept of c which is in $incr_c$.

Structure $incr_c$ avoids to test all attributes at each step. Indeed, the set of increments is reducing when the lattice is explored top-down. Therefore, $incr_c$ avoids to test a lot of irrelevant attributes, by storing relevant choice points from the previous step in the computation. In practice, concepts are represented by their extent, so that $incr_c$ is represented by a trie indexed by extents.

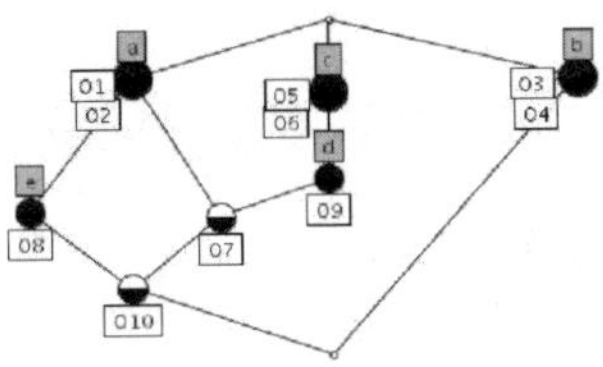

Fig. 1. Concept lattice without taxonomy

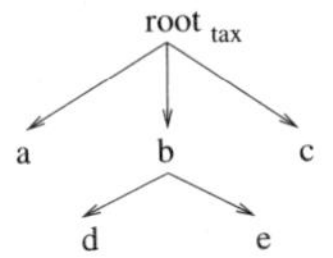

Fig. 2. Taxonomy of the example

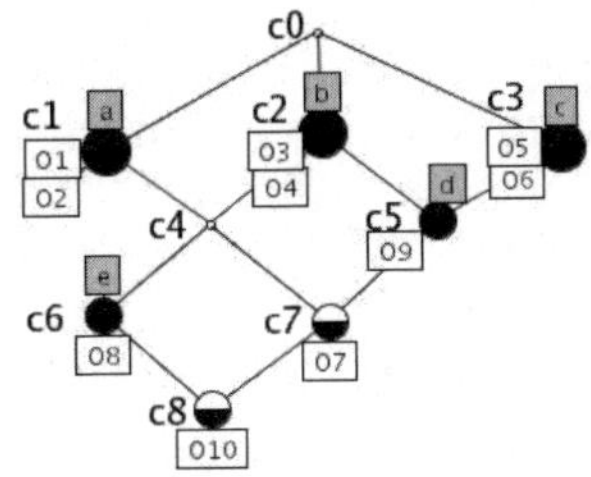

Fig. 3. Completed concept lattice

2.2 Algorithm

Algorithm Frequent_concepts computes all frequent concepts, exploring the concept lattice top-down. SOLUTION is initially empty (step 1). The top concept, labelled by the root of the taxonomy ($root_{tax}$), is put in EXPLORATION (step 2). At each iteration of the while loop (step 3), an element of EXPLORATION with the largest possible extent is selected: $(ext_s \rightarrow X, int_{g(s)}, incr_{g(s)})$ (step 4), where $(ext_s \rightarrow X)$ is an element of $incr_{g(s)}$.

First, the intent of s is computed by completing $(int_{g(s)} \cup X)$ (step 5) with successors of X in the taxonomy. $succ_{tax}(X)$ returns immediate successors of attributes of X in the taxonomy and $succ^{+}_{tax}$ is the transitive closure. This is here that the elimination of redundant attributes takes place, thanks to $\cup_{tax}$. $\cup_{tax}$ is the union of two sets of attributes with elimination of redundancies due to the taxonomy.

Second, the increments of s are computed by exploring the increments of $g(s)$ (step 6) and the successors of the attributes of X in the taxonomy (steps 7-10). Indeed, the first are still possible increments for s. For each candidate increment, the algorithm checks whether it actually leads to a frequent subconcept. Finally, EXPLORATION (steps 13-14) and SOLUTION (step 16) are updated.

The context and the taxonomy of an example are given in Figure 1 and Figure 2. Figure 3 shows the completed context, i.e. the explored lattice.

For this example, we assume *min_sup*=3, and we give the first 2 steps of computation. Initially, SOLUTION and EXPLORATION are:

- SOLUTION $= \emptyset$
- EXPLORATION $= \{((\mathcal{O} \rightarrow \{root_{tax}\}), \emptyset, \emptyset)\}$.

First step: the top of the lattice is explored, i.e. s=c_0. Increments of s are computed from the taxonomy only, as there is no generator concept:

- $incr_{c_0} = \{ (\{o_3, o_4, o_7, o_8, o_9, o_{10}\} \rightarrow \{\text{b}\}), (\{o_1, o_2, o_7, o_8, o_{10}\} \rightarrow \{\text{a}\}), (\{o_5, o_6, o_7, o_9, o_{10}\} \rightarrow \{\text{c}\})\}$
- SOLUTION $= \{c_0\}$
- EXPLORATION $= \{((\{o_3, o_4, o_7, o_8, o_9, o_{10}\} \rightarrow \{\text{b}\}), \emptyset, incr_{c_0}), ((\{o_1, o_2, o_7, o_8, o_{10}\} \rightarrow \{\text{a}\}), \emptyset, incr_{c_0}), ((\{o_5, o_6, o_7, o_9, o_{10}\} \rightarrow \{\text{c}\}), \emptyset, incr_{c_0})\}$.

Second step: an element of EXPLORATION with the largest possible extent is explored: $s = c_2$, $g(s) = c_0$. In order to compute $incr_{c_2}$, we have to consider the elements of $incr_{c_0}$ and the elements in the taxonomy.

- $incr_{c_2} = \{ (\{o_7, o_8, o_{10}\} \rightarrow \{a\}), (\{o_7, o_9, o_{10}\} \rightarrow \{c, d\}), \not(\not\{\not o_8, \not o_{10}\not\} \not\rightarrow \not\{\not e\not\}\not) \}$
- SOLUTION $= \{c_0, c_2\}$
- EXPLORATION $= \{((\{o_1, o_2, o_7, o_8, o_{10}\} \rightarrow \{a\}), \emptyset, incr_{c_0}), ((\{o_5, o_6, o_7, o_9, o_{10}\} \rightarrow \{c\}), \emptyset, incr_{c_0}), ((\{o_7, o_9, o_{10}\} \rightarrow \{c,d\}), \{b\}, incr_{c_2}), ((\{o_7, o_8, o_{10}\} \rightarrow \{a\}), \{b\}, incr_{c_2})\}$.

In the second step, attributes d and e are introduced as successors of attributes b, and attributes a and c are introduced as increments of c_0, the generator of c_2.

Increment {e} is eliminated because it leads to an infrequent concept. Attributes c and d are grouped into a single increment because they lead to the same subconcept. This ensures that computed intents are complete.

2.3 Properties

The algorithm has two properties: 1) it computes all frequent concepts; 2) all intents of computed concepts are maximal and without redundancy according to the taxonomy.

The first property is given by the fact that every frequent concept is a subconcept of a frequent concept (except top) [PBTL99], and the concepts in EXPLORATION are treated from the largest (with respect to the cardinal of the extent) to the smallest.

The second property (the intent of a computed concept is without redundancy), is given by the use of $\cup_{tax}$ which explicitly removes redundancy. In addition, when computing $incr_s$, the increments from $g(s)$ and from the taxonomy which lead to the same concept are grouped together.

3 Experiments

The algorithm is implemented in the CAML language inside LISFS [PR03]. In LISFS, attributes can be ordered to create a taxonomy (for more details see [PR03]). We ran experiments on an Intel(R) Pentium(R) M processor 2.00 GHz with Fedora Core release 4, 1GB of main memory.

We study a context with taxonomy about Java methods[1]. The context contains 5 526 objects which are the methods of java.awt. They are described by their input and output types, visibility modifiers, exceptions, keywords extracted from their identifiers, and keywords from their comments. The context has 1 624 properties. Due to the class inheritance, the context has a natural hierarchy, i.e. a taxonomy. There are 134 780 concepts but few of them are really frequent. For this context, the execution time is proportional to the number of found concepts. For example, with a threshold min_sup of 5%, 189 frequent concepts are

[1] Available on the web at http://lfs.irisa.fr/demo-area/awt-source/

computed in 8s and taking into account the taxonomy to compute intent allows to reduce 39% of irrelevant attributes.

In order to study the impact on the performance, of taking into account the taxonomy, we test the method on a context without taxonomy, using the mushroom benchmark[2]. The mushroom context has 8 416 objects and 127 different properties. The computation time is similar to the results of the algorithms Close and A-Close on the same data [Pas00], for example with a threshold min_sup of 10%, 4 793 concepts are computed in 76s. This shows that in practice, taking into account the taxonomy does not negatively impact the performance.

4 Conclusion

We have proposed an algorithm to compute all frequent concepts in a context with taxonomy. The main advantage of the presented algorithm is to avoid redundancies due to the taxonomy, in the intents of the computed frequent concepts. The resulting concepts are therefore more relevant. Experiments have shown that, in practice, taking a taxonomy into account does not negatively impact the performance.

References

[Bor86] Bordat, J.: Calcul pratique du treillis de Galois d'une correspondance. Mathématiques, Informatiques et Sciences Humaines 24(94), 31–47 (1986)

[FR04] Ferré, S., Ridoux, O.: An introduction to logical information systems. Information Processing & Management 40(3), 383–419 (2004)

[GW99] Ganter, B., Wille, R.: Formal Concept Analysis: Mathematical Foundations. Springer, Heidelberg (1999)

[KO02] Kuznetsov, S.O., Objedkov, S.A.: Comparing performances of algorithms for generating concept lattices. JETAI: Journal of Experimental & Theoretical Artificial Intelligence 14, 189–216 (2002)

[Pas00] Pasquier, N.: Data Mining: Algorithmes d'extraction et de réduction des règles d'association dans les bases de données. Computer science, Université Blaise Pascal - Clermont-Ferrand II (January 2000)

[PBTL99] Pasquier, N., Bastide, Y., Taouil, R., Lakhal, L.: Discovering Frequent Closed Itemsets for Association Rules. In: Beeri, C., Bruneman, P. (eds.) ICDT 1999. LNCS, vol. 1540, pp. 398–416. Springer, Heidelberg (1999)

[PR03] Padioleau, Y., Ridoux, O.: A logic file system. In: Proc. USENIX Annual Technical Conference (2003)

[STB+02] Stumme, G., Taouil, R., Bastide, Y., Pasquier, N., Lakhal, L.: Computing iceberg concept lattices with TITANIC. Data Knowl. Eng. 42(2), 189–222 (2002)

[2] Available at ftp://ftp.ics.uci.edu/pub/machine-learning-databases/mushroom/

Unique Factorization Theorem and Formal Concept Analysis

Peter Mihók[1,⋆] and Gabriel Semanišin[2,⋆⋆]

[1] Department of Applied Mathematics,
Technical University Košice, Faculty of Economics,
B.Němcovej 32, 040 01 Košice, Slovak Republic
and
Mathematical Institute, Slovak Academy of Sciences,
Grešákova 6, 040 01 Košice, Slovak Republic
peter.mihok@tuke.sk
[2] Institute of Computer Science,
P.J. Šafárik University, Faculty of Science,
Jesenná 5, 041 54 Košice, Slovak Republic
gabriel.semanisin@upjs.sk

Abstract. In the theory of generalised colourings of graphs, the Unique Factorization Theorem (UFT) for additive induced-hereditary properties of graphs provides an analogy of the well-known Fundamental Theorem of Arithmetics. The purpose of this paper is to present a new, less complicated, proof of this theorem that is based on Formal Concept Analysis. The method of the proof can be successfully applied even for more general mathematical structures known as relational structures.

1 Introduction and Motivation

Formal Concept Analysis (briefly FCA) is a theory of data analysis which identifies conceptual structures among data sets. It was introduced by R. Wille in 1982 and since then has grown rapidly (for a comprehensive overview see [12]). The mathematical lattices that are used in FCA can be interpreted as classification systems. Formalized classification systems can be analysed according to the consistency of their relations. Some extensions and modifications of FCA can be found e.g. in [16].

In this paper we provide a new proof of the Unique Factorization Theorem (UFT) for induced-hereditary additive properties of graphs. The problem of unique factorization of reducible hereditary properties of graphs into irreducible factors was formulated as Problem 17.9 in the book [15] of T.R. Jensen and B. Toft. Our proof is significantly shorter as the previous ones and it is based on FCA. Moreover, FCA allows us to work with concepts instead of graphs and the

⋆ Research supported in part by Slovak VEGA Grant 2/7141/27
⋆⋆ Research supported in part by Slovak VEGA Grant 1/3004/06 and Slovak APVV grant 20-004104.

S. Ben Yahia et al. (Eds.): CLA 2006, LNAI 4923, pp. 232–239, 2008.

reader can rather easily see that using this approach we can prove UFT even for properties of more general structures like hypergraphs, coloured hypergraphs, posets, etc. Such general mathematical object are very often called relational structures.

In general, we follow standard graph terminology (see e.g. [1]). In particular, we denote by $\mathbb{N}$ the set of positive integers and by $\mathcal{I}^\omega$, $\mathcal{I}$ and $\mathcal{I}^{conn}$ the class of all simple countable graphs, simple finite graphs and simple finite connected graphs, respectively and K_n stands for the complete graph of order n. For a positive integer k and a graph G, the notation $k.G$ is used for the union of k vertex disjoint copies of G. The *join* of graphs G, H is the graph obtained from the disjoint union G and H by joining all vertices of G with all the vertices of H.

All our considerations can be done for arbitrary infinite graphs, however, in order to avoid formal set-theoretical problems, we shall consider only countable infinite graphs. Moreover, we assume that the vertex set $V(G)$ of a graph G is a subset of a given countable set, say U. A graph property $\mathcal{P}$ is any isomorphism-closed nonempty subclass of $\mathcal{I}^\omega$. It means that investigating graph properties, in principle, we restrict our considerations to unlabeled graphs.

Let $\mathcal{P}_1, \mathcal{P}_2, \ldots, \mathcal{P}_n$ be graph properties. A vertex $(\mathcal{P}_1, \mathcal{P}_2, \ldots, \mathcal{P}_n)$-*colouring* (*partition*) of a graph $G = (V, E)$ is a partition $(V_1, V_2, \ldots, V_n)$ of $V(G)$ (every pair of V_i's has empty intersection and the union of V_i's forms V) such that each colour class V_i induces a subgraph $G[V_i]$ having property $\mathcal{P}_i$. For convenience, we allow empty partition classes in the partition sequence. An empty class induces the null graph $K_0 = (\emptyset, \emptyset)$. If each of the $\mathcal{P}_i$'s, $i = 1, 2, \ldots, n$, is the property $\mathcal{O}$ of being edgeless, we have the well-known proper vertex n-colouring. A graph G which have a $(\mathcal{P}_1, \mathcal{P}_2, \ldots, \mathcal{P}_n)$-colouring is called $(\mathcal{P}_1, \mathcal{P}_2, \ldots, \mathcal{P}_n)$-*colourable*, and in such a situation we say that G has property $\mathcal{P}_1 \circ \mathcal{P}_2 \circ \cdots \circ \mathcal{P}_n$. For more details concerning generalized graph colourings we refer the reader to [2,3,15].

In 1951, de Bruijn and Erdős proved that an infinite graph G is k-colourable if and only if every finite subgraph of G is k-colourable. An analogous compactness theorem for generalized colourings was proved in [7]. The key concept for the Vertex Colouring Compactness Theorem of [7] is that of a property being of *finite character*. Let $\mathcal{P}$ be a graph property, $\mathcal{P}$ is of *finite character* if a graph in $\mathcal{I}^\omega$ has property $\mathcal{P}$ if and only if each its finite induced subgraph has property $\mathcal{P}$. It is easy to see that if $\mathcal{P}$ is of finite character and a graph has property $\mathcal{P}$ then so does every induced subgraph. A property $\mathcal{P}$ is said to be *induced-hereditary* if $G \in \mathcal{P}$ and $H \leq G$ implies $H \in \mathcal{P}$, that is $\mathcal{P}$ is closed under taking induced subgraphs. Thus properties of finite character are induced-hereditary. However not all induced-hereditary properties are of finite character; for example the graph property $\mathcal{Q}$ of not containing a vertex of infinite degree is induced-hereditary but not of finite character. Let us also remark that every property which is hereditary with respect to every subgraph (we say simply *hereditary*) is induced-hereditary as well. The properties of being edgeless, of maximum degree at most k, K_n-free, acyclic, complete, perfect, etc. are properties of finite character. The compactness theorem for $(\mathcal{P}_1, \mathcal{P}_2, \ldots, \mathcal{P}_n)$-colourings, where the $\mathcal{P}_i$'s are of finite character, have been proved using Rado's Selection Lemma.

Theorem 1 (Vertex Colouring Compactness Theorem, [7]**).** *Let G be a graph in $\mathcal{I}^\omega$ and let $\mathcal{P}_1, \mathcal{P}_2, \ldots, \mathcal{P}_n$ be properties of graphs of finite character. Then G is $(\mathcal{P}_1, \mathcal{P}_2, \ldots, \mathcal{P}_n)$-colourable if every finite induced subgraph of G is $(\mathcal{P}_1, \mathcal{P}_2, \ldots, \mathcal{P}_n)$-colourable.*

Let us denote by $\mathcal{R} = \mathcal{P}_1 \circ \mathcal{P}_2 \circ \cdots \circ \mathcal{P}_n$, $n \geq 2$ the set of all $(\mathcal{P}_1, \mathcal{P}_2, \ldots, \mathcal{P}_n)$-colourable graphs. The binary operation $\circ$ is obviously commutative, associative on the class of graph properties and $\Theta = \{K_0\}$ is its neutral element. The properties $\Theta, \mathcal{I}$ and $\mathcal{I}^\omega$ are said to be trivial. A nontrivial graph property $\mathcal{P}$ is said to be *reducible* if there exist nontrivial graph properties $\mathcal{P}_1, \mathcal{P}_2$, such that $\mathcal{P} = \mathcal{P}_1 \circ \mathcal{P}_2$; otherwise $\mathcal{P}$ is called *irreducible.* In what follows each property is considered to be nontrivial.

The problem of unique factorization of a reducible induced-hereditary property into induced-hereditary factors was introduced in connection with the study of the existence of uniquely colourable graphs with respect to hereditary properties (see [2,3] and Problem 17.9. in the book [15]). In general, there are only few graph properties that have a unique factorization into irreducible ones (see [8,10]). However, for some important classes of graph properties the Unique Factorization Theorems can be proved. In [19] it is proved that every reducible property of finite graphs, which is closed under taking subgraphs and disjoint union of graphs (such properties are called *additive*) is uniquely factorisable into irreducible additive hereditary factors. An analogous result was obtained in [10,17] for additive induced-hereditary properties of finite graphs. Following [2] let us denote by $\mathbb{M}^a$ the set of all additive induced-hereditary properties of finite graphs. Then UFT can be stated as follows.

Theorem 2 (Unique Factorization Theorem - UFT, [10,17]**).** *Every additive induced-hereditary property of finite graphs is in $\mathbb{M}^a$ uniquely factorisable into a finite number of irreducible additive induced-hereditary properties, up to the order of factors.*

Let us remark, that using Theorem 1 we can prove UFT for the class $\mathbb{M}^{\omega a}$ of the additive properties of infinite (countable) graphs of finite character (see [13]). The proof of the Unique Factorization Theorem is rather complicated. The problems concerning the proof were discussed from different points of view in several papers [6,10,11,13,17] and in details in PhD thesis (see e.g. [8]). On the other hand, the Theorem 2 has several deep applications related to the existence of uniquely partitionable graphs (see [4,5]) and consequently the complexity of generalized colourings. A. Farrugia in [9] proved that if $\mathcal{P}$ and $\mathcal{Q}$ are additive induced-hereditary graph properties, then $(\mathcal{P}, \mathcal{Q})$-colouring is NP-hard, with the sole exception of graph 2-colouring (the case where both $\mathcal{P}$ and $\mathcal{Q}$ are the set $\mathcal{O}$ of finite edgeless graphs). Moreover, $(\mathcal{P}, \mathcal{Q})$-colouring is NP-complete if and only if $\mathcal{P}$- and $\mathcal{Q}$-recognition are both in NP. It shows that additive induced-hereditary properties are rather complicated mathematical structures.

The aim of this paper is to present a new method of the proof of the Unique Factorization Theorem, which will eliminate some technical difficulties in the previous proofs. Moreover it shows a new utilisation of the methods of FCA.

2 Hereditary Graph Properties in the Language of FCA

It is quite easy to prove that the sets $\mathbb{M}^a$ ($\mathbb{M}^{\omega a}$) of all additive and induced-hereditary graph properties of finite graphs (of finite character), partially ordered by set inclusion, forms a complete distributive lattice. The lattices of hereditary graph properties have been studied intensively, references may be found in [2,14,18]. In this section we will present a new approach to the lattice of additive induced-hereditary graph properties.

In order to proceed we need to introduce some concepts of FCA according to a fundamental book of B. Ganter and R. Wille [12].

Definition 1. *A* **formal context** $\mathbb{K} := (O, M, I)$ *consists of two sets O and M and a relation I between O and M. The elements of O are called the* **objects** *and the elements of M are called the* **attributes** *of the context.*

For a set $A \subseteq O$ of objects we define

$$A' := \{m \in M : gIm \text{ for all } g \in A\}.$$

Analogously, for a set B of attributes we define

$$B' := \{g \in O : gIm \text{ for all } \in B\}.$$

A **formal concept** *of the context (O, M, I) is a pair (A, B) with $A \subseteq O, B \subseteq M, A' = B$ and $B' = A$.*

We call A the **extent** *and B the* **intent** *of a concept (A, B). $\mathbb{L}(O, M, I)$ denotes the set of all concepts of the context (O, M, I).*

If (A_1, B_1) and (A_2, B_2) are concepts of a context and $A_1 \subseteq A_2$ (which is equivalent to $B_2 \subseteq B_1$), we write $(A_1, B_1) \leq (A_2, B_2)$.

For an object $g \in O$ we write $g' = \{m \in M | gIm\}$ and γg for the **object concept** *(g'', g'), where $g'' = \{\{g\}'\}'$.*

Let us mention that, by the Basic Theorem on Concept Lattices, the set $\mathbb{L}(O, M, I)$ of all concepts of the context (O, M, I) partially ordered by the relation $\leq$ (see Definition 1) is a complete lattice.

Let us present additive induced-hereditary graph properties as concepts in a given formal context. Using FCA we can proceed in the following way. Let us define a context (O, M, I) by setting objects to countable simple graphs, e.g. $O = \mathcal{I}^\omega$. For each connected finite simple graph $F \in \mathcal{I}$ let us consider an attribute m_F : "do not contain an induced-subgraph isomorphic to F". Thus GIm_F means that the graph G does not contain any induced subgraph isomorphic to F. We can immediately observe the following:

Lemma 1. *Let $O = \mathcal{I}^\omega$ and $M = \{m_F, F \in \mathcal{I}^{conn}\}$. Then the concepts of the context $\mathbb{K} = (O, M, I)$ are additive induced-hereditary graph properties of finite character and the concept lattice $(\mathbb{L}(O, M, I), \leq)$ is isomorphic to the lattice $(\mathbb{M}^{\omega a}, \subseteq)$. Moreover, for each concept $\mathcal{P} = (A, B)$ there is an object - a countable graph $G \in O$ such that $\mathcal{P} = \gamma G = (G'', G')$.*

Proof. It is easy to verify that the extent of any concept (A, B) of $\mathbb{K}$ forms an additive induced-hereditary property $\mathcal{P} = A$ of finite character. Obviously, each countable graph $G = (V, E)$ in the context $\mathbb{K}$ leads to an "object concept" $\gamma G = (G'', G')$. On the other hand, because of additivity, the disjoint union of all finite graphs having a given additive induced-hereditary property $\mathcal{P} \in \mathbb{M}^{\omega a}$ is a countable infinite graph K satisfying $\gamma K = (\mathcal{P}, \mathcal{I}^{conn} - \mathcal{P})$. □

In order to describe additive induced-hereditary properties of finite graphs, mainly two different approaches were used: a characterization by generating sets and/or by minimal forbidden subgraphs (see [2] and [11]). While the extent A of a concept $(A, B) \in \mathbb{L}(O, M, I)$ is related to a graph property $\mathcal{P}$, the intent B consists of forbidden connected subgraphs of $\mathcal{P}$. The set $\boldsymbol{F}(\mathcal{P})$ of *minimal forbidden subgraphs* for $\mathcal{P}$ consists of minimal elements of the poset $(B, \leq)$. For a given countable graph $G \in \mathcal{I}^{\omega}$ let us denote by $age(G)$ the class of all finite graphs isomorphic to finite induced-subgraph of G (see e.g. [20]). Scheinerman in [21] showed, that for each additive induced-hereditary property $\mathcal{P}$ of finite graphs, there is an infinite countable graph G such that $\mathcal{P} = age(G)$. This result corresponds to the proof of Lemma 1. On the other hand, it is worth to mention that $\gamma G = (\mathcal{P}, G')$ does not imply, in general, that $\mathcal{P} = age(G)$. Let us define a binary relation $\cong$ on $\mathcal{I}^{\omega}$ by $G_1 \cong G_2$ whenever $\gamma G_1 = \gamma G_2$ in the context $\mathbb{K}$, and we say that G_1 is congruent with G_2 with respect to $\mathbb{K}$. Obviously, $\cong$ is an equivalence relation on $\mathcal{I}^{\omega}$. The aim of the next section is to find appropriate representatives of congruence classes and to describe their properties.

3 Uniquely Decomposable Graphs

All the previous proofs of UFT are based on a construction of *uniquely $\mathcal{R}$-decomposable* graphs that are defined as follows.

Definition 2. *For given (finite or infinite) graphs $G_1, G_2, \ldots, G_n$, $n \geq 2$, denote by $G_1 * G_2 * \cdots * G_n$ the set of graphs*

$$\left\{ H \in \mathcal{I}^{\omega} : \bigcup_{i=1}^{n} G_i \subseteq H \subseteq \sum_{i=1}^{n} G_i \right\},$$

where $\bigcup_{i=1}^{n} G_i$ denotes the disjoint union and $\sum_{i=1}^{n} G_i$ the join of the graphs $G_1, G_2, \ldots, G_n$, respectively. For a graph G, $s \geq 2$, $s * G$ stands for the class $G * G * \cdots * G$, with s copies of G.

Let G be a graph and $\mathcal{R}$ be an additive induced-hereditary property of graphs. Then we put $dec_{\mathcal{R}}(G) = \max\{n :$ there exist a partition $\{V_1, V_2, \ldots, V_n\}$, $V_i \neq \emptyset$, of $V(G)$ such that for each $k \geq 1$, $k.G[V_1] * k.G[V_2] * \cdots * k.G[V_n] \subseteq \mathcal{R}\}$. If $G \notin \mathcal{R}$ we set $dec_{\mathcal{R}}(G)$ to zero.

A graph G is said to be **$\mathcal{R}$-decomposable** *if $dec_{\mathcal{R}}(G) \geq 2$; otherwise G is $\mathcal{R}$-indecomposable.*

A graph $G \in \mathcal{P}$ is called **$\mathcal{P}$-strict** *if $G * K_1 \nsubseteq \mathcal{P}$. The class of all $\mathcal{P}$-strict graphs is denoted by $S(\mathcal{P})$. Put $dec(\mathcal{R}) = \min\{dec_{\mathcal{R}}(G) : G \in S(\mathcal{R})\}$.*

A $\mathcal{R}$-strict graph G with $dec_{\mathcal{R}}(G) = dec(\mathcal{R}) = n \geq 2$ is said to be **uniquely $\mathcal{R}$-decomposable** *if there exists exactly one $\mathcal{R}$-partition $\{V_1, V_2, \dots, V_n\}$, $V_i \neq \emptyset$, such that for each $k \geq 1$, $k.G[V_1] * k.G[V_2] * \cdots * k.G[V_n] \subseteq \mathcal{R}$. We call the graphs $G[V_1], G[V_2], \dots, G[V_n]$* **ind-parts** *of the uniquely decomposable graph G.*

These notions are motivated by the following observation: Let us suppose that $G \in \mathcal{R} = \mathcal{P} \circ \mathcal{Q}$ and let (V_1, V_2) be a $(\mathcal{P}, \mathcal{Q})$-partition of G. Then by additivity of $\mathcal{P}$ and $\mathcal{Q}$ we have that $k.G[V_1] * k.G[V_2] \subseteq \mathcal{R}$ for every positive integer k. Thus, if the property $\mathcal{R}$ is reducible, every graph $G \in \mathcal{R}$ with at least two vertices is $\mathcal{R}$-decomposable.

We proved in [13,17] that for any reducible additive induced-hereditary property also the converse assertion holds:

Theorem 3. *An induced-hereditary additive property $\mathcal{R}$ is reducible if and only if all graphs in $\mathcal{R}$ with at least two vertices are $\mathcal{R}$-decomposable.*

Remark that almost all graphs in $\mathcal{R}$ are $\mathcal{R}$-strict and each graph $G \in \mathcal{R}$ is an induced subgraph of a $\mathcal{R}$-strict graph. To present our main result we need some notions from [10]:

Definition 3. *Let $d_0 = \{U_1, U_2, \dots, U_m\}$ be a $\mathcal{P}$-partition of a graph $G \in \mathcal{P}$. A $\mathcal{P}$-partition $d_1 = \{V_1, V_2, \dots, V_n\}$ of G* **respects** *d_0 if no V_i intersects two or more U_j's; that is each V_i is contained in some U_j. We say that the graph $G^* \in s * G$* **respects** *d_0 if $G^* \in s.G[U_1] * s.G[U_2] * \cdots * s.G[U_m]$. For a graph $G^* \in s * G$, denote the copies of G by $G^1, G^2, \dots, G^s$. Then we say that a $\mathcal{P}$-partition $d = \{V_1, V_2, \dots, V_n\}$ of G^** **respects** *d_0* **uniformly** *whenever for each V_i there is a U_j such that for every G^k, $V_i \cap V(G^k) \subseteq U_j$.*

If G is uniquely $\mathcal{R}$-decomposable, its ind-parts respect d_0 if its unique $\mathcal{R}$-partition respects d_0. If G^* is uniquely $\mathcal{R}$-decomposable, it ind-parts respect d_0 uniformly if for some s the graph $G^* \in s * G$ respects d_0 and the unique $\mathcal{R}$-partition of G^* respects d_0 uniformly.

Based on the construction given in [17] A. Farrugia and R.B. Richter proved:

Theorem 4. ([10,17]) *Let G be an $\mathcal{R}$-strict graph with $dec_{\mathcal{R}}(G) = dec(\mathcal{R}) = n \geq 2$ and let $d_0 = \{U_1, U_2, \dots, U_m\}$ be a fixed $\mathcal{R}$-partition of G. Then there is a uniquely $\mathcal{R}$-decomposable finite graph $G^* \in s * G$, for some s, that respects d_0, and moreover any $\mathcal{R}$-partition of G^* with n parts respects d_0 uniformly.*

Using Theorem 4 we can prove:

Theorem 5. *Let $\mathcal{R} \in \mathbb{M}^{\omega a}$ be a reducible graph property of finite character. Then there exists a uniquely $\mathcal{R}$-decomposable infinite countable graph H such that $\gamma H = (\mathcal{R}, H')$ and $age(H) = \mathcal{R} \cap \mathcal{I}$.*

Proof. Following E. Scheinerman [21], a *composition sequence* of a class $\mathcal{P}$ of finite graphs is a sequence of finite graphs $H_1, H_2, \dots, H_n, \dots$ such that $H_i \in \mathcal{P}, H_i < H_{i+1}$ for all $i \in \mathbb{N}$ and for all $G \in \mathcal{P}$ there exists a j such that $G \leq G^j$. According to Theorem 4, we can easily find a composition sequence $H_1, H_2, \dots, H_n, \dots$

of $\mathcal{R} \cap \mathcal{I}$ consisting of finite uniquely $\mathcal{R}$-decomposable graphs. Without loss of generality, we may assume that if $i < j$, $i, j \in \mathbb{N}$, then $V(H_i) \subset V(H_j)$. Let $V(H) = \bigcup_{i \in \mathbb{N}} V(H_i)$ and $\{u, v\} \in E(H)$ if and only if $\{u, v\} \in E(H_j)$ for some $j \in \mathbb{N}$. It is easy to see that $age(H) = \mathcal{R} \cap \mathcal{I}$, implying $\gamma H = (\mathcal{R}, H')$. Let us remark that, according to the Theorem 1, H is $\mathcal{R}$-decomposable if every finite induced subgraph of H is $\mathcal{R}$-decomposable. In order to verify, that H is uniquely $\mathcal{R}$-decomposable it is sufficient to verify that if $\{V_{j_1}, V_{j_2}, \dots, V_{j_n}\}$, $V_{j_i} \neq \emptyset$ is the unique $\mathcal{R}$-partition of H_j, $j \in \mathbb{N}$, then $\{U_1, U_2, \dots, U_n\}$, where $U_k = \bigcup_{j \in \mathbb{N}} V_{j_k}$, $k = 1, 2, \dots, n$, is the unique $\mathcal{R}$-partition of H. Indeed, this is because the existence of other $\mathcal{R}$-partition of H would imply the existence of other partition of some H_i and it provides a contradiction. □

4 Unique Factorization Theorem for Properties of Finite Character

In [13], based on Theorem 1 and Theorem 2 we proved:

Theorem 6. *Every reducible additive property $\mathcal{R}$ of finite character is uniquely factorisable into finite number of irreducible factors belonging to $\mathbb{M}^{\omega a}$.*

Here we present a new proof of the Theorem 6 based on the Theorem 5 in the context $\mathbb{K}$.

Proof. According to the Theorem 5, let H be a uniquely $\mathcal{R}$-decomposable infinite countable graph such that $\gamma H = (\mathcal{R}, H')$ and let $d_H = \{W_1, W_2, \dots, W_n\}$ be the unique $\mathcal{R}$-partition of H. Let $\mathcal{P}_i = \gamma H[W_i]$ for $i = 1, 2, \dots, n = dec(\mathcal{R})$. Then obviously we have $\mathcal{R} = \mathcal{P}_1 \circ \mathcal{P}_2 \circ \dots \circ \mathcal{P}_n$. If there would be some other factorization of $\mathcal{R}$ into n irreducible factors then obviously H would have another $\mathcal{R}$-partition, which contradicts to the fact that H is uniquely $\mathcal{R}$-decomposable. Since $dec(H) = dec(\mathcal{R}) = n$, there is no factorization of $\mathcal{R}$ into more then n factors. Thus to prove that $\mathcal{R} = \mathcal{P}_1 \circ \mathcal{P}_2 \circ \dots \circ \mathcal{P}_n$ is the unique factorization of $\mathcal{R}$. Further, let $\mathcal{R} = \mathcal{Q}_1 \circ \mathcal{Q}_2 \circ \dots \circ \mathcal{Q}_m, m < n$ and $d_0 = \{U_1, U_2, \dots U_m\}$ be a $(\mathcal{Q}_1, \mathcal{Q}_2, \dots, \mathcal{Q}_m)$-partition of H. Then, by Theorem 4, there is an $s \in \mathbb{N}$ such that $s * H$ respects d_0 uniformly. Thus, since $m < n$, there exists an index j such that $H[U_j] \in H[W_r] * H[W_s]$, implying $\mathcal{Q}_j$ is reducible. □

5 Conclusion

By a careful examination of the previous considerations and arguments, it is not very difficult to see, that for the presented method of the proof it is not important that we are dealing with simple graphs. Indeed, without any substantial change the presented proofs can be applied for directed graphs, hypergraphs or partially ordered sets. All these mathematical objects are examples of so-called relational structures. Thus we obtain a general UFT that is applicable for additive properties of finite character for different objects, with various applications in computer science. For other details we refer the reader to [6].

References

1. Bondy, J.A., Murty, U.S.R.: Graph Theory with Applications. Elsevier North-Holland (1976)
2. Borowiecki, M., Broere, I., Frick, M., Mihók, P., Semanišin, G.: Survey of hereditary properties of graphs. Discuss. Math. - Graph Theory 17, 5–50 (1997)
3. Borowiecki, M., Mihók, P.: Hereditary properties of graphs. In: Kulli, V.R. (ed.) Advances in Graph Theory Vishwa International Publication, Gulbarga, pp. 42–69 (1991)
4. Broere, I., Bucko, J.: Divisibility in additive hereditary properties and uniquely partitionable graphs. Tatra Mt. Math. Publ. 18, 79–87 (1999)
5. Broere, I., Bucko, J., Mihók, P.: Criteria for the existence of uniquely partitionable graphs with respect to additive induced-hereditary properties. Discuss. Math. - Graph Theory 22, 31–37 (2002)
6. Bucko, J., Mihók, P.: On uniquely partitionable systems of objects. Discuss. Math. - Graph Theory 26, 281–289 (2006)
7. Cowen, R., Hechler, S.H., Mihók, P.: Graph coloring compactness theorems equivalent to BPI. Scientia Math. Japonicae 56, 171–180 (2002)
8. Farrugia, A.: Uniqueness and complexity in generalised colouring. Ph.D. thesis, University of Waterloo (April 2003), available at: `http://etheses.uwaterloo.ca`
9. Farrugia, A.: Vertex-partitioning into fixed additive induced-hereditary properties is NP-hard. Elect. J. Combin. 11, R46, 9 (2004) (electronic)
10. Farrugia, A., Richter, R.B.: Unique factorization of additive induced-hereditary properties. Discuss. Math. - Graph Theory 24, 319–343 (2004)
11. Farrugia, A., Mihók, P., Richter, R.B., Semanišin, G.: Factorizations and Characterizations of Induced-Hereditary and Compositive Properties. J. Graph Theory 49, 11–27 (2005)
12. Ganter, B., Wille, R.: Formal Concept Analysis - Mathematical Foundation. Springer, Berlin Heidelberg (1999)
13. Imrich, W., Mihók, P., Semanišin, G.: A note on the unique factorization theorem for properties of infinite graphs. Stud. Univ. Žilina, Math. Ser. 16, 51–54 (2003)
14. Jakubík, J.: On the lattice of additive hereditary properties of finite graphs. Math. - General Algebra and Applications 22, 73–86 (2002)
15. Jensen, T.R., Toft, B.: Graph colouring problems. Wiley–Interscience Publications, New York (1995)
16. Krajči, S.: A generalized concept lattice. Logic Journal of IGPL 13(5), 543–550 (2005)
17. Mihók, P.: Unique Factorization Theorem. Discuss. Math. - Graph Theory 20, 143–153 (2000)
18. Mihók, P.: On the lattice of additive hereditary properties of object systems. Tatra Mt. Math. Publ. 30, 155–161 (2005)
19. Mihók, P., Semanišin, G., Vasky, R.: Additive and hereditary properties of graphs are uniquely factorizable into irreducible factors. J. Graph Theory 33, 44–53 (2000)
20. Sauer, N.W.: Canonical Vertex Partitions. Combinatorics, Probability and Computing 12, 671–704 (2003)
21. Scheinerman, E.R.: On the Structure of Hereditary Classes of Graphs. J. Graph Theory 10, 545–551 (1986)

Towards Concise Representation for Taxonomies of Epistemic Communities

Camille Roth[1], Sergei Obiedkov[2], and Derrick Kourie[3]

[1] CIRESS/LEREPS, University of Toulouse, France
[2] Department of Applied Mathematics, Higher School of Economics, Moscow, Russia
[3] Department of Computer Science, University of Pretoria, South Africa
camille.roth@polytechnique.edu, sergei.obj@gmail.com,
dkourie@cs.up.ac.za

Abstract. We present an application of formal concept analysis aimed at representing a meaningful structure of knowledge communities in the form of a lattice-based taxonomy. The taxonomy groups together agents (community members) who interact and/or develop a set of notions—i.e. cognitive properties of group members. In the absence of appropriate constraints on how it is built, a knowledge community taxonomy is in danger of becoming extremely complex, and thus difficult to comprehend. We consider two approaches to building a concise representation that respects the underlying structural relationships, while hiding uninteresting and/or superfluous information. The first is a pruning strategy that is based on the notion of concept stability, and the second is a representational improvement based on nested line diagrams. We illustrate the method with a small sample of a community of embryologists.

1 Introduction

A knowledge community is a group of agents who produce and exchange knowledge within a given knowledge field, achieving a widespread social cognition task in a rather decentralized, collectively interactive, and networked fashion. The study of such communities is frequent topic in social epistemology as well as in scientometrics and political science (refer, *inter alia*, to [1,2,3]).

In particular, a traditional concern relates to the description of the structure of knowledge communities [4], generally organized in several subcommunities. In contrast to the limited, subjective, and implicit representation that agents have of their own global community—a folk taxonomy [5]—epistemologists typically use expert-made taxonomies, which are somewhat more reliable but which still fall short in terms of precision, objectivity, and comprehensiveness.

We describe here an application of formal concept analysis (FCA) aimed at representing a meaningful structure of a given knowledge community in the form of a lattice-based taxonomy which is built upon groups of agents who jointly manipulate sets of notions. Formal concepts in this case relate loosely to the sociological idea of "structural equivalence" [6].

This work is a development of the approach presented in [7,8], where it was shown how to use FCA to identify the main fields in a scientific community and describe their

S. Ben Yahia et al. (Eds.): CLA 2006, LNAI 4923, pp. 240–255, 2008.

taxonomy with several levels of detail. Section 2 gives an overview of the approach. In section 3, we concentrate on how to make lattice-based taxonomies concise and intelligible. Concept lattices faithfully represent all features of data, including those due to noise. Therefore, we need tools that would allow us to abstract from insignificant and noisy features. To this end, we suggest a pruning technique based on stability indices of concepts [9] and apply it on its own and in combination with nested line diagrams [10]. The latter allows for representing the community structure at various levels of precision, depending on which subcommunities are most interesting to the user of the taxonomy. The techniques described in section 3 admit modifications, which are a subject for further research and experiment. Some possible directions and open questions are listed in section 4.

2 A Formal Concept Analysis Approach in Applied Epistemology

2.1 Framework

Representing taxonomies of knowledge communities has routinely been an issue for applied epistemology and scientometrics [2], addressed notably by describing community partitions with trees or two-dimensional maps of agents and topics. Various quantitative methods have been used, often based on categorization techniques and data describing links between authors, papers, and/or notions—such as co-citation [4], co-authorship [11], or co-occurrence data [12].

The lattice-based taxonomies discussed here allow overlapping category building, with agents possibly belonging to several communities at once. They render a finer and more accurate structure of knowledge fields by representing various kinds of interrelationships. Our notion of a community is both looser and more general than the sociological notion of structural equivalence [6] in that we identify maximal groups of agents linked jointly to various sets of notions instead of *exactly* the same notions.

A similar problem of identifying communities exists in the area of social networks. Lattices have also been used there [13,14,15], but in that context, groups of actors are generally considered to be disjoint and a lattice is merely a first step in their construction. Besides, social network researchers are interested in social aspects of the community structure (who the leaders are, how they influence peripheral members, how actors cooperate within their own group and between different groups, etc.), whereas we rather try to discover a structure of a scientific field (and are not particularly concerned with individuals). Because of these differences in emphasis, social network lattices are typically based on data describing interaction and relations between actors, while our data, as will be seen later, describes actors in terms of the domain for which we want to build a taxonomy.

Before proceeding, we briefly recall the FCA terminology [10]. Given a *(formal) context* $\mathbb{K} = (G, M, I)$, where G is called a set of *objects*, M is called a set of *attributes*, and the binary relation $I \subseteq G \times M$ specifies which objects have which attributes, the derivation operators $(\cdot)^I$ are defined for $A \subseteq G$ and $B \subseteq M$ as follows:

$$A^I = \{m \in M \mid \forall g \in A : gIm\} \qquad B^I = \{g \in G \mid \forall m \in B : gIm\}$$

In words, A^I is the set of attributes common to all objects of A and B^I is the set of objects sharing all attributes of B.

If this does not result in ambiguity, $(\cdot)'$ is used instead of $(\cdot)^I$. The double application of $(\cdot)'$ is a closure operator, i.e., $(\cdot)''$ is extensive, idempotent, and monotonous. Therefore, sets A'' and B'' are said to be *closed*.

A *(formal) concept* of the context (G, M, I) is a pair (A, B), where $A \subseteq G$, $B \subseteq M$, $A = B'$, and $B = A'$. In this case, we also have $A = A''$ and $B = B''$. The set A is called the *extent* and B is called the *intent* of the concept (A, B).

A concept (A, B) is a *subconcept* of (C, D) if $A \subseteq C$ (equivalently, $D \subseteq B$). In this case, (C, D) is called a *superconcept* of (A, B). We write $(A, B) \leq (C, D)$ and define the relations $\geq$, $<$, and $>$ as usual. If $(A, B) < (C, D)$ and there is no (E, F) such that $(A, B) < (E, F) < (C, D)$, then (A, B) is a *lower neighbor* of (C, D) and (C, D) is an *upper neighbor* of (A, B); notation: $(A, B) \prec (C, D)$ and $(C, D) \succ (A, B)$.

The set of all concepts ordered by $\leq$ forms a lattice, which is denoted by $\mathfrak{B}(\mathbb{K})$ and called the *concept lattice* of the context $\mathbb{K}$. The relation $\prec$ defines edges in the *covering graph* of $\mathfrak{B}(\mathbb{K})$. The meet and join in the lattice are denoted by $\wedge$ and $\vee$, respectively.

An expression $B \to D$, where $B \subseteq M$ and $D \subseteq M$, is called an *(attribute) implication*. An implication $B \to D$ *holds* in (G, M, I) if all objects from G having all attributes from B also have all attributes from D, i.e., $B' \subseteq D'$ (equivalently, $D'' \subseteq B''$). The set of all implications is summarized by the Duquenne–Guigues basis [16].

Epistemic community taxonomy. Our primary data consists of scientific papers dealing with a certain (relatively broad) topic, from which we construct a set G of authors and a set M of terms and notions used in these papers. Thus, we have a context (G, M, I), where I describes which author uses which term in one of his or her papers: gIm iff g uses m. Then, for a group of authors $A \subseteq G$, A' represents notions being used by every author $a \in A$, while, for a set of notions $B \subseteq M$, B' is the set of authors using every notion $b \in B$. Thus, we see notions as cognitive *properties* of authors who use them (skills in scientific fields).

The intent of a concept in this context is a subtopic and the extent is the set of all authors active in this subtopic. Thus, formal concepts provide a solid formalization of the notion of *epistemic community* (EC) traditionally defined as a group of agents dealing with a common set of issues and aiming towards a common goal of knowledge creation [3]. By EC, we understand henceforth a field within a given knowledge community together with authors working in this field irrespective of their affiliation or personal interactions, i.e., neither a department nor a research project. The concept lattice represents the structure of a given knowledge community as a taxonomy of ECs, with more populated and less specific subtopics closer to the top [7].

2.2 Empirical Example and Protocol

We focus on a bibliographical database of `MedLine` abstracts coming from the fast-growing community of embryologists working on the zebrafish during the period

1998–2003.[1] We build up a context describing which author used which notion during the whole period, where the notion set is made of a limited dictionary of about 70 lemmatized words selected by the expert [17] among the most frequent yet significant words of the community, i.e., excluding rhetorical and paradigmatic words such as "*is*", "*with*", "*study*", "*biology*", "*develop*", etc. At first, we thus should say that each term appearing in an article is a notion, which is a classical assumption in scientometrics [12,18]. In other words, we extract the semantics from article contents rather than from their metadata. As such, scientific fields will be defined by notion sets describing EC intents. Then, we extract a random sample context of 25 authors and 18 words, which we use to illustrate the techniques described in the paper. The concept lattice of this context is shown in Fig. 1 (only attribute labels are shown); it contains 69 formal concepts or epistemic communities.[2]

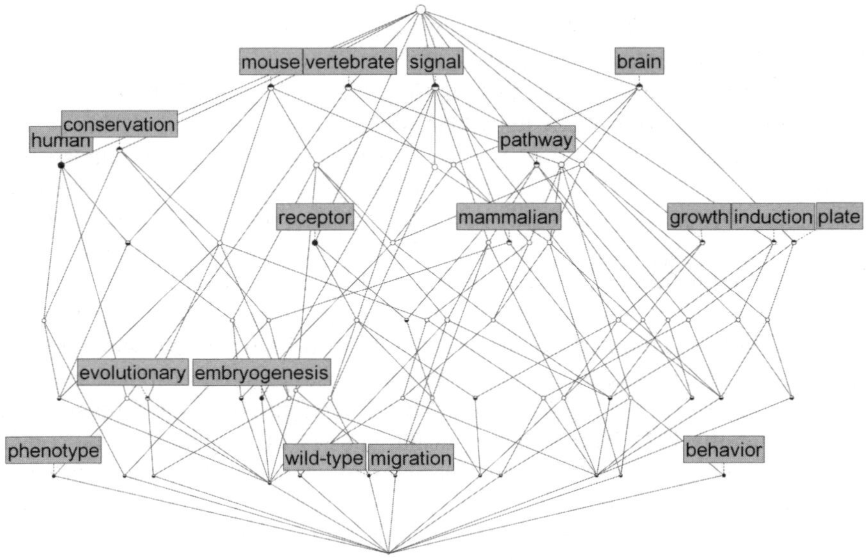

Fig. 1. The concept lattice of a sample zebrafish context

We use an expert-based description of the zebrafish community taxonomy as a benchmark for our procedure [17,19,20]. Three major subfields are to be distinguished according to the description by the expert. First, an important part of the community focuses on biochemical signaling mechanisms, involving pathways and receptor, which are crucial in understanding embryo growth processes. A second field includes comparative studies: the zebrafish, as a model animal, may show similarities with other close

[1] Data is obtained from a query on article abstracts containing the term "*zebrafish*" at http://www.pubmed.com. Using a precise term is likely to delimit properly the community, in contrast to global terms such as "*molecular biology*".

[2] Diagrams are produced with ConExp (http://sourceforge.net/projects/conexp) and ToscanaJ (http://sourceforge.net/projects/toscanaj).

vertebrate species, in particular, with mice and humans. Finally, another significant area of interest relates to the brain and the nervous system, notably in association with signaling in brain development.

3 Concise Representation

3.1 Rationale

The concept lattice in Fig. 1 might, at first sight, appear adequately to identify and organize ECs. However, the number of ECs is rather large and the diagram in Fig. 1 is indeed rather complicated, even though it is derived from a fairly small context. This is a well-known risk when using concept lattices. To quote [10], "even carefully constructed line diagrams lose their readability from a certain size up, as a rule from around 50 elements up". Unfortunately, there is no hope that lattices built from real-size data will be limited to anything close to fifty elements.

Moreover, various ECs turn out to be irrelevant for the purposes of deriving a practical taxonomy of the knowledge field. One solution is to compute only an upper part of the lattice (an order filter), e.g., concepts covering at least $n\%$ of authors. In this case, we get an "iceberg lattice" [21]. Here, one should be careful not to overlook small but interesting groups, for example: a group not yet supported by a large number of followers and that represents a new research trend; or a group that contains individuals who are not members of any other group. To take account of such groups, one should also compute all lower neighbors (proper subgroups) of "large" ECs (satisfying the $n\%$ threshold). Top-down lattice construction algorithms are particularly suitable for this approach [22]. Alternatively, one may look at algorithms designed specifically for constructing iceberg lattices [21] and other algorithms from the frequent itemset mining community [23]. The reduction in the number of concepts can be considerable; however, though computationally feasible, this would still be unsatisfying from the standpoint of manual analysis.

Clearly, the size of the concept lattice is not only a computational problem. The lattice may contain nodes that are just too similar to each other because of noise in data or real minor differences yet irrelevant to our purposes. In this case, taking an upper part of the lattice does not solve the problem, since this part may well contain such similar nodes. Besides, it is also of interest to distinguish major trends from minor subfields, perhaps, with a representation allowing for different levels of precision.

In this section, we consider two approaches to improve the readability of line diagrams: pruning and nesting. When pruning, we assume that some concepts are irrelevant: we filter out those that do not satisfy specified constraints of a certain kind. In a previous attempt to use concept lattices to represent EC taxonomies [7,8], heuristics combining various criteria—such as extent size, the shortest distance from the top, the number of lower neighbors, etc.—were used to score ECs and keep only the n best ones. The resulting pictures were meaningful taxonomies, but required *a posteriori* manual analysis, while it was unclear whether it could be possible to go further than a rough representation. Here, we focus on a particular pruning strategy based on the notion of the stability of a concept [9].

Nested line diagrams [10], on the other hand, provide no reduction and, hence, do not incur any loss of information. Rather, they rearrange the concepts in such a way that the entire structure becomes more readable; they provide the user with a partial view, which can then be extended to a full view if so desired. Thus, nested line diagrams offer a useful technique for representing complex structures. Yet, because they preserve all details of the lattice, in order to remove (many) irrelevant details we combine nesting and pruning in section 3.4. We thus try to get a representation that respects the original taxonomy while hiding at the same time uninteresting and superfluous information; our aim is a compromise between the noise level, the number of details, and readability.

3.2 Stability-Based Pruning

Our structures are complex, but, in fact, they are more complex than they should be, since our data is fairly noisy. As a result, many concepts do not correspond to real communities, and some pruning seems unavoidable. For instance, an author might use a term accidentally (e.g., discussing related work), or there may be different names for the same thing (e.g., "Galois lattice" and "concept lattice"). In the latter case, even if it is not obvious that people preferring one term should be grouped under the exact same field as people preferring another, at least there ought to be a super-field uniting them (especially since we are interested in the taxonomy of knowledge fields rather than in social networks within academia).

The pruning technique we describe here is based on the notion of stability, first introduced in [24] in relation to hypotheses generated from positive and negative examples. It can be easily extended to formal concepts of a context [9]. The definition we use is slightly different from the original one, but the difference is irrelevant to our discussion.

Definition 1. *Let* $\mathbb{K} = (G, M, I)$ *be a formal context and* (A, B) *be a formal concept of* $\mathbb{K}$. *The* stability index, σ, *of* (A, B) *is defined as follows:*

$$\sigma(A, B) = \frac{|\{C \subseteq A \mid C' = B\}|}{2^{|A|}}. \tag{1}$$

The stability index of a concept indicates how much the concept intent depends on particular objects of the extent. A stable intent is probably "real" even if the description of some objects is "noisy". In application to our data, the stability index shows how likely we are to still observe a field if we ignore several authors. Apart from noise-resistance, a stable field does not collapse (e.g., merge with a different field, split into several independent subfields) when a few members stop being active or switch to another topic. The following proposition describing the stability index of a concept (A, B) as a ratio between the number of subcontexts of $\mathbb{K}$ where B is an intent and the total number of subcontexts of $\mathbb{K}$ makes the idea behind stability more explicit:

Proposition 1. *Let* $\mathbb{K} = (G, M, I)$ *be a formal context and* (A, B) *be a formal concept of* $\mathbb{K}$. *For a set* $H \subseteq G$, *let* $I_H = I \cap (H \times M)$ *and* $\mathbb{K}_H = (H, M, I_H)$. *Then,*

$$\sigma(A, B) = \frac{|\{\mathbb{K}_H \mid H \subseteq G \text{ and } B = B^{I_H I_H}\}|}{2^{|G|}}. \tag{2}$$

Proof. Every $C \subseteq A$ defines a family of contexts:

$$\mathfrak{F}_C(\mathbb{K}) = \{\mathbb{K}_H \mid C \subseteq H \subseteq G \text{ and } A \cap H = C\}. \qquad (3)$$

Obviously, $\mathfrak{F}_C(\mathbb{K}) \cap \mathfrak{F}_D(\mathbb{K}) = \emptyset$ if $C \neq D$. In fact, the sets $\mathfrak{F}_C(\mathbb{K})$ form a partition of subcontexts of $\mathbb{K}$ (with the same attribute set M). It is easy to see that all sets $\mathfrak{F}_C(\mathbb{K})$ (with $C \subseteq A$) have the same size: $|\mathfrak{F}_C(\mathbb{K})| = 2^{|G|-|A|}$. Note also that, for $\mathbb{K}_H \in \mathfrak{F}_C(\mathbb{K})$, we have $B^{I_H I_H} = C^{I_H} = C'$; hence, B is closed in the context $\mathbb{K}_H \in \mathfrak{F}_C(\mathbb{K})$ if and only if $C' = B$. Therefore,

$$|\{\mathbb{K}_H \mid H \subseteq G \text{ and } B = B^{I_H I_H}\}| = \frac{2^{|G|}|\{C \subseteq A \mid C' = B\}|}{2^{|A|}}, \qquad (4)$$

which proves the proposition. In other words, the stability of a concept is the probability of preserving its intent after leaving out an arbitrary subset of objects from the context. This is the idea of cross-validation [25] carried to its extreme: stable intents are those generated by a large number of subsets of the data. In the case of cross-validation, it is more common to consider only (some) subsets of a fixed size. Indeed, one may argue that subcontexts of different sizes should have different effect on the stability: the smaller the subcontext is, the further it is from the initial—observed—context, and, hence, the smaller should be its contribution to the instability of a concept. However, we leave these matters for further research and use the definition of the stability given above.

Computing stability. In [9], it is shown that, given a formal context and one of its concepts, the problem of computing the stability index of this concept is #P-complete. Below, we present a simple algorithm that takes the covering graph of a concept lattice $\mathfrak{B}(\mathbb{K})$ and computes the stability indices for every concept of the lattice. The algorithm is meant only as an illustration of a general strategy for computing the stability; therefore, we leave out any possible optimizations.

```
Algorithm ComputeStability
  Concepts := 𝔅(𝕂)
  for each (A, B) in Concepts
    Count[(A, B)] := the number of lower neighbors of (A, B)
    Subsets[(A, B)] := 2^|A|
  end for
  while Concepts is not empty
    let (C, D) be any concept from Concepts with Count[(C,D)] = 0
    Stability[(C, D)] := Subsets[(C, D)] / 2^|C|
    remove (C, D) from Concepts
    for each (A, B) > (C, D)
      Subsets[(A, B)] := Subsets[(A, B)] − Subsets[(C, D)]
      if (A, B) ≻ (C, D)
        Count[(A, B)] := Count[(A, B)] − 1
      end if
    end for
  end while
  return Stability
```

To determine the stability index $\sigma(A, B)$, we compute the number of subsets $E \subseteq A$ that generate the intersection B (i.e., for which $E' = B$) and store it in `Subsets`. The index $\sigma(A, B)$ is simply the number of such subsets divided by the number of all subsets of A, that is, by $2^{|A|}$. Once computed, $\sigma(A, B)$ is stored in `Stability`, which is the output of the algorithm.

The algorithm traverses the covering graph from the bottom concept upwards. A concept is processed only after the stability indices of all its subconcepts have been computed; the `Count` variable is used to keep track of concepts that become eligible for processing. In the beginning of the algorithm, `Count[(A, B)]` is initialized to the number of lower neighbors of (A, B). When the stability index is computed for some lower neighbor of (A, B), we decrement `Count[(A, B)]`. By the time `Count[(A, B)]` reaches zero, we have computed the stability indices for all lower neighbors of (A, B) and, consequently, for all subconcepts of (A, B). Then, it is possible to determine the stability index of (A, B).

Initially, `Subsets[(A, B)]` is set to the number of all subsets of A, that is, $2^{|A|}$. Before processing (A, B), we process all subconcepts (C, D) of (A, B) and decrement `Subsets[(A, B)]` by the number of subsets of C generating the intersection D. By doing so, we actually subtract from $2^{|A|}$ the number of subsets of A which do not generate B: indeed, every subset of A generates either B or the intent of a subconcept of (A, B). Thus, the value of `Subsets[(A, B)]` eventually becomes equal to the number of subsets of A generating B.

Applying stability. The basic stability-based pruning method is to remove all concepts with stability below a fixed threshold. We computed the stability indices for concepts of our example from Fig. 1. There are 17 concepts with stability index above 0.5. Coincidentally, they are closed under intersection of intents. Hence, they form a lattice, which is shown in Fig. 2.

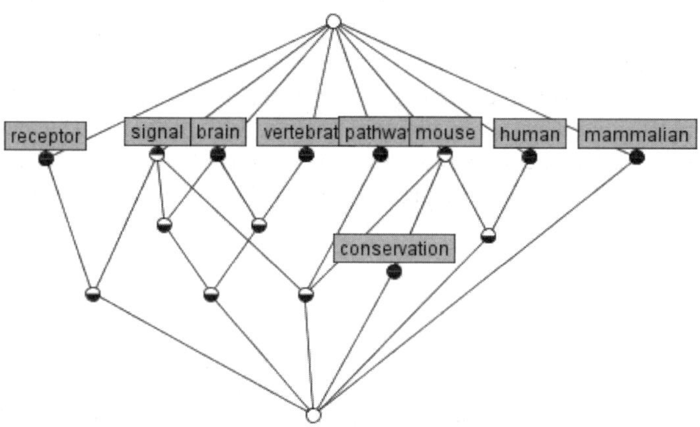

Fig. 2. The pruned lattice of Fig.1, with stability threshold strictly above 0.5

Naturally, removing an unstable node from a line diagram requires that upper and lower neighbors of the remaining stable concepts (i.e. those satisfying the chosen stability threshold) need to be re-established. The resulting structure of stable concepts need not *necessarily* form a lattice. This may or may not be a problem. If all we need is a directly observable taxonomy of scientific fields, there seems to be no reason to require that this taxonomy should be a lattice. In other contexts, however, a lattice may be required in order to apply further lattice-based analysis techniques. Although a full elaboration of this issue is beyond the scope of the present paper; we nonetheless suggest some possible strategies in section 4.2.

Apart from the obvious compression—we kept 17 concepts out of 69—Fig. 2 presents a more readable epistemic taxonomy representation, displaying the major fields of the community along with some meaningful joint communities (such as "*mouse*" and "*human*", as well as "*signal, receptor*"). However, some less important communities, like "*mouse, conservation*", "*mammalian*", or "*signal, pathway, mouse*", are also shown, and it is not clear from the picture that they are less important. Raising the stability threshold would eliminate these communities. Conversely, the stability threshold chosen for Fig. 2 has already eliminated the concept "*signal, receptor, growth, pathway*", even though it is of interest, according to the expert-based description of the field (see section 2.2). Instead of summarily throwing concepts out of a representation, it seems preferable to have a multi-level representation in which certain communities are not entirely eliminated, but rendered instead at a deeper level. In this respect, nested line diagrams would appear to provide a handy representation that distinguishes between various levels of importance of notions.

3.3 Nested Line Diagrams

Nested line diagrams are a well-established tool in formal concept analysis that makes it possible to distribute representation details across several levels [10]. The main idea is to divide the attribute set of the context into two (or more) parts, construct the concept lattices for the generated subcontexts, and draw the diagram of one lattice inside each node of the other lattice. In the case of two parts, an inner concept (A, B) enclosed within an outer concept (C, D) corresponds to a pair $(A \cap C, B \cup D)$. Not every such pair is a concept of the original context. Only inner nodes that correspond to concepts are represented by circles; such nodes are said to be "realized". The outer diagram structures the data along one attribute subset, while the diagram inside an outer concept describes its structure in terms of the remaining attributes. For more details, see [10].

Partitioning the attribute set. The first step in constructing a nested line diagram is to split the attribute set into several parts. These parts do not have to be disjoint, but they will be in our case; hence, we are looking for a partition of the attribute set. As we seek to improve readability, we should display foremost the most significant attributes; therefore, we should assign major notions to higher levels, leaving minor distinctions for lower levels. To this end, words can be partitioned according to a "preference function", which ranges from the simple (e.g., word frequency within the corpus) to more complicated designs.

One could consider a minimal set of notions covering all authors, i.e., find an irredundant cover set, as words from such a set could be expected to play a key role in describing the community. Clearly, this minimal set is not always unique. In practice, we use the algorithm from [26]. We apply it iteratively: the first subcontext contains notions forming an irredundant cover set for the whole author set; the second subcontext includes notions not occurring in the first subcontext, while covering the set of authors excluding those that use only notions from the first subcontext, etc. The last level contains the remaining notions. Denoting by $\mathcal{IC}(\mathbb{K})$ an irredundant cover set of a context $\mathbb{K}$, we start with the context $\mathbb{K}_0 = (G_0, M_0, I_0)$ and, for $k > 0$, recursively define the context $\mathbb{K}_k = (G_k, M_k, I_k)$, where $M_k = M_{k-1} \setminus \mathcal{IC}(\mathbb{K}_{k-1})$, $G_k = \bigcup_{m \in M_k} \{m\}'$, and $I_k = I \cap (G_k \times M_k)$. The sequence $(\mathcal{IC}(\mathbb{K}_0), \mathcal{IC}(\mathbb{K}_1), \ldots, \mathcal{IC}(\mathbb{K}_n), M_0 \setminus M_n)$ for some $n > 0$ defines a partition of the attribute set M_0 to be used for nesting.

In our example, "*receptor*", "*growth*", "*signal*", "*brain*", "*mouse*", and "*human*" cover the whole set of authors and constitute the outer-level subcontext notions. As we use only two levels, the inner-level subcontext is made of the remaining terms: "*embryogenesis*", "*evolutionary*", "*conservation*", "*mammalian*", "*behavior*", "*vertebrate*", "*plate*", "*pathway*", "*induction*", "*phenotype*", "*wild-type*" and "*migration*".

The resulting diagrams are shown in Fig. 3. Yet, while nesting makes it possible to distinguish between various levels of precision, both the outer and inner diagrams are still too large and recall the jumbled picture of Fig. 1. Stability-based pruning will address this problem; combining both procedures should yield a concise hierarchical representation.

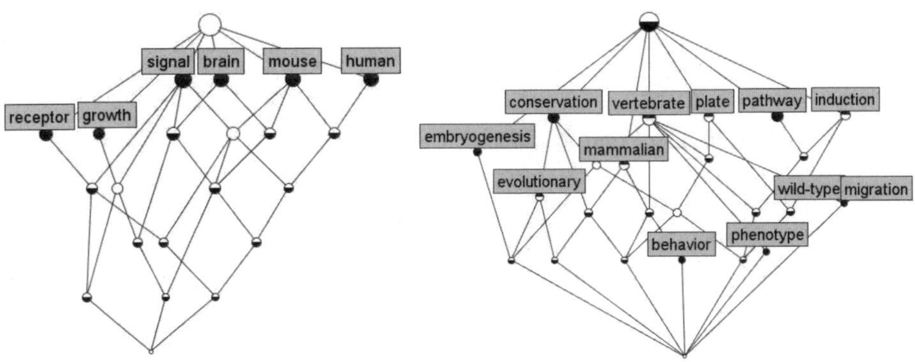

Fig. 3. Outer and inner diagrams for the nested line diagram of the zebrafish context

3.4 Combining Nesting and Stability-Based Pruning

After partitioning the set of words and building lattices for individual parts, we prune each lattice using the stability criterion. We can use different thresholds for different parts depending on the number of concepts we are comfortable to work with. For our example, we get the two diagrams shown in Fig. 4. Many attributes of the inner diagram are not shown in the picture, as they are not contained in any stable intent.

We proceed by drawing one diagram inside the other and interpret the picture as usual. Again, only inner nodes corresponding to concepts of the full context are represented by circles. Figure 5 shows the resulting structure for our context.

This approach may also help in reducing the computational complexity. Generally, computing inner concepts is the same as computing the lattice for the whole context,

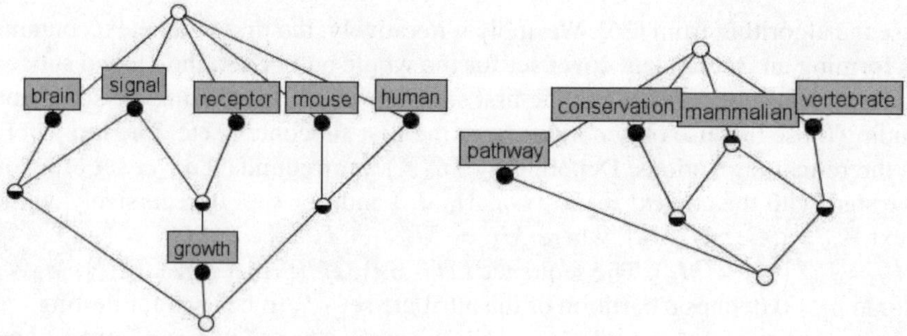

Fig. 4. The pruned outer and inner lattices from Fig. 3 (resp. thresholds 0.70 and 0.54)

but, combining nesting and pruning, we compute inner nodes only for relevant (that is, non-pruned) outer nodes.

Let us denote by $\underline{\mathfrak{B}}_p(\mathbb{K})$ the set of concepts of $\mathbb{K}$ satisfying the chosen pruning criteria and ordered in the usual way (one may regard p as an indicator of a specific pruning strategy). Assume that contexts $\mathbb{K}_1 = (G, M_1, I_1)$ and $\mathbb{K}_2 = (G, M_2, I_2)$ are subcontexts of $\mathbb{K} = (G, M, I)$ such that $M = M_1 \cup M_2$ and $I = I_1 \cup I_2$. We define the set of concepts corresponding to nodes of the nested line diagram of the pruned concept sets $\underline{\mathfrak{B}}_p(\mathbb{K}_1)$ and $\underline{\mathfrak{B}}_p(\mathbb{K}_2)$:

$$\underline{\mathfrak{B}}_p(G, M_1, M_2, I) = \{(A, B) \in \underline{\mathfrak{B}}(\mathbb{K}) \mid \forall i \in \{1, 2\} : ((B \cap M_i)', B \cap M_i) \in \underline{\mathfrak{B}}_p(\mathbb{K}_i)\}. \tag{5}$$

Proposition 2. *If $\underline{\mathfrak{B}}_p(\mathbb{K}_1)$ and $\underline{\mathfrak{B}}_p(\mathbb{K}_2)$ are $\bigvee$-subsemilattices of $\underline{\mathfrak{B}}(\mathbb{K}_1)$ and $\underline{\mathfrak{B}}(\mathbb{K}_2)$, respectively, then $\underline{\mathfrak{B}}_p(G, M_1, M_2, I)$ is a $\bigvee$-subsemilattice of $\underline{\mathfrak{B}}(\mathbb{K})$ and the map*

$$(A, B) \mapsto (((B \cap M_1)', B \cap M_1), ((B \cap M_2)', B \cap M_2)) \tag{6}$$

is a $\bigvee$-preserving order embedding of $\underline{\mathfrak{B}}_p(G, M_1, M_2, I)$ in the direct product of $\underline{\mathfrak{B}}_p(\mathbb{K}_1)$ and $\underline{\mathfrak{B}}_p(\mathbb{K}_2)$.

Proof. Let $(A, B), (C, D) \in \underline{\mathfrak{B}}_p(G, M_1, M_2, I)$. Then, we have $(A, B) \vee (C, D) = ((B \cap D)', B \cap D) \in \underline{\mathfrak{B}}_p(G, M_1, M_2, I)$, since $B \cap D \cap M_i = (B \cap M_i) \cap (D \cap M_i)$ is the intent of a concept in $\underline{\mathfrak{B}}_p(\mathbb{K}_i)$ for $i \in \{1, 2\}$. Hence, $\underline{\mathfrak{B}}_p(G, M_1, M_2, I)$ is indeed a $\bigvee$-subsemilattice of $\underline{\mathfrak{B}}(\mathbb{K})$. To see that the above-mentioned mapping is $\bigvee$-preserving, note that it maps the intent $B \cap D$ to the pair of intents $(B \cap D \cap M_1, B \cap D \cap M_2)$, and $B \cap D \cap M_i$ is the intent of the join of concepts with intents $B \cap M_i$ and $D \cap M_i$.

Unlike in standard nesting [10], the component maps $(A, B) \mapsto ((B \cap M_i)', B \cap M_i)$ are not necessarily surjective on $\underline{\mathfrak{B}}(\mathbb{K}_i)$. Hence, some outer nodes in our nested line diagram may be empty, i.e., contain no realized inner nodes, and some nodes of the inner diagram may never be realized.

Back to our example, the pruned outer diagram embraces most of the expert-based description outlined in section 2.2 within a readable structure: it shows a joint focus on "*human*" and "*mouse*" (comparative studies); features several subfields made of "*signal*", "*receptor*", and "*growth*"; and displays brain studies, also in connection with signaling issues. The nested line diagram allows a deeper insight into the substructure of

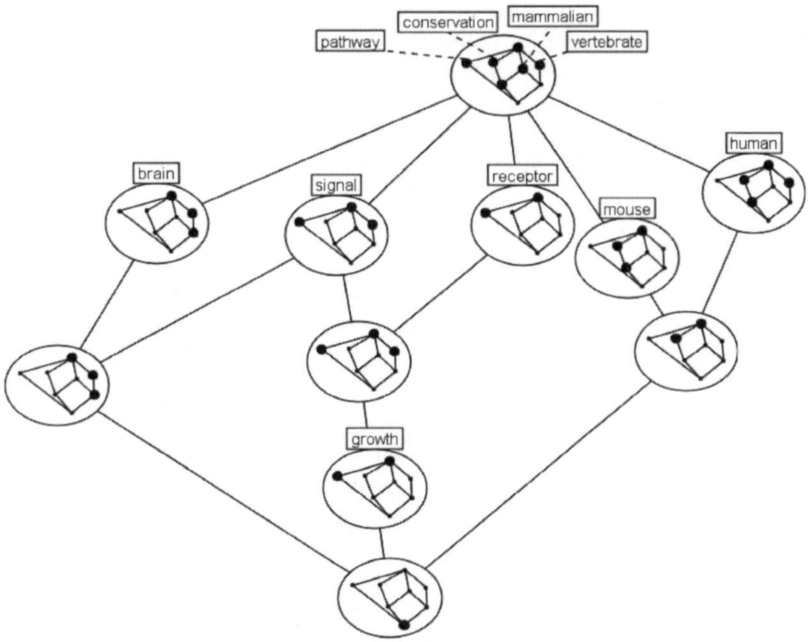

Fig. 5. Nested line diagram of pruned lattices from Fig. 4

particular fields embedded within the pruned outer diagram. One may notice that outer nodes involving "*human*" and "*mouse*" show "*conservation*" in their inner diagrams (also together with "*mammalian*"), while outer nodes involving "*signal*" and "*receptor*" display "*pathway*". This is consistent with the real state of affairs.

4 Further Work

4.1 Variants of Stability

The stability index σ as in Def. 1 and [9] refers to the stability of an intent; we call it *intensional*. The *extensional stability index* of a concept (A, B) can be defined similarly:

$$\sigma_e(A, B) = \frac{|\{D \subseteq B \mid D' = A\}|}{2^{|B|}}. \tag{7}$$

The extensional stability of a concept is the probability of preserving its extent after leaving out an arbitrary number of attributes, and a proposition similar to Proposition 1 holds. Extensional stability relates to the social aspect of the concept, measuring how much the community as a group of people depends on a particular topic. It also allows one to fight noisy words—a community based on a noisy word (or, e.g., a homograph used differently within different communities) will be extensionally unstable.

Proposition 1 suggests how the *general stability index* of a concept (A, B) could be defined—as the ratio between the number of subcontexts of $\mathbb{K} = (G, M, I)$ preserving the concept up to the omitted objects and attributes and the total number of subcontexts:

$$\frac{|\{(H, N, J) \mid H \subseteq G, N \subseteq M, J = I \cap (H \times N), A_H^J = B_N, B_N^J = A_H\}|}{2^{|G|+|M|}} \tag{8}$$

where $A_H = A \cap H$ and $B_N = B \cap N$. As of now, we are not aware of any realistic method for computing the general stability; thus, it is only of theoretical interest. On the other hand, limited versions of stability (e.g., computed over subsets of a certain size) and various combinations of extensional and intensional stability, are worth trying.

4.2 Strategies for Pruning

Other techniques aiming at reducing the number of concepts should be tested and, perhaps, some of them can be combined with stability for better results—notably pruning based on monotonous criteria like extent/intent size. Another method is given by attribute-dependency formulas [27], involving an expert-specified hierarchy on the attribute set (e.g., "*human*" and "*mouse*" are subtypes of "*vertebrate*").

As noted in section 3.2, pruning may not necessarily yield a lattice. We can handle this situation in several ways: for example, enlarging the pruned structure by including all intersections of stable intents; or reducing the structure by eliminating some stable intents. We may prefer to merge an unstable concept (A, B) with one of its subconcept (C, D), rather than simply drop (A, B). However, it is not immediately clear how to choose (C, D)—only that it should be "close" to (A, B) in some or other sense. Merging can be done by assuming that all objects from A have all attributes from D and replacing the context relation I by $I \cup A \times D$ (cf. [28]). However, the modified context may have intents that are absent from the initial context, which is probably undesirable. Alternatively, one could add $B \rightarrow D$ to the implication system of the context. The lattice of attribute subsets generated by this augmented implication system will be different from the original lattice only in that B and possibly some of its previously closed supersets are not in the new lattice.

A different approach would involve merging based on partial implications (or association rules): compute all partial implications for the given confidence threshold and add them to the implication system of the context. It is a matter of further experimentation to see which strategies are suitable for our goals.

4.3 Nesting

Nested line diagrams are not limited to two levels, although it still has to be investigated whether multi-level diagrams remain readable and interpretable. Various techniques for partitioning attribute sets should be explored. One strategy specific to our application is to partition words according to their type: as a verb, noun, adjective; or as a method, object, property, etc. This can be combined with other feature selection algorithms.

It should be noted that nesting seems to have more potential if used in interactive software tools that allow the user to zoom in and out on particular communities instead of having to deal with the entire picture. The fact that one need not compute everything at once provides an additional computational advantage.

4.4 Dynamic Monitoring

Modeling changes of the community structure should be particularly useful to describe the evolution of fields historically, either longitudinally or dynamically. The longitudinal approach means establishing a relation between community structures corresponding to different time points, e.g., identifying cases when several communities have merged into one or a community has divided into several sub-communities. FCA offers some methods for comparing two lattices built from identical objects and/or attributes (e.g., see [29]). Yet the relevance of such methods is likely to be application-dependent, and they should certainly be adapted for the reduced lattice-based structures we work with. One possibility in line with the static approach is to use nested line diagrams by nesting diagrams of contexts corresponding to successive time points. It also seems worth exploring whether temporal concept analysis [30] has anything to offer in this regard.

A more ambitious dynamic approach to modeling changes assumes that any elementary change in the database (any modification of G, M, or I: a new author, a new word, an author using a particular word for the first time, or removal of authors due to their inactivity, etc.) should correspond to a concrete change in the representation of communities. Although not every such change will have effect on the structure of the communities, it should always be possible to trace a change in the structure to a sequence of elementary changes in the database.

5 Conclusion

The approach discussed in this paper is based on the assumption that community structure in knowledge-based social networks should be dealt with more deeply than by simply relying on single-mode characterizations, as is often the case. In previous work [7,8], it was shown how concept lattices can be used to build knowledge taxonomies from data describing authors by sets of terms they use in their papers. As frequently happens with concept lattices derived from real data, such taxonomies tend to be huge and, therefore, hard to compute and analyze. The computational complexity can be partially addressed by reducing the number of agents, since a taxonomy centered on knowledge fields rather than individuals justifies the use of a random representative sample of authors.

However, the interpretability of results requires a more serious effort. In this paper, we proposed a pruning method based on the stability indices of formal concepts [9]. We think that this method does not merely reduce the concept lattice to a somewhat rougher structure; it also helps to combat noise in data, so that the resulting structure might even be more accurate in describing the knowledge community than the original lattice is.

We suggested that this method could also be applied to constituent parts of a nested line diagram to achieve an optimal relationship between the readability of the taxonomy and the level of detail in it. This is beneficial from the stance of computational complexity, too: it is easier to compute the lattices of subcontexts used in nesting and then prune each of them individually than to compute the lattice of the entire context and prune it. Besides, nested line diagrams admit "lazy" computation: within an interactive software tool the user can choose which outer nodes to explore. As a result, inner diagrams corresponding to neglected outer nodes need not be computed, unless required.

We have illustrated the proposed techniques with a small example. Of course, wider experiments are needed to see how this works. There are open questions: how to efficiently compute stability, how exactly stability-based criteria should be formulated and applied, how other compression techniques perform against stability-based pruning, etc. (see section 4). Thus, this paper is merely a first step towards a consistent methodology for creating concise knowledge taxonomies based on concept lattices.

References

1. Schmitt, F. (ed.): Socializing Epistemology: The Social Dimensions of Knowledge. Lanham, MD: Rowman & Littlefield (1995)
2. Leydesdorff, L.: In search of epistemic networks. Social Studies of Science 21, 75–110 (1991)
3. Haas, P.: Introduction: Epistemic communities and international policy coordination. International Organization 46(1), 1–35 (1992)
4. McCain, K.W.: Cocited author mapping as a valid representation of intellectual structure. J. Am. Society for Information Science 37(3), 111–122 (1986)
5. Atran, S.: Folk biology and the anthropology of science: Cognitive universals and cognitive particulars. Behavioral and Brain Sciences 21, 547–609 (1998)
6. Lorrain, F., White, H.C.: Structural equivalence of individuals in social networks. Journal of Mathematical Sociology 1, 49–80 (1971)
7. Roth, C., Bourgine, P.: Epistemic communities: Description and hierarchic categorization. Mathematical Population Studies 12(2), 107–130 (2005)
8. Roth, C., Bourgine, P.: Lattice-based dynamic and overlapping taxonomies: The case of epistemic communities. Scientometrics 69(2) (2006)
9. Kuznetsov, S.O.: On stability of a formal concept. In: SanJuan, E. (ed.) JIM, Metz, France (2003)
10. Ganter, B., Wille, R.: Formal Concept Analysis: Mathematical Foundations. Springer, Berlin (1999)
11. Girvan, M., Newman, M.E.J.: Community structure in social and biological networks. PNAS 99, 7821–7826 (2002)
12. Noyons, E.C.M., van Raan, A.F.J.: Monitoring scientific developments from a dynamic perspective: self-organized structuring to map neural network research. Journal of the American Society for Information Science 49(1), 68–81 (1998)
13. White, D.R., Duquenne, V.: Social network & discrete structure analysis: Introduction to a special issue. Social Networks 18, 169–172 (1996)
14. Freeman, L.: Cliques, Galois lattices, and the structure of human social groups. Social Networks 18, 173–187 (1996)
15. Falzon, L.: Determining groups from the clique structure in large social networks. Social Networks 22, 159–172 (2000)
16. Guigues, J.L., Duquenne, V.: Familles minimales d'implications informatives resultant d'un tableau de données binaires. Math. Sci. Humaines 95, 5–18 (1986)
17. Peyriéras, N.: Personal communication (2005)

18. McCain, K.W., Verner, J.M., Hislop, G.W., Evanco, W., Cole, V.: The use of bibliometric and Knowledge Elicitation techniques to map a knowledge domain: Software Engineering in the 1990s. Scientometrics 65(1), 131–144 (2005)
19. Grunwald, D.J., Eisen, J.S.: Headwaters of the zebrafish – emergence of a new model vertebrate. Nature Rev. Genetics 3(9), 717–724 (2002)
20. Bradbury, J.: Small fish, big science. PLoS Biology 2(5), 568–572 (2004)
21. Stumme, G., Taouil, R., Bastide, Y., Pasquier, N., Lakhal, L.: Computing iceberg concept lattices with TITANIC. Data & Knowledge Engineering 42, 189–222 (2002)
22. Kuznetsov, S.O., Obiedkov, S.: Comparing performance of algorithms for generating concept lattices. J. Expt. Theor. Artif. Intell. 14(2/3), 189–216 (2002)
23. Bayardo, J. R., Goethals, B., Zaki, M. (eds.): Proc. of the IEEE ICDM Workshop on Frequent Itemset Mining Implementations (FIMI 2004), CEUR-WS.org (2004)
24. Kuznetsov, S.O.: Stability as an estimate of the degree of substantiation of hypotheses derived on the basis of operational similarity. Nauchn. Tekh. Inf., Ser.2 (Automat. Document. Math. Linguist.) (12), 21–29 (1990)
25. Kohavi, R.: A study of cross-validation and bootstrap for accuracy estimation and model selection. In: IJCAI, pp. 1137–1145 (1995)
26. Batni, R.P., Russell, J.D., Kime, C.R.: An efficient algorithm for finding an irredundant set cover. J. Ass. for Comp. Machinery 21(3), 351–355 (1974)
27. Belohlávek, R., Sklenar, V.: Formal concept analysis constrained by attribute-dependency formulas. In: Ganter, B., Godin, R. (eds.) ICFCA 2005. LNCS (LNAI), vol. 3403, pp. 176–191. Springer, Heidelberg (2005)
28. Rome, J.E., Haralick, R.M.: Towards a formal concept analysis approach to exploring communities on the world wide web. In: Ganter, B., Godin, R. (eds.) ICFCA 2005. LNCS (LNAI), vol. 3403, pp. 33–48. Springer, Heidelberg (2005)
29. Wille, R.: Conceptual structures of multicontexts. In: Eklund, P.W., Mann, G.A., Ellis, G. (eds.) ICCS 1996. LNCS, vol. 1115, pp. 23–29. Springer, Heidelberg (1996)
30. Wolff, K.E.: Temporal concept analysis. In: Mephu Nguifo, E., et al. (eds.) ICCS-2001 Intl. Workshop on Concept Lattices-Based Theory, Methods and Tools for Knowledge Discovery in Databases, Palo Alto (CA), Stanford Univ, pp. 91–107 (2001)

Towards an Iterative Classification Based on Concept Lattice

Stéphanie Guillas, Karell Bertet, and Jean-Marc Ogier

L3I laboratory, Université de La Rochelle
av M. Crépeau, 17042 La Rochelle Cedex 1, France
{sguillas,kbertet,jmogier}@univ-lr.fr

Abstract. In this paper, we propose a generic description of the concept lattice as classifier in an iterative recognition process. We also present the development of a new structural signature adapted to noise context. The experimentation is realized on the noised symbols of GREC database [4]. Our experimentation presents a comparison with the two classical numerical classifiers that are the bayesian classifier and the nearest neighbors classifier and some comparison elements for an iterative process.

1 Introduction

The work presented in this paper takes place in the field of automatic retro-conversion of technical documents and proposes to use concept lattice to recognize graphic objects, and more precisely to classify noised symbols images. This graph issued from Formal Concept Analysis [14], has often been used in data mining. A recent study [9] gives a comparison of several classification methods based on concept lattice, and clearly shows the interest of its use in classification.

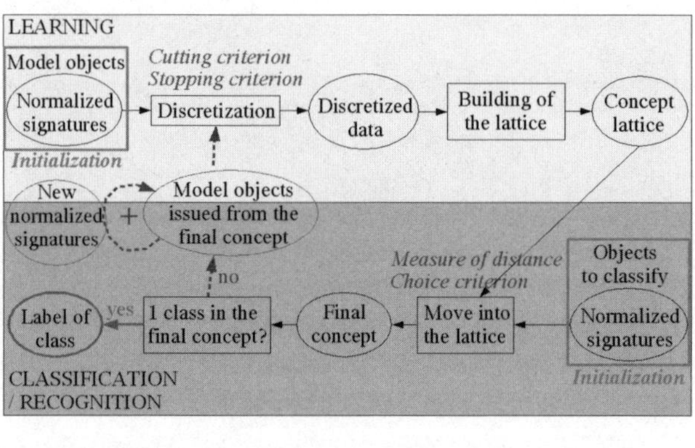

Fig. 1. The iterative recognition stages

In study [6], we showed that concept lattice has a structure which looks like the decision tree, and that its bigger size gives more robustness to the noise than

S. Ben Yahia et al. (Eds.): CLA 2006, LNAI 4923, pp. 256–262, 2008.

the decision tree. We also highlighted the recognition parameters and use the concept lattice as a classifier in a one-step process.

Here, we present an iterative process (Fig. 1), in which we repeat the recognition process with selection of new attributes (or characteristics) in the signatures at each iteration. In the field of symbols recognition, an iterative process is attractive because various techniques (structural, statistical) enable to extract new data from images. In our one-step process, we used a statistical signature which gave good results of recognition, and for the iterative process, we have chosen to complete the description by a structural signature adapted to the context of noise. In our process, discretization and particularly selection of attributes are necessary to reduce the context size. Moreover, we chose to build the concept lattice because the graph allows to navigate and to progressively validate attributes to classify the noised data.

Recognition process (Fig. 1) is usually composed of the *learning* stage and the *classification* stage (section 2). In part 2.2, we describe the data learning which data are discretized and the concept lattice is built. Classification and especially navigation in the concept lattice is described in part 2.3. Part 3 proposes a comparison in cross-validation with the bayesian classifier and the nearest neighbors classifier and to finish, conclusion and extensions are presented in part 4.

2 Process

The iterative recognition process follows a coarse-to-fine strategy with selection of new attributes at each iteration. The recognition process (Fig. 1) is composed of learning and classification. We first have a set of *model* objects (classes are known) and a set of objects to classify. Classification aims to attribute a class label to each object. After each iteration, we propose a *final concept* (defined in part 2.3) which contains one or several classes. When it contains only one class, the process is finished, otherwise, the signatures don't discriminate enough the classes, and another selection of attributes is needed to determine the class label.

2.1 Signatures

In our case, objects to recognize are graphic images described by equal size normalized numerical signatures [13]. We chose to use the two following types of signatures : *statistical* and *structural.*

Statistical. These kind of signatures describe the pixels distribution on the images. In our previous work [6], experimental results showed that the statistical signature based on the Radon transform [12] seems to be the most appropriate for symbols description. This signature is composed of 50 values and describes the pixels organization in several orientations.

Structural. Symbol images are particular and can be characterized by their spatial organization. A structural signature aims to describe the topology of the

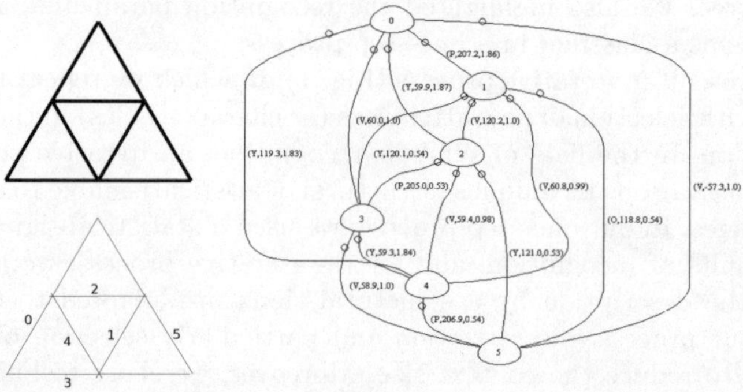

Fig. 2. Example of symbol, its extracted segments (left), and its topological graph (right)

primitive elements which composed the symbols. We want to discriminate the symbols by searching in their structure, specific forms as rectangles, triangles, ..., but in a generic way, without define the forms we are looking for.

First, we extract these primitives elements, as segments, using the robust Hough transform. We treat segments by pairs, and extract their relative position in a triplet of information : <*relation type, relation value, length ratio*>. Relation type corresponds to the visual interpretation of the intersection of the pair (X, Y, V, P for parallel, and O for others), relation value is the relative angle between the pair for X, Y, V and O, and the relative distance for P, and length ratio is the relative length of the pair of segments. These triplets are introduced in a topological graph and its corresponding adjacency matrix. They describe the whole relations which compose the symbol (see Fig. 2).

Then, our signature is based on Geibel work [3] : compute the number of occurrences of various paths in the topological graph. Paths represent generic forms which characterize the structure of the symbols. Depending on the length of the searched paths, we can determine the precision level of the symbols description. The final signature will be composed of the paths and their corresponding occurrences in the different symbols.

2.2 Learning

The learning stage consists in organizing a concept lattice data issued from a set of objects. It is composed of: *discretization of data* and *building of the lattice* (Fig. 1).

Discretization. This stage [11] consists in organizing the signatures $p = (p_i)_{i \leq n}$ issued from the objects set O, in intervals, that characterize each class of objects. At each step of discretization, an interval is selected to be cut. This selection depends on a *cutting criterion*, and the cutting process is repeated until a *stopping criterion* is validated. In study [6], we selected the maximal distance as non supervised criterion and the Hotelling's coefficient as supervised criterion.

Here are some stopping criteria: $crit_{class\ separated}$ is *"to stop when classes are separated"*; $crit_{nb\ steps}$ is *"to stop when the discretization steps number equals a constant nb"*; $crit_{nb\ classes\ max}$ means that the final concept contains at most nb classes; and $crit_{cutting\ min}$ limits the cutting criterion above a minimal value.

When discretization is performed, objects $p \in O$ are characterized by intervals $I = I_1 \times I_2 \times \ldots \times I_n$ with I_i the intervals set of each attribute $i = 1 \ldots n$, and the membership relation $\mathcal{R}$ between objects and intervals can be deduced.

Building of the concept lattice. This stage immediately follows the discretization stage and is totally determined by the membership relation $\mathcal{R}$ between objects and intervals without criterion or parameter.

A concept lattice is composed of a set of *concepts* ordered by inclusion, which forms a graph (that has the lattice properties [1]). We associate to a set of objects $A \subseteq O$, the set $f(A)$ of intervals in relation $\mathcal{R}$ with A: $f(A) = \{x \in I \mid p\mathcal{R}x\ \forall\ p \in A\}$. Dually, for a set of intervals $B \subseteq I$, we define the set $g(B)$ of objects in relation $\mathcal{R}$ with B: $g(B) = \{p \in O \mid p\mathcal{R}x\ \forall\ x \in B\}$.

A *formal concept* is a pair objects-intervals (A, B) with $A \subseteq O$, $B \subseteq I$, $f(A) = B$ and $g(B) = A$. Two concepts (A_1, B_1) and (A_2, B_2) are in relation in the concept lattice if they verify the inclusion property: $(A_1, B_1) \leq (A_2, B_2) \Leftrightarrow A_1 \supseteq A_2$ equivalent to $B_1 \subseteq B_2$. Let $\prec$ be the transitive reduction associated to $\leq$. The minimal concept $(O, f(O))$ according to the relation $\leq$ contains the whole objects O and the set $f(O)$. Note that $f(O) = \emptyset$ when intervals shared by all the objects are removed. Dually, the maximal concept is $(g(I), I)$. For more information about Galois connection and concept lattice, see [1].

Our algorithm is based on Bordat [2] and Morvan and Nourine [10] ones. We choose it for its implementation simplicity and the possibility to generate on demand the required concepts of the lattice, instead of building the whole graph. It is a real advantage because the main limit of concept lattice is its cost in time and space. Indeed, its size is bounded by $2^{|S|}$ in the worst case, and by $|S|$ in the best case. The main advantage of this graph is its good readability because it is easy to interpret.

2.3 Classification

Concept lattice can be seen as a search space in which we move by validation of the intervals issued from the discretization stage. The signature $s = (s_1, \ldots, s_n)$ of the object to recognize is introduced in the concept lattice starting from the *minimal concept*: $(O, f(O))$ meaning that the whole classes of objects are *candidates* to recognition and no interval is validated. We progress step by step in the concept lattice by validation of new intervals and consequently by reduction of the objects set and their corresponding classes, until we reach a *final concept*.

A concept is a *final concept* when it is the last concept in the classification progress containing objects of some class. A final concept (A, B) corresponds to the sup-irreducibles of the lattice. (see [1]) and is characterized by:

$$|GetClasses((A, B))|! = \sum_{(A',B') \succ (A,B)} |GetClasses((A', B'))|$$

From a current concept, an elementary step of classification consists in selecting an interval from a set of intervals S, to progress to a new concept. More precisely, S is a family of intervals obtained from the n immediate successors $(A_1, B_1), \ldots, (A_n, B_n)$ of the current concept (A, B) and defined by: $S = \bigcup_{i=1}^{n} B_i \backslash B = \{X_1, \ldots, X_n\}$. Thus, the *choice criterion* parameter consists in *choosing a subset* X_i *of intervals among* S using a fuzzy distance measure d.

In our experiments, symbols are noised and thus signature values can be modified. To make supple the boundaries intervals, we use a fuzzy distance measure $d(s_i, x) = \mu_A(x)$, with μ the *likelihood degree* of $x \in A$, and A a fuzzy set.

3 Experimental Results

Our previous work [6] showed that concept lattice is more appropriate to the classification of noised graphic objects than the decision tree. Moreover, experimental results showed that the Radon signature, the Hotelling's cutting criterion seem to be the most appropriate. So we used them in these new tests.

3.1 Tests with Separation of Classes

In this experiment, concept lattice is compared to bayesian classifier and nearest neighbors classifier (k-NN). For the concept lattice, we use $crit_{class\ separated}$ as stopping criterion, so one iteration is required to obtain a label of class. Our data consist of 2 sets of 10 classes of symbols of GREC2003 [4] (namely cl1-10 and cl11-20), in which each class contains 1 "model" symbol and 90 symbols (Fig. 3) noised by the Kanungo method [7]. We use another data set composed of 25 classes (namely cl1-25) of GREC2005 database. This symbols set is more noised than those of GREC2003, and is composed of 175 symbols.

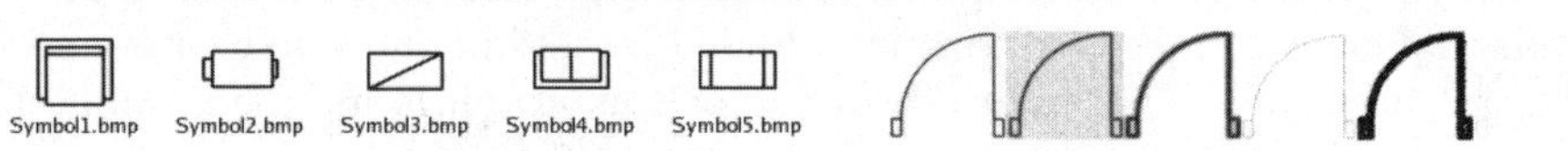

Fig. 3. 5 "model" symbols of GREC2003 database (left) and 6 noised symbols of GREC2005 database (right)

We test the 3 classifiers by the cross-validation technique [8]. The test result is the average of the n recognition rates. On GREC2003 symbols, we try: 5 blocks of 182 symbols (test 1), 10 blocks of 91 symbols (test 2) and 26 blocks of 35 symbols (test 3). On GREC2005 symbols, we try 5 blocks of 35 symbols (test 4). Recognition rates are shown in Figure 4.

For test 4, results are really low due to the high level of noise. From these results, we deduced that k-NN classifier gives the best rates, and bayesian classifier gives better rates than the concept lattice only for big sizes of the learning set (tests 1 and 2). Notice that concept lattice only needs between 6 and 15 attributes of the Radon signature among the 50 values, on the contrary to the bayesian and the k-NN classifiers. The relatively good results of these tests indicate that an iterative process is an interesting way to explore.

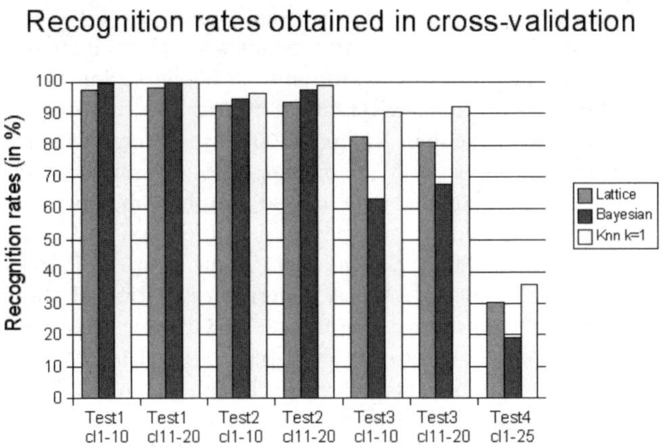

Fig. 4. Results of cross-validation for the 3 classifiers

3.2 Tests without Separation of Classes

In order to set up an iterative process, we need to define a stopping criterion of discretization. In paper [5], we study the recognition potential of 4 stopping criteria. Results were encouraging and showed that the improvement of the recognition rates was possible with an iterative process. However, we would like to find a converging stopping criterion, but none of studied criteria respect this property. We decided to use a validation set of symbols to determine the concepts which produce classification errors and stop the progression in the graph for these concepts. Then, from this first step of recognition, it is possible to pursuit the process for the concerned symbols with a new signature.

4 Conclusion

The experimentations show that concept lattice gives relatively close recognition rates than the famous k-NN classifier. Moreover, the iterative recognition approach described here is interesting to handle big learning sets, what was relatively costly, and the first results are promising. Moreover, this iterative system could be useful when classes are few separable. Indeed, to characterize these classes, we would like to use our new structural signature with the most appropriate description level, and to complete with the information given by the statistical signature.

References

1. Barbut, M., Monjardet, B.: Ordres et classifications: Algèbre et combinatoire (tome II). In: Hachette, Paris (1970)
2. Bordat, J.: Calcul pratique du treillis de Galois d'une correspondance. Math. Sci. Hum. 96, 31–47 (1986)

3. Geibel, P., Wysotzki, F.: Learning relational concepts with decision trees. In: Saitta, L. (ed.) Machine Learning: Proceedings of the Thirteenth International Conference, pp. 166–174. Morgan Kaufmann Publishers, San Francisco (1996)
4. GREC. Symbol images database GREC 2003 (Graphics RECognition), Last access 09/10/2007 (2003), www.cvc.uab.es/grec2003/symreccontest/index.htm
5. Guillas, S., Bertet, K., Ogier, J.-M.: Concept lattice classifier: A first step towards an iterative process of recognition of noised graphic objects. In: Ben Yahia, S., Mephu Nguifo, E. (eds.) Fourth International Conference on Concept Lattices and their Applications (CLA 2006), Hammamet, Tunisia, pp. 257–263 (2006)
6. Guillas, S., Bertet, K., Ogier, J.-M.: A generic description of the concept lattices classifier: Application to symbol recognition. In: Graphics Recognition: Ten Years Review and Future Perspectives - Selected papers from GREC 2005, Hong Kong, China, August 2005. LNCS, vol. 3926, pp. 47–60. Springer, Berlin / Heidelberg (2006) (Revised and extended version of paper first presented at Sixth IAPR International Workshop on Graphics Recognition (GREC 2005))
7. Kanungo, T., et al.: Document degradation models: Parameter estimation and model validation. In: IAPR Workshop on machine vision applications, Kawasaki (Japan), pp. 552–557 (1994)
8. Krus, D., Fuller, E.: Computer assisted multicrossvalidation in regression analysis. Educational and Psychological Measurement 42, 187–193 (1982)
9. Mephu Nguifo, E., Njiwoua, P.: Treillis des concepts et classification supervisée. In: Technique et Science Informatiques, RSTI, Hermès - Lavoisier, Paris, France, vol. 24(4), pp. 449–488 (2005)
10. Morvan, M., Nourine, L.: Simplicial elimination shemes, extremal lattices and maximal antichains lattice. Order 13(2), 159–173 (1996)
11. Rakotomalala, R.: Graphes d'induction. PhD thesis, Université Claude Bernard, Lyon I, Décembre (1997)
12. Tabbone, S., Wendling, L.: Adaptation de la transformée de Radon pour la recherche d'objets à niveaux de gris et de couleurs. In: Technique et Science Informatiques, RSTI, Hermès - Lavoisier, Paris, France, vol. 22(9), pp. 1139–1166 (2003)
13. Teague, M.: Image analysis via the general theory of moments. Journal of Optical Society of America (JOSA) 70, 920–930 (2003)
14. Wille, R.: Restructuring lattice theory: An approach based on hierarchies of concepts. In: Rival, I. (ed.) Ordered sets, pp. 445–470. Reidel, Dordrecht-Boston (1982)

Using Formal Concept Analysis for Mining and Interpreting Patient Flows within a Healthcare Network

Nicolas Jay[1,2], François Kohler[2], and Amedeo Napoli[1]

[1] Équipe Orpailleur, LORIA, Vandoeuvre-lès-Nancy, France
[2] Laboratoire SPI-EAO, Faculté de Médecine, Vandoeuvre-lès-Nancy, France

Abstract. This paper presents an original experiment based on frequent itemset search and lattice based classification. This work focuses on the ability of iceberg-lattices to discover and represent flows of patient within a healthcare network. We give examples of analysis of real medical data showing how Formal Concept Analysis techniques can be helpful in the interpretation step of the knowledge discovery in databases process. This combined approach has been successfully used to assist public health managers in designing healthcare networks and planning medical resources.

Keywords: Formal Concept Analysis, frequent itemsets, network.

1 Introduction

Knowledge Discovery in Databases (KDD) is an iterative and interactive process for identifying valid, novel, and potentially useful patterns in data [1]. KDD is usually divided into three main steps: data preparation, data mining, and interpretation of the extracted units. Data mining, often considered as the central step in this process, is still an active field of research. The success key in KDD practice relies also on ability of easily producing units understandable as knowledge units. One way of achieving such a goal relies on an adapted visualization of the extracted units.

In this paper, we present an original experiment based on both frequent itemset search and lattice-based classification. This experiment holds on medical data and is aimed at showing the interactions and collaborations between hospitals in the French Region of Lorraine. This experiment may be regarded from two points of view: on the one hand, it is based on frequent itemset search on a medico-economic database, and on the other hand, the visualization of extracted units is based on Formal Concept Analysis (FCA) techniques [2], organizing the extracted units into a lattice for medical analysis and interpretation. At our knowledge, this is an original combination of data mining and FCA techniques that has been rarely carried on until now. Indeed, this is one of the main feature of this paper to show how FCA techniques can be very helpful in the interpretation step of KDD process. The results of this experiment have been used by healthcare administration in Lorraine for planning and evaluation purposes [3].

S. Ben Yahia et al. (Eds.): CLA 2006, LNAI 4923, pp. 263–268, 2008.

2 Health Networks and Collaborations

Healthcare networks are sets of healthcare actors working in cooperation, sharing information, and providing care for the same patients. In France, some networks are formally structured but others are still in an implicit existence. Thus, healthcare policy should be based on this current state of things to plan new networks or optimize existing ones. However, for both structured and implicit networks, knowledge on the degree of collaboration between hospitals is poor, because no information system is dedicated to this type of monitoring. Such an information system could help measuring collaboration by analyzing the flow of patients being treated in more than one hospital.

This issue is close to the problem of cartographying a communication network [4]. A healthcare network can be represented by an undirected graph where hospitals are the nodes, and edges represent patients flows, i.e. sets of patients shared by two hospitals. In our context, healthcare networks can involve hundreds of hospitals and tens of thousands of patients. It is a challenge to visualize a network with such a volume of data.

Furthermore, this problem goes beyond simple cartography. Patient flows depend on several constraints: geography, location of high technology devices and specialized medical teams, personal affinities between physicians, regulations, type of disease... According to these constraints, hospitals do not have the same role within a healthcare network. There exists high level relations that cannot be represented in usual network maps. In the domain of social network analysis, Freeman [5] has proposed to use FCA to produce useful insights about structural properties of relationships between social actors. This approach could be extended to our problem. Nevertheless, a lattice-based representation does not always support the size of data. A way to deal with that issue is to only represent the most significant flows.

The analysis of patient flows can also be seen as a consumer behavior problem. Consumer behavior and market basket analysis are well-known problems in data mining and can be solved using frequent itemset search and association rule extraction [6]. In our application domain, a formal context can be built with patients as objects and the hospitals in which they have been treated as attributes. Discovering significant flows of patient between hospitals can be achieved by mining this context for searching for frequent itemsets of hospitals sharing the same patients. However, it may be difficult to exploit the results because of the large number of extracted units, and because of the lack of visualization support.

The links between the frequent itemset search and FCA have been studied by several research groups [7,8,9]. Stumme [10] has introduced iceberg lattices, which are concept lattices of frequent closed itemsets. The approach combining visualization and frequent itemset search is a feature of first importance in our research work. Firstly, it is a top-down method for gradually discovering and representing significant patient flows. Secondly, it provides easily understandable results, especially for novice users. In a similar way, Duquenne [11] has studied associations of psychological handicaps of children. Using filters on a weighted lattice, he has shown the ability of FCA to describe profiles of patients.

Furthermore, due to their ability to encode dualities [12], concept lattices can provide two points of view for interpreting patient flows: an intensional one in which flows result from interaction and collaboration of healthcare providers within a network, and an extensional one where flows can be regarded as groups of patients sharing a common medical profile.

3 Iceberg-Lattices

Let $\mathbb{K} := (G, M, I)$ be a formal context where G is a set of objects, M a set of attributes and I a binary relation between G an M.

Definition 1. *Let $B \subseteq M$ and let minsupp be a threshold $\in [0, 1]$. The support count of the attribute set B in $\mathbb{K}$ is $supp(B) := \frac{|B'|}{|G|}$. B is said to be a frequent attribute set if $supp(B) \geqslant minsupp$.*

A concept is called frequent concept if its intent is frequent. The set of all frequent concepts of a context $\mathbb{K}$ is called iceberg lattice of the context $\mathbb{K}$.

4 Discovery Process of Patient Flows

In France, the PMSI[1] database is a national information system used to describe hospital activity with both an economical and medical point of view. We have worked on two years of PMSI data of the Lorraine Region in France. Data preparation consists in building a formal context where objects are patients and attributes are hospitals lying in the database. A patient is related to a hospital whenever the patient has been treated in that hospital. An iceberg-lattice is then built from this context using the Titanic algorithm [10] implemented in Galicia 3.0 [13]. Hasse diagrams are drawn with the Graphiz [14] tools.

5 Results

We present here an example of cancer network analysis. In this experiment, the formal context holds 28009 patients and 158 hospitals. Figure 1 shows the resulting iceberg for a minsupp=0.017, i.e. 50 patients. For clarity, $\bot$ was removed and right and leftmost part of the lattice are not drawn. A first comment can be made about its general shape. It is more wide than deep because the context is sparse and data are poorly correlated. This means that patient flows are most of the time tightly partitioned, and that patients are rarely hospitalized in more than two hospitals. The intent of co-atoms, i.e. immediate descendants of $\top$, is always a singleton. This means a hospital never shares all of its patients with another one, or if it is so, less than 50 patients are involved in the interaction. The intent of atoms, i.e. the immediate ascendant of $\bot$, is always a pair. The extent of atoms gives an idea of the strength of the collaboration between the

[1] Programme de Médicalisation des Systèmes d'information.

two hospitals: the larger is the cardinal of the extent, the higher is the strength of the collaboration(i.e. the more patients are shared between the two hospitals). The iceberg can be divided in two parts:

- on the right, concepts that are both atoms and co-atoms. They represent institutions that share a few patients with others. This is that either they treat a few patients, or they work in a relative autonomy, or collaboration is split with many other hospitals.
- on the left, concepts that have at least a sub-concept (different from $\bot$). They represent a hospital receiving a significant number of patients, and having collaborations with at least one other establishment.

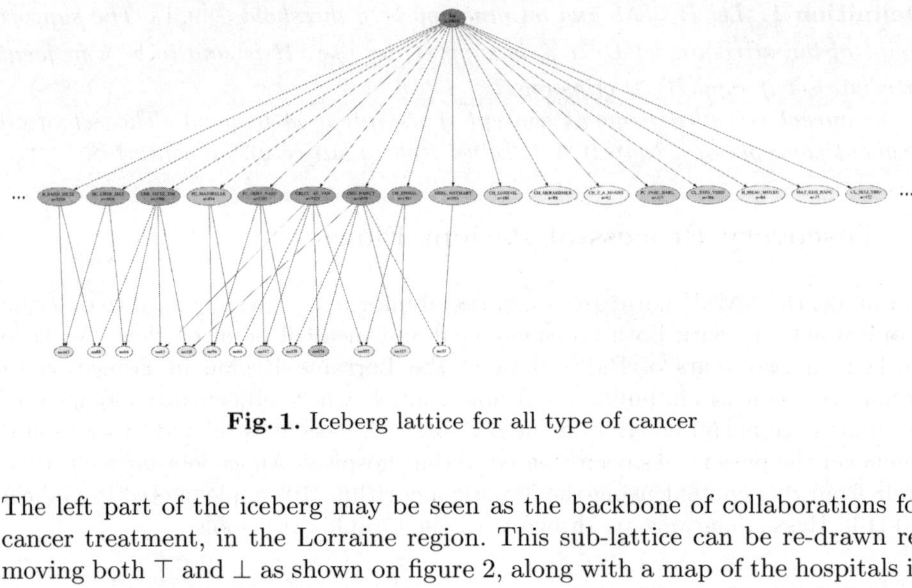

Fig. 1. Iceberg lattice for all type of cancer

The left part of the iceberg may be seen as the backbone of collaborations for cancer treatment, in the Lorraine region. This sub-lattice can be re-drawn removing both $\top$ and $\bot$ as shown on figure 2, along with a map of the hospitals in the Lorraine region. Co-atoms are then represented by ellipses. Their label shows the name of the hospital in their intent and the size of their extent. Diamonds are the second rank concepts (i.e. the atoms). Their label shows the size of their extent. Arrows represent the super-concept/sub-concept relation. These diamonds can be seen as cooperation between several hospitals. For example,CHU_NANCY and CRLCC_AV_VAN hospitals share 624 patients.

This figure contains a lot of information for the domain expert. First of all, three concepts have a large number of patients and many sub-concepts: CHU-NANCY, CHR-METZ-THI, and CRLCC-AV-VAN. They are located in Nancy and Metz, the two largest cities in Lorraine . The large number of sub-concepts related to these institutions precisely shows that they are reference centers. They employ highly skilled and specialized personnel. Treatments given there rely on state-of-art technology. Furthermore, they actively participate in anti-cancer research programs.

The aspect of the lattice reflects geographical constraints that shapes the network structure. An East-West separation line clearly appears. Many flows are

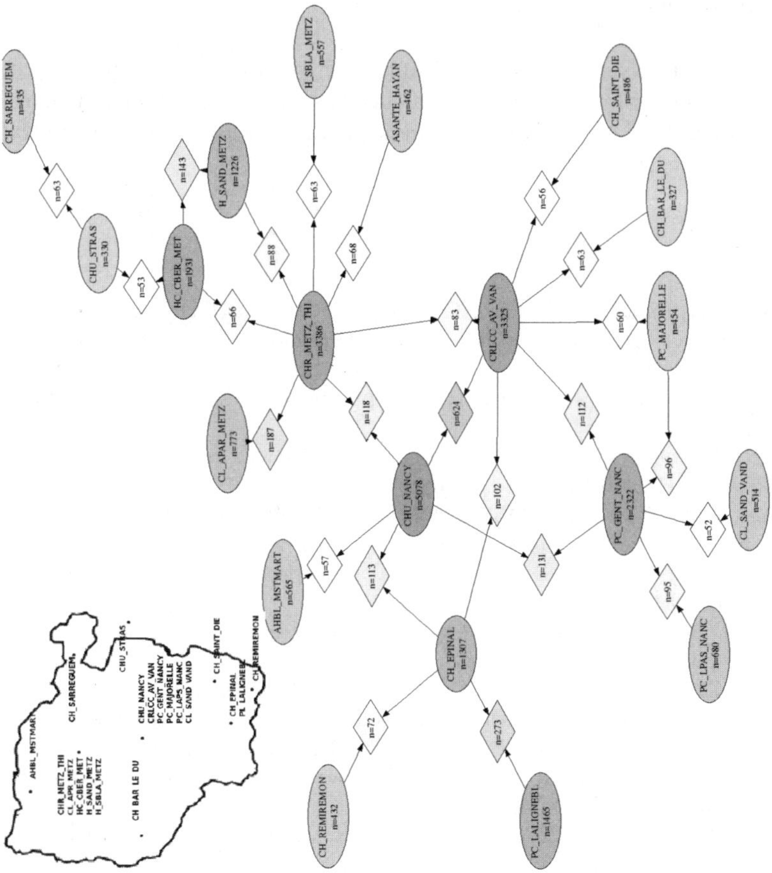

Fig. 2. Cooperation between hospitals for the treatment of cancer

concentrated in the north around the CHR-METZ-TH hospital. By contrast, concepts sharing sub-concepts with CHU-NANCY and CRLCC-AV-VAN concepts concern most of the time hospitals located in the southern Lorraine. Let us also notice that the CH-SARRGUEMIN hospital on the top right of the figure has a trans-border cooperation with the CHU-STRAS hospital in the next region of Alsace.

The lattice also illustrates the influence of statutory constraints The PC-GENT-NANCY concept has common sub-concepts with PC-LPAS-NANCY, CL-SAND-VAND and PC-MAJORELLE. This makes a sub-network gathering private hospitals in the city of Nancy.

Finally, the lattice allows to visualize two important sets of knowledge units: one is on the most important centers for the treatment of cancer, the other on the geographical locations of centers and the patient flows between these locations. Indeed, this can be seen as the concrete result of a working KDD system.

6 Conclusion

We have presented here a combined approach relying on data mining and FCA for representing patient flows in a healthcare system. This method takes advantage of iceberg-lattices to discover and to display in a simple way the backbone of healthcare networks.

References

1. Fayyad, U., Piatetsky-Shapiro, G., Smyth, P.: The KDD Process for Extracting Useful Knowledge from Volumes of Data. Communication of the ACM 29–11, 27–34 (1996)
2. Ganter, B., Wille, R.: Formal Concept Analysis: mathematical foundations. Springer, Heidelberg (1999)
3. Jay, N., Napoli, A., Kohler, F.: Cancer Patient Flows Discovery in DRG Databases. In: Proc. MIE 2006 Conf. (to appear)
4. Becker, R.A., Eick, S.G., Wilks, A.R.: Visualizing Network Data IEEE Transactions on Visualization and Computer Graphics, vol. 1, pp. 16–28 (1995)
5. Freeman, L.C., White, D.R.: Using Galois Lattices to Represent Network Data. Sociological Methodology 23, 127–146 (1993)
6. Agrawal, R., Imielinski, T., Swami, A.: Mining association rules between sets of items in large databases. In: Proc. of the ACM SIGMOD Conference on Management of Data, Washington, D.C, pp. 207–216 (May 1993)
7. Pasquier, N., Bastide, Y., Taouil, R., Lakhal, L.: Closed set based discovery of small covers for association rules. In: Proc. BDA conf., pp. 361–381 (1999)
8. Stumme, G.: Conceptual Knowledge Discovery with Frequent Concept Lattices. FB4-Preprint 2043, TU Darmstadt (1999)
9. Zaki, M.J., Hsiao, C.: CHARM: An Efficient Algorithm for Closed Itemset Mining. In: SDM (2002)
10. Stumme, G., Taouil, R., Bastide, Y., Pasquier, N., Lakhal, L.: Computing iceberg concept lattices with TITANIC Data Knowl. Eng., vol. 42, pp. 189–222. Elsevier Science Publishers B. V, Amsterdam (2002)
11. Duquenne, V.: Lattice analysis and the representation of handicap associations. Social Networks 18, 217–230 (1996)
12. Duquenne, V.: Latticial structures in data analysis. Theorical Computer Science 217, 407–436 (1999)
13. Valtchev, P., Grosser, D., Roume, C., Hacene, M.R.: GALICIA: An open platform for lattices. In: Using Conceptual Structures: Contributions to the 11th Intl. Conference on Conceptual Structures (ICCS 2003), Dresden, Germany, vol. 25, pp. 241–254. Shaker, Ithaca (2003)
14. Gansner, E.R., North, S.C.: An open graph visualization system and its applications to software engineering. Softw. Pract. Exper. 30(11), 1203–1233 (2000)

Using FCA to Suggest Refactorings to Correct Design Defects

Naouel Moha, Jihene Rezgui, Yann-Gaël Guéhéneuc,
Petko Valtchev, and Ghizlane El Boussaidi

GEODES, Department of Informatics and Operations Research
University of Montreal, Quebec, Canada
{mohanaou,rezguiji,guehene,valtchev,elboussg}@iro.umontreal.ca

Abstract. Design defects are poor design choices resulting in a hard-to-maintain software, hence their detection and correction are key steps of a disciplined software process aimed at yielding high-quality software artifacts. While modern structure- and metric-based techniques enable precise detection of design defects, the correction of the discovered defects, e.g., by means of refactorings, remains a manual, hence error-prone, activity. As many of the refactorings amount to re-distributing class members over a (possibly extended) set of classes, formal concept analysis (FCA) has been successfully applied in the past as a formal framework for refactoring exploration. Here we propose a novel approach for defect removal in object-oriented programs that combines the effectiveness of metrics with the theoretical strength of FCA. A case study of a specific defect, the *Blob*, drawn from the Azureus project illustrates our approach.

Keywords: Design Defects, Formal Concept Analysis, Refactoring.

1 Introduction

Design defects are bad solutions to recurring design problems in object-oriented programming. The activities of detection and correction of design defects are essential to improve the quality of programs and to ease their maintenance and evolution. Indeed, design defects have a strong negative impact on quality characteristics such as evolvability and maintainability [4]. A program without design defects is easier to understand and change and thus has lower maintenance costs.

However, the detection and correction of design defects are time-consuming and error-prone activities because of lack of (semi-)automated techniques and tools. Although approaches exist to detect design defects, using metrics [7] for example, to the best of our knowledge, no approach attempts to correct design defects (semi-)automatically. Huchard and Leblanc [6] use formal concept analysis (FCA) to suggest restructurations of class hierarchies to maximise the sharing of data structure and code through fields and methods and remove code smells from the program. Arévalo *et al.* applied FCA to identify implicit dependencies among classes in program models [1]. They build models from source code and

S. Ben Yahia et al. (Eds.): CLA 2006, LNAI 4923, pp. 269–275, 2008.

extract contexts from the models. Concepts and lattices generated from the contexts with the ConAn engine are filtered out to build a set of views at different levels of abstraction. These two approaches provide interesting results but none attempts *to suggest refactorings to correct design defects.*

We propose to apply FCA on a suitable representation of a program to suggest appropriate refactorings for certain design defects. A refactoring is a change performed on the source code of a program to improve its internal structure without changing its external behaviour [4]. In particular, we examine the benefits of FCA and concept lattices for the correction of a very common design defect, the Blob [2, p. 73–83]. It is generally accepted that a Blob reflects procedural thinking during the design of an object-oriented program. It manifests through a large class monopolising the computation, surrounded by a number of smaller data classes, which embed a lot of attributes and few or no methods.

Design defects are the results of *bad* practices that transgress *good* object-oriented principles. Thus, we use the degree of satisfaction of those principles before and after the refactorings as a measure of progress. Technically speaking, we rely on quantification of *coupling* and *cohesion*, which are among the most widely acknowledged software quality characteristics, key for the target maintainability factor. The cohesion of a class reflects *how closely the methods are related to the instance variables in the class* [3]. A low cohesion score witnesses a cohesive class whereas a value close to 1 indicates a lack of cohesion and suggests the class might better be split into parts. The coupling of a class is defined as the degree of its reliance on services provided by other classes [3], i.e. it counts the classes to which a class is coupled. A well-designed program exhibits *high* average cohesion and *low* average coupling, but it is well known that these criteria are antinomic hence a trade-off is usually sought.

Our intuition is that design defects resulting in high coupling and low cohesion could be improved by redistributing class members among existing classes (with possibly new classes) to increase cohesion and–or decrease coupling. FCA provides a particularly suitable framework for helping in redistributing class members because it can discover strongly related sets of individuals wrt. shared properties and hence supports the search of cohesive subsets of class members.

2 Combining Metrics and FCA to Correct Design Defects

2.1 Running Example

We illustrate our approach using Azureus version 2.3.0.6, a peer-to-peer program [8] that contains 143 Blobs for 1,449 classes (191,963 lines of code) and show that FCA can suggest relevant refactorings to improve the program. We choose Azureus because it has been heavily changed and maintained since its first release in July 2003. The addition of new features, optimisations, and bugs fixes have introduced design defects. We choose to illustrate our approach with the Blob because it impacts negatively the two important quality characteristics: such classes show low cohesion and high coupling. We notice that the underlying classes that constitute the Blobs in Azureus are difficult to understand, maintain, and reuse because they have a large number of fields and methods. For example, the class `DHTTransportUDPImpl` in the package `com.aelitis.azureus.core.dht.transport.udp.impl`, which implements a distributed sloppy hash table (DHT) for storing peer contact information over UDP, has an atypically large size. It declares 52 fields and 71 methods for 3,211 lines of code. It has a medium-to-high cohesion of 0.542 and a high coupling of 41 (8th highest value among 1,449 classes). The data classes that surround this large class are: `Average`, `HashWrapper` in package `org.gudy.azureus2.core3.util` and `IpFilterManagerFactory` in package `org.gudy.azureus2.core3.ipfilter`.

2.2 Our Approach in a Nutshell

Figure 1 depicts our approach for the identification of refactorings to correct design defects in general and the Blob in particular. The diagram shows the activities of detection of design defects and correction of user-validated defects.

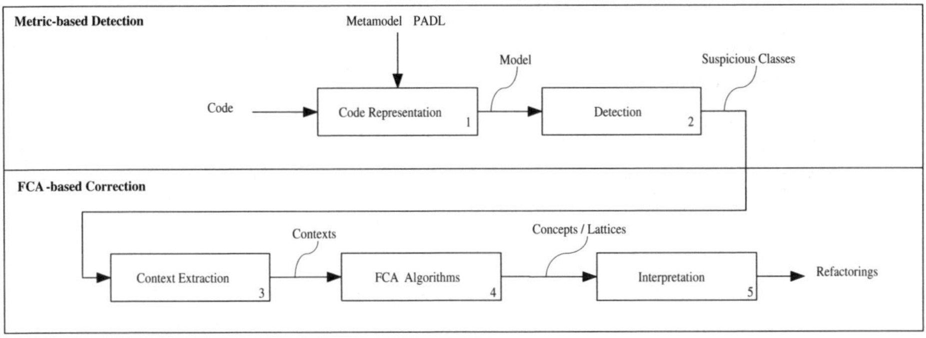

Fig. 1. Detection and FCA-based Correction of Design Defects

First, we build a model of the source code which is simpler to manipulate than the raw source code and therefore eases the subsequent activities of detection and correction. The model is instantiated from a meta-model to describe object-oriented programs.

Next, we apply well-known algorithms based on metrics and–or structural data on this model to single out suspicious classes having potential design defects. We automatically extract contexts related to these classes (and to their methods and fields) from the model of the source code. These contexts are built to enable the detection of related methods, fields, and classes (see Section 2.3).

Then, the contexts are fed into a FCA engine which generates concept lattices. We explore the lattice structure (order) and interpret the discovered concepts to clarify the relationships among members of the suspect classes and their links to the rest of the program. Both concepts and order are analysed to suggest refactorings to recreate the discovered related sets of elements.

2.3 Encoding Blobs into Formal Contexts

To correct Blob design defects, we need to identify cohesive sets of methods and, possibly, fields with respect to three criteria: usage of fields, calls to other methods, and reliance on data classes. Hence, our individuals can be either methods or fields, our properties are substitutable to fields, methods, or data classes and our incidence relations represent associations, method invocations, or use-relationships.

Table 1. Context $\mathcal{K}_1$ linking methods of the large class to fields of the class

	(a0) alien_average	(a1) alien_fv_average	(a2) bad_ip_bloom_filter	(a3) bootstrap_node	(a4) external_address	(a5) last_address_change	(a6) last_alien_count	(a7) last_alien_fv_count	(a8) listeners	(a9) local_contact	(a10) logger	(a11) other_non_routable_total	(a12) other_routable_total	(a13) packet_handler	(a14) reachable	(a15) reachable_accurate	(a16) recent_reports	(a17) request_handler	(a18) request_timeout	(a19) stats	(a20) stats_start_time	(a21) store_timeout	(a22) STATS_INIT_PERIOD	(a23) STATS_PERIOD
(m0) checkAddress()			×																					
(m1) externalAddressChange()						×																		
(m2) getAddressChange()					×																			
(m3) process()				×						×	×			×			×	×		×				
(m4) sendFindNode()										×				×					×	×				
(m5) sendFindValue()										×				×					×	×				
(m6) sendStore()										×				×						×		×		
(m7) setRequestHandler()																		×						
(m8) testInstanceIDChange()										×														
(m9) testTransportIDChange()					×					×														
(m10) updateContactStatus()												×	×											
(m11) updateStats()	×	×					×	×	×						×	×				×	×		×	×

Context 1. In the first formal context, $\mathcal{K}_1$, individuals are methods of a suspect large class and properties are fields of that class. The incidence relation is the *method-uses-field* relationship. The context aims at identifying methods using the same sets of fields and fields used by cohesive sets of methods. It allows to assess the cohesion of a class because methods sharing the same fields are, by definition, cohesive.

Table 1 illustrates the context drawn from the large class `DHTTransportUDPImpl` in Azureus. It shows the methods (individuals in rows) and their use-relationship links with fields (properties in columns) of the large class. Codes are provided that are used when presenting lattices in the next paragraphs.

We defined three contexts. In the first formal context, $\mathcal{K}_1$, individuals are methods of a suspect large class and properties are fields of that class. The incidence relation is the *method-uses-field* relationship. The context aims at identifying methods using the same sets of fields and fields used by cohesive sets of methods. It allows to assess the cohesion of a class because methods sharing the same fields are, by definition, cohesive. In the second formal context, $\mathcal{K}_2$, both individuals and properties are methods of the suspect large class, while the incidence is the *method-invocation* relationship. This context highlights subsets of cohesive methods, because methods invoking the same set of other methods are highly cohesive. In the third formal context, $\mathcal{K}_3$, individuals are methods and fields of the large class and properties are the surrounding data classes. This context represents the *use-relationship* and allows to assess the coupling between the large class and its data classes. We can identify which methods and fields of the large class should be moved together to some data class.

2.4 Interpretation of Lattice Structure

We build lattices from the contexts $\mathcal{K}_1$, $\mathcal{K}_2$, and $\mathcal{K}_3$, respectively. We use these lattices to interpret the inner structure of the large class and then to suggest refactorings. More specifically, we look for specific configurations of concepts that reflect the presence of cohesive and (un)coupled sets. Intuitively, shared usages of fields and calls of methods is a sign of cohesion whereas coupling is directly expressed by the reliance of a class member on a data class. We define the following interpretation rules.

Rule 1. *[Collection of cohesive and independent subsets.]* If a set of concepts has only the lattice supremum (top) as a successor and only the infimum (bottom) as a predecessor (*pancake lattice*), then they all represent cohesive and disjoint subsets of the individuals. For instance, in Figure 2, we interpret the concepts in the area 2 (on the right of the oblique line) as sets of elements that, whenever put together, form a low-cohesion group. Indeed, there is no collaboration (*i.e.*, no shared fields) between the individuals in different concepts.

Rule 2. *[A large cohesive subset.]* If a sub-structure of the lattice has many concepts that form a network with all their meets and joins (different from the supremum and the infimum of the lattice), then that structure represents a cohesive set of individuals. Such a situation is depicted in Figure 2, on the left of the oblique line (zone 1).

Lattice 1. Recall that the lattice in Figure 2 represents the method-uses-attribute relationship. By applying **Rules 1,2**, we obtain the following four concept sets representing cohesive subsets of methods and fields in the large class:

Combination of lattices. Following the interpretation of the lattices, we split the large class into two ways. First, we move disjoint and cohesive subsets of methods and–or fields that are related to a data class in that data class. Second, we organise cohesive subsets that are not related to data classes in separate classes.

Refactorings. Before the refactorings, class `DHTTransportUDPImpl` had a cohesion of 0.542 and a coupling of 41. After the refactorings, the cohesion of the classes is maximum of 0.278 and the coupling has reduced to 34, which shows a better compliance to good object-oriented design principles and highlights the interest of our approach.

Implementation. We use PADL to model source code and GALICIA to construct and visualize the lattices. PADL is the meta-model at the heart of the

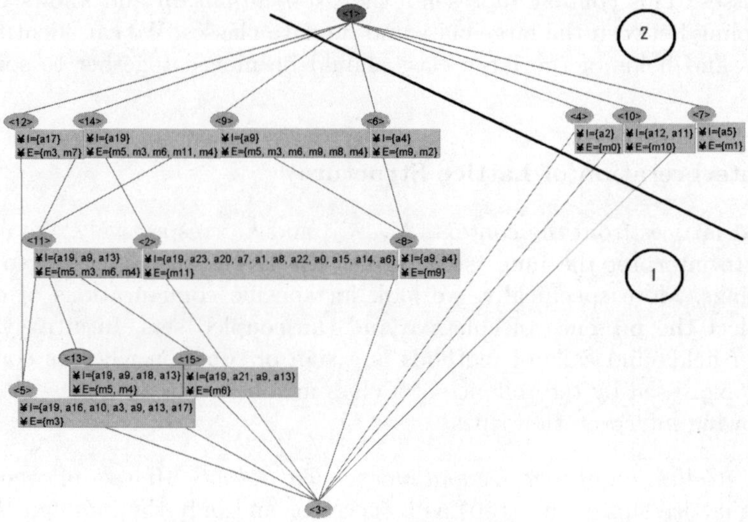

Fig. 2. Concept lattice of the methods x fields context

PTIDEJ open-source tool suite (*Pattern Trace Identification, Detection, and Enhancement in Java*) [5]. GALICIA is a multi-tool open-source platform for creating, visualizing, and storing lattices. Both tools communicate by means of XML files describing data and results. Thus, an add-on to PTIDEJ generates contexts in the XML format of GALICIA, which are then transformed by the tool into lattices and shown on screen for exploration.

3 Conclusion

Design defects are the results of bad practices that transgress good object-oriented design principles. A low coupling and a high cohesion are among the most recognised design principles to assess the quality of programs, in particular their maintainability and evolvability.

We propose an approach based on the joint use of metrics and FCA to suggest corrections to design defects in object-oriented programs. FCA provides a sketch of the target design by grouping methods and fields into cohesive sets which, once turned into separate classes, represent a better trade-off between coupling and cohesion. Our approach can be systematically generalised to other design defects characterised by a high coupling and a low cohesion.

In the long term, we plan to develop fully automatic correction mechanisms and to propose an integrated tool platform to support FCA-based refactorings. A refinement of the proposed rules for lattice structure interpretation will also be developed, allowing for more subtle, possibly numerical, decision criteria for cohesive sets of concepts.

References

1. Nierstrasz, O., Ducasse, S., Arévalo, G.: Lessons Learned in Applying Formal Concept Analysis to Reverse Engineering. In: Ganter, B., Godin, R. (eds.) ICFCA 2005. LNCS (LNAI), vol. 3403, pp. 95–112. Springer, Heidelberg (2005)
2. Brown, W.J., Malveau, R.C., Brown, W.H., McCormick III, H.W., Mowbray, T.J.: Anti Patterns: Refactoring Software, Architectures, and Projects in Crisis, 1st edn. John Wiley and Sons, Chichester (1998)
3. Fenton, N., Pfleeger, S.L.: Software metrics: A rigorous and practical approach, 2nd edn. PWS Publishing Co., Boston (1997)
4. Fowler, M.: Refactoring – Improving the Design of Existing Code, 1st edn. Addison-Wesley, Reading (1999)
5. Guéhéneuc, Y.-G.: A reverse engineering tool for precise class diagrams. In: Singer, J., Lutfiyya, H. (eds.) Proceedings of the 14th IBM Centers for Advanced Studies Conference, pp. 28–41. ACM Press, New York (2004)
6. Huchard, M., Leblanc, H.: Computing interfaces in java. In: ASE, pp. 317–320 (2000)
7. Marinescu, R.: Detection strategies: Metrics-based rules for detecting design flaws. In: Proceedings of the 20th International Conference on Software Maintenance, pp. 350–359. IEEE Computer Society Press, Los Alamitos (2004)
8. Open source project. Azureus (June 2003)

Type Signature Induction with FCA$_{\text{Type}}$

Wiebke Petersen

Institute of Language and Information
Heinrich-Heine-Universität Düsseldorf
petersew@uni-duesseldorf.de

Abstract. Type signatures are common in modern linguistic theories. Their construction and maintenance is intricate, and therefore, an automatic induction method is desirable. In the present paper we present FCA$_{\text{Type}}$, a module of our system FCA$_{\text{Ling}}$, that automatically induces type signatures from sets of untyped feature structures. The induction procedure is based on so-called *decomposition semilattices* which serve as a basis for initial type signatures. These signatures can be folded up to result in compact and restrictive type signatures which adequately specify the input structures.

1 Introduction

The primary task in grammar engineering is to construct a grammar which generates exactly those phrases which are well-formed in the target language. The purpose of the lexicon is to provide the basic units of the language. Modern linguistic theories tend to express more and more grammatical information in the lexicon. Hence, "lexical entries have evolved from simple pairings of phonological forms with grammatical categories into elaborate information structures, in which phonological forms are now paired with more articulated feature structure descriptions.", [1, p.173]. *Feature structures* (FSs) are recursive attribute-value structures which are known as frames in other disciplines, e.g. [2].[1] An example lexicon with small 'toy' FSs taken from [4] is shown in Fig. 1.

As depicted in Fig. 1, FSs can be written as recursive *attribute-value matrices* (AVMs). The AVMs are constructed as follows: FSs are enclosed in square brackets. Each first-level attribute is followed by a colon and its value. The values are either atomic (i.e., not specified by further attributes) or complex FSs. Restricting a FS to one of its paths yields the value of the path in the FS, e.g.,

$$\begin{bmatrix} \text{CAT}: & \mathsf{np} \\ \text{HEAD}: & \begin{bmatrix} \text{AGR}: \begin{bmatrix} \text{PERS}: \mathsf{third} \\ \text{NUM}: \mathsf{sing} \end{bmatrix} \end{bmatrix} \end{bmatrix} @\text{HEAD AGR} = \begin{bmatrix} \text{PERS}: \mathsf{third} \\ \text{NUM}: \mathsf{sing} \end{bmatrix}.$$

Ordered by subsumption the FSs form a semilattice where the *generalization* of two FSs is the most specific FS which subsumes both FSs.

[1] Due to space limits we decided to omit all formal definitions and to concentrate on why FCA$_{\text{Type}}$ is useful and how it works in principal. A detailed description of FCA$_{\text{Type}}$ and a formal proof that the described procedures are well-defined can be found in [3]. FCA$_{\text{Type}}$ can be obtained from the author on request.

S. Ben Yahia et al. (Eds.): CLA 2006, LNAI 4923, pp. 276–281, 2008.

$$\text{Uther} = \begin{bmatrix} \text{CAT}: & \mathsf{np} \\ \text{HEAD}: & \begin{bmatrix} \text{AGR}: & \begin{bmatrix} \text{PERS}: \mathsf{third} \\ \text{NUM}: \mathsf{sing} \end{bmatrix} \end{bmatrix} \end{bmatrix} \qquad \text{sleeps} = \begin{bmatrix} \text{CAT}: & \mathsf{vp} \\ \text{HEAD}: & \begin{bmatrix} \text{FORM}: & \mathsf{finite} \\ \text{SUBJ} & \begin{bmatrix} \text{CAT}: & \mathsf{np} \\ \text{HEAD}: & \begin{bmatrix} \text{AGR}: & \begin{bmatrix} \text{PERS}: \mathsf{third} \\ \text{NUM}: \mathsf{sing} \end{bmatrix} \end{bmatrix} \end{bmatrix} \end{bmatrix} \end{bmatrix}$$

$$\text{knights} = \begin{bmatrix} \text{CAT}: & \mathsf{np} \\ \text{HEAD}: & \begin{bmatrix} \text{AGR}: & \begin{bmatrix} \text{PERS}: \mathsf{third} \\ \text{NUM}: \mathsf{plur} \end{bmatrix} \end{bmatrix} \end{bmatrix} \qquad \text{sleep} = \begin{bmatrix} \text{CAT}: & \mathsf{vp} \\ \text{HEAD}: & \begin{bmatrix} \text{FORM}: & \mathsf{finite} \\ \text{SUBJ} & \begin{bmatrix} \text{CAT}: & \mathsf{np} \\ \text{HEAD}: & \begin{bmatrix} \text{AGR}: & \begin{bmatrix} \text{PERS}: \mathsf{third} \\ \text{NUM}: \mathsf{plur} \end{bmatrix} \end{bmatrix} \end{bmatrix} \end{bmatrix} \end{bmatrix}$$

Fig. 1. Example lexicon with small untyped feature structures

In order to organize the lexicon, avoid redundancy, and capture generalizations, a strict type discipline has been developed [5]. Types are assigned to FSs and their restrictions and they are organized in a *type hierarchy*, that is, in a finite semilattice. In the AVM representation of a FS types are represented as small indices. In order to restrict the class of admissible FSs, plain type hierarchies are typically enriched by appropriateness conditions [5,6]. They regulate which features are appropriate for FSs of a special type and restrict the values of the appropriate features. A type hierarchy enriched by appropriateness conditions is called a *type signature*. Fig. 3 (top) shows a small type signature. The appropriateness condition 'CAT : np' at type t_3 means that the attribute CAT is appropriate for structures of type t_3 and its value is restricted to structures of type np or subtypes of np. Appropriateness conditions are inherited downwards. Hence, the subtype t_4 of t_3 inherits the condition 'CAT : np' from t_3. It also inherits the condition 'HEAD : t_9' from t_3, but tightens it up to 'HEAD : t_{10}'.

We can consider a type signature as a *specification* of a set of FSs, namely the set of its totally well-typed FSs. We call a FS *totally well-typed* with respect to a type signature if all its attributes are licensed by the type signature and their values are at least as specific as demanded by the appropriateness conditions. Additionally, all attributes which are prescribed by the appropriateness conditions need to be present. For example, $\begin{bmatrix} \text{CAT}: & \mathsf{np} \\ \text{HEAD}: & \begin{bmatrix} \text{AGR}: & \begin{bmatrix} \text{PERS}: \mathsf{third} \\ \text{NUM}: \mathsf{sing} \end{bmatrix}_{\mathsf{t}_{14}} \end{bmatrix}_{\mathsf{t}_{11}} \end{bmatrix}_{\mathsf{t}_5}$ is totally well-typed w.r.t. the type signature in Fig. 3 (top), but neither $\begin{bmatrix} \text{PERS}: \mathsf{third} \\ \text{NUM}: \mathsf{num} \end{bmatrix}_{\mathsf{t}_{14}}$ nor $\begin{bmatrix} \text{CAT}: \mathsf{np} \end{bmatrix}_{\mathsf{t}_5}$ are.

FCAType is a system for the automatic induction of a type signature from a set of untyped FSs. Generally, in the grammar engineering process, the type signature is constructed simultaneously with the rest of the grammar starting with a small grammar covering only a few linguistic phenomena. However, FSs which encode all the necessary phonological, morphological, syntactic, and semantic information of a lexical entry are huge, and type signatures which cover generalizations about such FSs become so complex that a purely manual construction and maintenance is intricate. Therefore, an induction of type signatures which is at least semiautomatic would be most welcome.

The following three grammar engineering tasks are particularly supported by our induction method: (1) corpus-driven grammar development, (2) reuse of grammar resources, and (3) grammar maintenance: Today, we are in the lucky position that we are provided with huge, corpus-extracted lexica. Usually, the entries of these lexica can be seen as untyped FSs, but the manual hierarchical organization of them is not feasible and must be automated. For economical reasons, the reusability of grammatical resources is desirable [7] and should be supported by automatic systems like $\mathsf{FCA}_{\mathsf{Type}}$. Transferring a grammar from one formalism into another one may also unveil new theoretical insights, cf. [8,9]. Finally, [10] discusses how error and consistency checking of a large scale untyped grammar can be facilitated by adding an appropriate type signature. However, constructing an appropriate type signature for an already existing grammar is usually a difficult task which requires deep insight into the structural design of the grammar. Therefore, we propose that an expert should intellectually investigate the automatically induced type signature in order to detect errors and inconsistencies in the given grammar which can be an easier task than to build up an appropriate type signature from scratch by hand.

The key idea of our system for the induction of type signatures is to construct the *decomposition semilattice* (DSL) from the untyped input FSs which can be seen as a *featureless* type signature [6]. A similar method is used by Sporleder [11] for a different task, namely for the automatic induction of lexical inheritance hierarchies, i.e. hierarchies of untyped FSs: she reduces the task to a classification problem where the search space is defined by a concept lattice.

2 $\mathsf{FCA}_{\mathsf{Type}}$ Approach

Our aim is to automatically induce an *adequate* type signature from a set of untyped FSs. The type signature is adequate if it specifies the input data, i.e., if for each untyped input structure a totally well-typed, typed version exists (a *typed version* of an untyped FS is identical to the untyped structure, except that types are assigned to each restriction). Therefore, we need types for the input structures themselves and for each restriction of them. Since it is required that the type signature expresses generalizations, we also need types for all possible generalizations about these structures. Moreover, we ask that the generalizations about restrictions of the input structures can be naturally order embedded (via their typed versions) into the ordered set of totally well-typed FSs of the induced type signature.

Further quality criteria for the induced type signatures are *restrictiveness* and *compactness*. Improving the restrictiveness of an induced type signature means reducing the set of totally well-typed FSs, and improving the compactness means reducing the number of types in the type hierarchy.

$\mathsf{FCA}_{\mathsf{Type}}$ is based on the construction of a DSL from a set of untyped FSs. The DSL consists of (1) the FSs themselves, (2) their restrictions, and (3) all

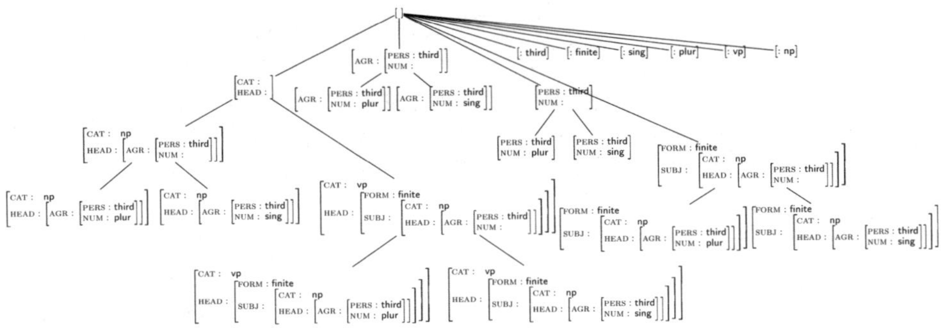

Fig. 2. The decomposition semilattice for the structures of Fig. 1

generalizations about structures from (1) and (2). All those structures are partially ordered by subsumption as in Fig. 2.[2]

By assigning a type to each element of the DSL one gains a type hierarchy which provides the required types. That a DSL can be straightforwardly transformed into a well-formed, adequate type signature by inferring adequate appropriateness conditions from the DSL can be seen by comparing the DSL in Fig. 2 with the inferred type signature in Fig. 3 (top) (for details see [3]).[3]

However, the type signature in Fig. 3 (top) still has two undesirable properties: First, the type signature is not very *compact* since some types are unnecessary (the set of totally well-typed FSs would not change substantially if the types t_4,t_5,t_7,t_8,t_{10},t_{11},t_{16}, and t_{17} were deleted). Second, the induced appropriateness conditions are not *restrictive* enough (e.g., the appropriateness condition 'NUM:t_1' permits that the attribute NUM takes a complex FS of type t_2 as value). The first problem is solved by *folding up* the signature and the second one by adding additional types to control the values:

We *fold up* a signature at a type t by deleting all proper subtypes of t under the condition that this deletion does not affect the set of well-typed FSs of our signature (up to typing). Hence, folding up a type signature results in a more compact type signature (the terminology of *folding* is taken from [6]). In

[2] Actually, instead of computing the set of all generalizations about restrictions of the input structures and ordering them by subsumption, FCAType implements an alternative approach: It generates the *decomposition context* of the input structures and computes its concept lattice which is isomorphic to the DSL (except for the bottom element). Its object set corresponds to the set of restrictions and its attribute set encodes information about paths, path equations, and path values (for details see [3]). This approach enables us on the one hand to use our FCA-submodule which is used by other modules of FCALing, too. On the other hand, we found it useful to have direct access to the decomposed properties of the input structures when it comes to infer appropriateness conditions and to determine folding opportunities. Another alternative would have been to use pattern concepts as described in [12].

[3] If decomposition contexts are employed (see footnote 2), the required appropriateness conditions can be immediately read off from the attribute concepts.

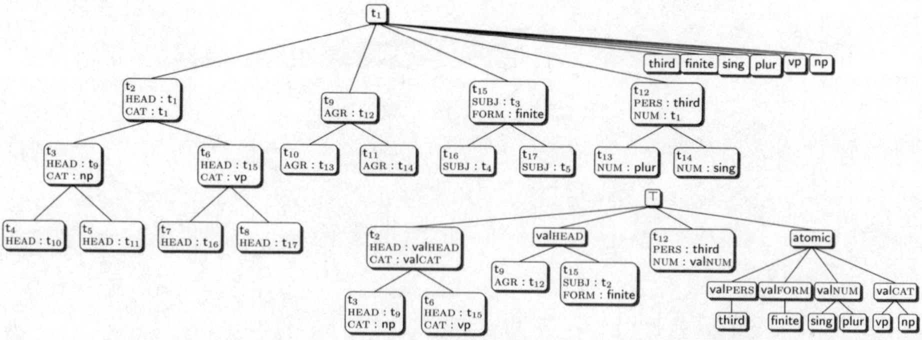

Fig. 3. Unfolded type signature (top) and maximally folded, value-controlled type signature (bottom) for the structures of Fig. 1 (each box shows a type label in the first line followed by the appropriateness conditions)

principle we have to consider two different folding opportunities. *Atomic folding opportunities* result from the distribution of the atomic types in the FSs. All folding opportunities of the type signature in Fig. 3 (top) are atomic. *Structural folding opportunities* are rare and therefore not discussed here (for details see [3]). In FCAType, we have chosen an easy way to take advantage of all atomic folding opportunities. The key idea is that manually constructed type signatures mainly encode information about the general structure of the lexical FSs. Hence, we have chosen to simplify the input structures in a first step by replacing each atomic value with a generic marker *av_*. Starting from the DSL of these simplified structures, we construct the corresponding type signature. It covers all structural aspects of the untyped FSs, and it lays the foundation for our target signature. In the next step, the atomic values are taken into account and the appropriateness conditions are tightened up, wherever possible. Finally, all atomic values are arranged under a type 'atomic' and meanwhile ordered by the features they can be values of. In the resulting type signature, no atomic folding opportunities are left, thanks to the preceding simplification of the input structure. The fact that the rigorous simplification of the input structures can theoretically result in type signatures which are too heavily folded up can be captured by additionally induced *feature co-occurrence restrictions* [3,9].

The deficient restrictiveness of type signatures of DSLs is caused by appropriateness conditions which do restrict values of an attribute to structures of the most common type. In such cases, the type signature has at least one recursive type and thus the set of totally well-typed FSs is infinite [6]. Therefore, we have decided to introduce an artificial type whenever such a situation would occur and to adjust the affected appropriateness conditions. By inserting those artificial types our type signatures become more restrictive.

Fig. 3 (bottom) shows the maximally folded, value-controlled type signature induced by FCAType from the input data of Fig. 1. A detailed description of the induction process is given in [3].

3 Conclusion

It would be interesting to combine our approach with that of Sporleder and to use our DSLs as input for her classification problem, since they encode much more detail than her lattices, and they can be used as basis for the construction of type signatures.

But also from a theoretical point of view, our observations are interesting: The set of typed FSs corresponding to a type signature is well understood [5,6]. However, a lot of work has still to be done to answer the question which type signature models a set of untyped FSs best. In our opinion, a closer investigation of DSLs and the related type signatures can provide answers. In [3] these questions are discussed in greater detail: A number of alternative type signatures induced from DSLs are presented and their properties are compared. Additionally, type constraints are induced which either restrict the admissible path-equation relations or express feature co-occurrence restrictions.

References

1. Sag, I., Wasow, T.A.: Syntactic Theory: A Formal Introduction. In: CSLI, Stanford (1999)
2. Minsky, M.: A framework for representing knowledge. In: Winston, P.H. (ed.) The Psychology of Computer Vision, pp. 211–277. McGraw-Hill, New York (1975)
3. Petersen, W.: Induktion von Typsignaturen mit Mitteln der Formalen Begriffsanalyse. PhD thesis, University of Düsseldorf (in submission process)
4. Shieber, S.M.: An Introduction to Unification-Based Approaches to Grammar. In: CSLI, Stanford (1986)
5. Carpenter, B.: The Logic of Typed Feature Structures. In: Cambridge Tracts in Theoretical Computer Science 32. CUP (1992)
6. Penn, G.: The Algebraic Structure of Attributed Type Signatures. PhD thesis, School of Computer Science, Carnegie Mellon University (2000)
7. Arnold, D.J., Badia, T., van Genabith, J., Markantonatou, S., Momma, S., Sadler, L., Schmidt, P.: Experiments in reusability of grammatical resources. In: Proceedings of 6th EACL, pp. 12–20 (1993)
8. Kilbury, J., Petersen, W., Rumpf, C.: Inheritance-based models of the lexicon. In: Wunderlich, D. (ed.) Advances in the Theory of the Lexicon, pp. 429–477. Mouton de Gruyter, Berlin (2006)
9. Petersen, W., Kilbury, J.: What feature co-occurrence restrictions have to do with type signatures. In: Proceedings of FG/MOL-2005, Edinburgh, pp. 125–139 (2005)
10. Wintner, S., Sarkar, A.: A note on typing feature structures. Computational Linguistics 28(3), 389–397 (2002)
11. Sporleder, C.: Discovering Lexical Generalisations. A Supervised Machine Learning Approach to Inheritance Hierarchy Construction. PhD thesis, School of Informatics, University of Edinburgh (2003)
12. Ganter, B., Kuznetsov, S.O.: Pattern Structures and Their Projections. In: Delugach, H.S., Stumme, G. (eds.) ICCS 2001. LNCS (LNAI), vol. 2120, Springer, Heidelberg (2001)

Author Index

Printing: Mercedes-Druck, Berlin
Binding: Stein + Lehmann, Berlin

Lecture Notes in Artificial Intelligence (LNAI)

Vol. 4953: N.T. Nguyen, G.S. Jo, R.J. Howlett, L.C. Jain (Eds.), Agent and Multi-Agent Systems: Technologies and Applications. XX, 909 pages. 2008.

Vol. 4938: T. Tokunaga, A. Ortega (Eds.), Large-Scale Knowledge Resources. IX, 367 pages. 2008.

Vol. 4933: R. Medina, S. Obiedkov (Eds.), Formal Concept Analysis. XII, 325 pages. 2008.

Vol. 4929: M. Helmert, Understanding Planning Tasks. XIV, 270 pages. 2008.

Vol. 4924: D. Riaño (Ed.), Knowledge Management for Health Care Procedures. X, 161 pages. 2008.

Vol. 4923: S. Ben Yahia, E. Mephu Nguifo, R. Belohlavek (Eds.), Concept Lattices and Their Applications. XII, 283 pages. 2008.

Vol. 4914: K. Satoh, A. Inokuchi, K. Nagao, T. Kawamura (Eds.), New Frontiers in Artificial Intelligence. X, 404 pages. 2008.

Vol. 4911: L. De Raedt, P. Frasconi, K. Kersting, S. Muggleton (Eds.), Probabilistic Inductive Logic Programming. VIII, 341 pages. 2008.

Vol. 4898: M. Kolp, B. Henderson-Sellers, H. Mouratidis, A. Garcia, A.K. Ghose, P. Bresciani (Eds.), Agent-Oriented Information Systems IV. X, 292 pages. 2008.

Vol. 4897: M. Baldoni, T.C. Son, M.B. van Riemsdijk, M. Winikoff (Eds.), Declarative Agent Languages and Technologies V. X, 245 pages. 2008.

Vol. 4894: H. Blockeel, J. Ramon, J. Shavlik, P. Tadepalli (Eds.), Inductive Logic Programming. XI, 307 pages. 2008.

Vol. 4885: M. Chetouani, A. Hussain, B. Gas, M. Milgram, J.-L. Zarader (Eds.), Advances in Nonlinear Speech Processing. XI, 284 pages. 2007.

Vol. 4874: J. Neves, M.F. Santos, J.M. Machado (Eds.), Progress in Artificial Intelligence. XVIII, 704 pages. 2007.

Vol. 4869: F. Botana, T. Recio (Eds.), Automated Deduction in Geometry. X, 213 pages. 2007.

Vol. 4865: K. Tuyls, A. Nowe, Z. Guessoum, D. Kudenko (Eds.), Adaptive Agents and Multi-Agent Systems III. VIII, 255 pages. 2008.

Vol. 4850: M. Lungarella, F. Iida, J.C. Bongard, R. Pfeifer (Eds.), 50 Years of Artificial Intelligence. X, 399 pages. 2007.

Vol. 4845: N. Zhong, J. Liu, Y. Yao, J. Wu, S. Lu, K. Li (Eds.), Web Intelligence Meets Brain Informatics. XI, 516 pages. 2007.

Vol. 4840: L. Paletta, E. Rome (Eds.), Attention in Cognitive Systems. XI, 497 pages. 2007.

Vol. 4830: M.A. Orgun, J. Thornton (Eds.), AI 2007: Advances in Artificial Intelligence. XIX, 841 pages. 2007.

Vol. 4828: M. Randall, H.A. Abbass, J. Wiles (Eds.), Progress in Artificial Life. XII, 402 pages. 2007.

Vol. 4827: A. Gelbukh, Á.F. Kuri Morales (Eds.), MICAI 2007: Advances in Artificial Intelligence. XXIV, 1234 pages. 2007.

Vol. 4826: P. Perner, O. Salvetti (Eds.), Advances in Mass Data Analysis of Signals and Images in Medicine, Biotechnology and Chemistry. X, 183 pages. 2007.

Vol. 4819: T. Washio, Z.-H. Zhou, J.Z. Huang, X. Hu, J. Li, C. Xie, J. He, D. Zou, K.-C. Li, M.M. Freire (Eds.), Emerging Technologies in Knowledge Discovery and Data Mining. XIV, 675 pages. 2007.

Vol. 4811: O. Nasraoui, M. Spiliopoulou, J. Srivastava, B. Mobasher, B. Masand (Eds.), Advances in Web Mining and Web Usage Analysis. XII, 247 pages. 2007.

Vol. 4798: Z. Zhang, J.H. Siekmann (Eds.), Knowledge Science, Engineering and Management. XVI, 669 pages. 2007.

Vol. 4795: F. Schilder, G. Katz, J. Pustejovsky (Eds.), Annotating, Extracting and Reasoning about Time and Events. VII, 141 pages. 2007.

Vol. 4790: N. Dershowitz, A. Voronkov (Eds.), Logic for Programming, Artificial Intelligence, and Reasoning. XIII, 562 pages. 2007.

Vol. 4788: D. Borrajo, L. Castillo, J.M. Corchado (Eds.), Current Topics in Artificial Intelligence. XI, 280 pages. 2007.

Vol. 4775: A. Esposito, M. Faundez-Zanuy, E. Keller, M. Marinaro (Eds.), Verbal and Nonverbal Communication Behaviours. XII, 325 pages. 2007.

Vol. 4772: H. Prade, V.S. Subrahmanian (Eds.), Scalable Uncertainty Management. X, 277 pages. 2007.

Vol. 4766: N. Maudet, S. Parsons, I. Rahwan (Eds.), Argumentation in Multi-Agent Systems. XII, 211 pages. 2007.

Vol. 4760: E. Rome, J. Hertzberg, G. Dorffner (Eds.), Towards Affordance-Based Robot Control. IX, 211 pages. 2008.

Vol. 4755: V. Corruble, M. Takeda, E. Suzuki (Eds.), Discovery Science. XI, 298 pages. 2007.

Vol. 4754: M. Hutter, R.A. Servedio, E. Takimoto (Eds.), Algorithmic Learning Theory. XI, 403 pages. 2007.

Vol. 4737: B. Berendt, A. Hotho, D. Mladenic, G. Semeraro (Eds.), From Web to Social Web: Discovering and Deploying User and Content Profiles. XI, 161 pages. 2007.

Vol. 4733: R. Basili, M.T. Pazienza (Eds.), AI*IA 2007: Artificial Intelligence and Human-Oriented Computing. XVII, 858 pages. 2007.

Vol. 4724: K. Mellouli (Ed.), Symbolic and Quantitative Approaches to Reasoning with Uncertainty. XV, 914 pages. 2007.

Vol. 4722: C. Pelachaud, J.-C. Martin, E. André, G. Chollet, K. Karpouzis, D. Pelé (Eds.), Intelligent Virtual Agents. XV, 425 pages. 2007.

Vol. 4720: B. Konev, F. Wolter (Eds.), Frontiers of Combining Systems. X, 283 pages. 2007.

Vol. 4702: J.N. Kok, J. Koronacki, R. Lopez de Mantaras, S. Matwin, D. Mladenič, A. Skowron (Eds.), Knowledge Discovery in Databases: PKDD 2007. XXIV, 640 pages. 2007.

Vol. 4701: J.N. Kok, J. Koronacki, R. Lopez de Mantaras, S. Matwin, D. Mladenič, A. Skowron (Eds.), Machine Learning: ECML 2007. XXII, 809 pages. 2007.

Vol. 4696: H.-D. Burkhard, G. Lindemann, R. Verbrugge, L.Z. Varga (Eds.), Multi-Agent Systems and Applications V. XIII, 350 pages. 2007.

Vol. 4694: B. Apolloni, R.J. Howlett, L. Jain (Eds.), Knowledge-Based Intelligent Information and Engineering Systems, Part III. XXIX, 1126 pages. 2007.

Vol. 4693: B. Apolloni, R.J. Howlett, L. Jain (Eds.), Knowledge-Based Intelligent Information and Engineering Systems, Part II. XXXII, 1380 pages. 2007.

Vol. 4692: B. Apolloni, R.J. Howlett, L. Jain (Eds.), Knowledge-Based Intelligent Information and Engineering Systems, Part I. LV, 882 pages. 2007.

Vol. 4687: P. Petta, J.P. Müller, M. Klusch, M. Georgeff (Eds.), Multiagent System Technologies. X, 207 pages. 2007.

Vol. 4682: D.-S. Huang, L. Heutte, M. Loog (Eds.), Advanced Intelligent Computing Theories and Applications. XXVII, 1373 pages. 2007.

Vol. 4676: M. Klusch, K.V. Hindriks, M.P. Papazoglou, L. Sterling (Eds.), Cooperative Information Agents XI. XI, 361 pages. 2007.

Vol. 4667: J. Hertzberg, M. Beetz, R. Englert (Eds.), KI 2007: Advances in Artificial Intelligence. IX, 516 pages. 2007.

Vol. 4660: S. Džeroski, L. Todorovski (Eds.), Computational Discovery of Scientific Knowledge. X, 327 pages. 2007.

Vol. 4659: V. Mařík, V. Vyatkin, A.W. Colombo (Eds.), Holonic and Multi-Agent Systems for Manufacturing. VIII, 456 pages. 2007.

Vol. 4651: F. Azevedo, P. Barahona, F. Fages, F. Rossi (Eds.), Recent Advances in Constraints. VIII, 185 pages. 2007.

Vol. 4648: F. Almeida e Costa, L.M. Rocha, E. Costa, I. Harvey, A. Coutinho (Eds.), Advances in Artificial Life. XVIII, 1215 pages. 2007.

Vol. 4635: B. Kokinov, D.C. Richardson, T.R. Roth-Berghofer, L. Vieu (Eds.), Modeling and Using Context. XIV, 574 pages. 2007.

Vol. 4632: R. Alhajj, H. Gao, X. Li, J. Li, O.R. Zaïane (Eds.), Advanced Data Mining and Applications. XV, 634 pages. 2007.

Vol. 4629: V. Matoušek, P. Mautner (Eds.), Text, Speech and Dialogue. XVII, 663 pages. 2007.

Vol. 4626: R.O. Weber, M.M. Richter (Eds.), Case-Based Reasoning Research and Development. XIII, 534 pages. 2007.

Vol. 4617: V. Torra, Y. Narukawa, Y. Yoshida (Eds.), Modeling Decisions for Artificial Intelligence. XII, 502 pages. 2007.

Vol. 4612: I. Miguel, W. Ruml (Eds.), Abstraction, Reformulation, and Approximation. XI, 418 pages. 2007.

Vol. 4604: U. Priss, S. Polovina, R. Hill (Eds.), Conceptual Structures: Knowledge Architectures for Smart Applications. XII, 514 pages. 2007.

Vol. 4603: F. Pfenning (Ed.), Automated Deduction – CADE-21. XII, 522 pages. 2007.

Vol. 4597: P. Perner (Ed.), Advances in Data Mining. XI, 353 pages. 2007.

Vol. 4594: R. Bellazzi, A. Abu-Hanna, J. Hunter (Eds.), Artificial Intelligence in Medicine. XVI, 509 pages. 2007.

Vol. 4585: M. Kryszkiewicz, J.F. Peters, H. Rybinski, A. Skowron (Eds.), Rough Sets and Intelligent Systems Paradigms. XIX, 836 pages. 2007.

Vol. 4578: F. Masulli, S. Mitra, G. Pasi (Eds.), Applications of Fuzzy Sets Theory. XVIII, 693 pages. 2007.

Vol. 4573: M. Kauers, M. Kerber, R. Miner, W. Windsteiger (Eds.), Towards Mechanized Mathematical Assistants. XIII, 407 pages. 2007.

Vol. 4571: P. Perner (Ed.), Machine Learning and Data Mining in Pattern Recognition. XIV, 913 pages. 2007.

Vol. 4570: H.G. Okuno, M. Ali (Eds.), New Trends in Applied Artificial Intelligence. XXI, 1194 pages. 2007.

Vol. 4565: D.D. Schmorrow, L.M. Reeves (Eds.), Foundations of Augmented Cognition. XIX, 450 pages. 2007.

Vol. 4562: D. Harris (Ed.), Engineering Psychology and Cognitive Ergonomics. XXIII, 879 pages. 2007.

Vol. 4548: N. Olivetti (Ed.), Automated Reasoning with Analytic Tableaux and Related Methods. X, 245 pages. 2007.

Vol. 4539: N.H. Bshouty, C. Gentile (Eds.), Learning Theory. XII, 634 pages. 2007.

Vol. 4529: P. Melin, O. Castillo, L.T. Aguilar, J. Kacprzyk, W. Pedrycz (Eds.), Foundations of Fuzzy Logic and Soft Computing. XIX, 830 pages. 2007.

Vol. 4520: M.V. Butz, O. Sigaud, G. Pezzulo, G. Baldassarre (Eds.), Anticipatory Behavior in Adaptive Learning Systems. X, 379 pages. 2007.

Vol. 4511: C. Conati, K. McCoy, G. Paliouras (Eds.), User Modeling 2007. XVI, 487 pages. 2007.

Vol. 4509: Z. Kobti, D. Wu (Eds.), Advances in Artificial Intelligence. XII, 552 pages. 2007.

Vol. 4496: N.T. Nguyen, A. Grzech, R.J. Howlett, L.C. Jain (Eds.), Agent and Multi-Agent Systems: Technologies and Applications. XXI, 1046 pages. 2007.